CHEMISTRY IN THE LABORATORY

A STUDY OF CHEMICAL AND PHYSICAL CHANGES

J. A. BERAN
TEXAS A&I UNIVERSITY

John Wiley & Sons

New York   Chichester   Brisbane   Toronto   Singapore

PHOTO CREDITS

PREFACE TO THE STUDENT

The chemistry laboratory will be one of the most interesting and rewarding academic experiences you will encounter as a college student. It will be one of those experiences that will enable you to apply the material presented in the textbook and in the lecture to observations made in the laboratory environment. The data that are collected in the laboratory can be observed, discussed, and interpreted in terms of these learned basic chemical principles. Predictions will be made as a result of the assimilation of the data. Solving for "unknowns" will be gratifying when you realize that your accumulated chemical knowledge is correctly used in your analysis of the observations.

The chemistry laboratory is also a social experience. Working with others while attempting to gain an understanding of your laboratory observations allows you to perceive the conceptual understanding of basic chemical principles from various viewpoints. More simply, sharing the rewards and frustrations of a laboratory assignment promotes lasting friendships. Recollections of the "lab" at class reunions or college social events vary from "the acid holes in the new pair of jeans" to "the percent antacid in Rolaids or the percent vitamin C in a vitamin tablet"—memorable events seem to happen in the laboratory. Its fun, challenging, rewarding, and memorable—enjoy it!

This manual covers a full year of general chemistry in which you may expect to spend an average of three hours per week in the laboratory. Advance preparation for each experiment may decrease this time; no preparation may extend this time to four or five hours! Although the manual parallels the material in Brady and Holum's, *Chemistry: The Study of Matter and Its Changes*, the experiments are chosen and written so they may easily be adapted to any general chemistry text.

These experiments were chosen for you to gain an appreciation and an understanding of chemistry at a level that is necessary for your chosen major field of study, provided it is science related. You need *not* be a chemistry major to understand the material in this manual.

An introductory paragraph to each experiment attempts to place the importance of the experiment in the realm of the real world. An accompanying photograph/line drawing again assists in making the experiment more attractive and interesting.

Each experiment has six major divisions: **Objectives, Principles, Techniques, Procedure, Lab Preview,** and **Data Sheet:**

•**Objectives.** Stated at the beginning of each experiment, the objectives focus on the purpose of the experiment and just what you are to learn and accomplish from the experiment.

•**Principles.** Several paragraphs describe the nature of the experiment and the chemical principles used in the interpretation and the understanding of the data. Appropriate equations, tests, colors, etc., are presented and/or illustrated. A running glossary helps to clarify terms and symbols.

•**Techniques.** Eighteen basic laboratory techniques are described on pages 1–20. Techniques that are appropriate for the successful completion of the experiment are listed. The technique icons are identified and serve as reminders in each experiment. You will need to refer to these quite often.

•**Procedure.** A detailed procedure for the experiment includes figures for the construction of the apparatus, icon reminders to use the proper techniques during the experiment, Cautions, and Disposal Information.

•**Lab Preview.** You will need to read and investigate *thoroughly* the Principles, Procedure, and Techniques sections of the experiment before the laboratory and answer corresponding questions *before* coming to the laboratory session. Its careful completion reduces the "waste of time and boredom" that the poorly prepared student experiences.

•**Data Sheet**. A detailed data sheet provides an outline for you to organize your data and to complete the necessary calculations. Questions reviewing the experiment and probing your interpretations appear at the end of the Data Sheet. You will learn to depend on your data collection and analysis of the data for these answers.

Several features of the manual will help make your experience more meaningful:

•Proper laboratory technique is necessary for the collection of reliable and reproducible data. The *Techniques Section* (pps. 1–20) illustrate the proper way to perform various laboratory operations. Icons for each of the 20 techniques have been drawn and are used as reminders in the Procedure of each experiment.

•A *running glossary* in the margin of each experiment is used to clarify terms, symbols, and interpretations.

•Line drawings and photographs are used extensively to illustrate the construction of the proper laboratory apparatus and the plotting of data.

•Careful attention to safety is an essential component in the design of each experiment. While all potentially hazardous chemicals cannot be eliminated from the laboratory, their numbers are minimal in this manual. Safety is a major concern—a safety icon in the margin cites potential danger spots in the Procedure and a **Caution** citation in the text depicts the actual danger of the chemical. In addition, **Disposal Information** for each experiment suggests the method of discarding your test solutions. Ultimately, it is your responsibility to practice safe laboratory guidelines, not only for your personal safety but for your friends and neighbors as well.

The author truly hopes you find the general chemistry laboratory an enjoyable/memorable experience. Take the time to critically analyze your observations and to clearly and conscientiously complete your calculations. Realize that references to the textbook or discussions with your laboratory instructor may be necessary to effectively complete and understand the experiment. It is inadvisable to think of the laboratory as only a three hour segment of your academic week.

Good luck! Have fun!

PREFACE TO THE LABORATORY INSTRUCTOR

This manual covers a full year of general chemistry for students who have chosen careers in the sciences or engineering. Although the manual parallels the material in Brady and Holum's, *Chemistry: The Study of Matter and Its Changes,*, the experiments are chosen and written so they may easily be adapted to any general chemistry text.

Each experiment begins with a friendly *Introduction* to help students relate the importance of the experiment to the real world. An accompanying photograph/line drawing again assists in making the experiment more attractive and interesting.

A *running glossary* in the margin of each experiment helps to clarify many terms, symbols, or interpretations that are known to give students difficulty. In the class testing of these experiments, students found this information to be very helpful.

The development of good *laboratory techniques* is a feature of the manual. The Techniques Section, pages 1–20, covers 18 basic laboratory techniques with 47 photographs and line drawings. Each technique is identified with an icon. The icons are then used in the Procedure of each experiment as reminders to practice the indicated technique at that point. The microscale technique has been adapted for many experiments, not only for costs related to the purchase and disposal of chemicals, but also for the convenience of time and for the more evident qualitative observations that are made. Wherever appropriate, a line drawing or a photograph illustrates the proper apparatus required for the experiment.

All of the experiments are considered "**safe**". Emphasis on *laboratory safety* is evident on nearly every page: a **Caution** icon identifies a part of the Procedure wherever warning is made or a potentially harmful chemical is used. Within the context of the Procedure, a caution statement clearly identifies the nature of the caution or the consequence of improperly handling the chemical. At the end of each laboratory procedure, **Disposal Information** is given for proper disposal of laboratory chemicals. The scale-down of many experiments (microscale) has reduced the volumes of chemicals and wastes, thus cutting purchasing and disposal costs. Ultimately, however, you and the students have the joint responsibility of maintaining a safe laboratory environment. Do not underestimate the value of a safe laboratory!

Lists of the common and special *laboratory equipment* required for the use of this manual are at the front of the manual. Balances with ±0.001 g sensitivity, visible spectrophotometers, and a slide projector are needed on occasion. A **special note** to adopters of the manual: the spectral slides for Experiment 12 are available from John Wiley & Sons, Publishers, Inc. upon request.

Each experiment requires *basic chemicals and apparatus*. An **Instructor's Manual** is available upon request from the publisher. An effective teaching schedule is outlined for each experiment, including: a suggested lecture outline, cautions, representative or expected data, common student questions about the experiment, answers to the Lab Preview and Data Sheet Questions, the chemicals and special equipment that are required, a detailed preparation of all reagent solutions, and the safety rules.

Several experiments were modified from those appearing elsewhere. The author thanks the following for their permission to include them, in principle, here.

James E. Brady, St. John's University, N.Y., suggestions appear throughout the manual
John R. Amend, Montana State University, Experiment 12
Jerry Mills, Texas Tech University, Experiment 22
John Holum, Augsburg College, Experiment 40

The valuable suggestions provided by the following reviewers were greatly appreciated:

Joan Reeder, Eastern Kentucky University
Robert Kowerski, College of San Mateo

Special recognition is given to Joan Kalkut, Associate Editor with Wiley, for her guidance, patience, valuable suggestions, and personal perceptions of what a laboratory manual should be about during the course of the project, and the general chemistry students, laboratory instructors, and my colleagues at Texas A&I University.

The author invites corrections and suggestions for improvements of this manual from colleagues and students.

J. A. Beran
Box 161, Department of Chemistry
Texas A&I University
Kingsville, Texas 78363

TABLE OF CONTENTS

LABORATORY TECHNIQUES

Glassware and Chemicals

Liquid Reagents and Solutions

Solid Reagents

Gases

Handling Data

EXPERIMENTS

Laboratory Skills and Analysis

Chemicals and Chemical Reactions

Atoms, Molecules and Their Structures

Gases

Solutions (Acid, Base, and Redox Systems)

Equilibria (Principles and Analysis)

Electrochemistry

Kinetics

A Summary of Inorganic Chemical Principles

Organic Chemistry and Biochemistry

APPENDICES

COMMON LABORATORY DESK EQUIPMENT

No.	Quantity	Size	Item	First Term		Second Term		Third Term	
				In	Out	In	Out	In	Out
1	1	10 mL	graduated cylinder						
2	1	50 mL	graduated cylinder						
3	5	—	beakers						
4	2	—	stirring rods						
5	1	500 mL	wash bottle						
6	1	75 mm, 60°	funnel						
7	1	125 mL	Erlenmeyer flask						
8	1	250 mL	Erlenmeyer flask						
9	2	25 x 200 mm	test tubes						
10	6	18 x 150 mm	test tubes						
11	8	10 x 75 mm	test tubes						
12	1	large	test tube rack						
13	1	small	test tube rack						
14	1	—	glass plate						
15	1	—	wire gauze						
16	1	—	crucible tongs						
17	1	—	spatula						
18	2	—	litmus, red and blue						
19	2	90 mm	watch glasses						
20	1	75 mm	evaporating dish						
21	4	—	medicine droppers						
22	1	—	test tube holder						
23	1	large	test tube brush						
24	1	small	test tube brush						

COMMON LABORATORY
DESK EQUIPMENT

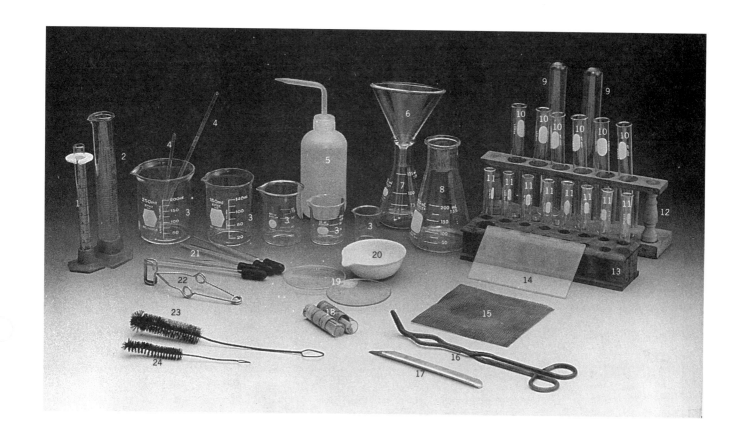

SPECIAL LABORATORY EQUIPMENT

Number	Item	Number	Item
1	reagent bottles	16	crucible and cover
2	condenser	17	mortar and pestle
3	500 mL Erlenmeyer flask	18	glass bottle
4	1000 mL beaker	19	pipets
5	Petri dish	20	ring stands
6	Büchner funnel	21	buret clamp
7	Büchner flask	22	double buret clamp
8	volumetric flasks	23	Bunsen burner
9	500 mL Florence flask	24	buret brush
10	110°C thermometer	25	clay triangle
11	100 mL graduated cylinder	26	rubber stoppers
12	50 mL buret	27	wire loop for flame test
13	glass tubing	28	pneumatic trough
14	U-tube	29	rubber pipet bulb
15	porous cup	30	iron ring

SPECIAL LABORATORY EQUIPMENT

LABORATORY TECHNIQUES

GLASSWARE AND CHEMICALS

1. Inserting Glass Tubing Through a Rubber Stopper

Firepolish: to round the edges of a sharp cut section of glass with heat.

The procedure for inserting glass tubing through a rubber stopper, when performed incorrectly, causes more serious injuries than any other single operation in the general chemistry laboratory. Please read this technique *CAREFULLY* when working with glass tubing.

Using water or glycerol (glycerol works best) moisten the glass tubing, that has been previously firepolished, *and* the hole in the stopper. Place your hand on the tubing no more than 2 to 3 cm (1 inch) from the stopper. Protect your hands with a towel (Figure T.1). Simultaneously *twist* and *push, slowly* and *carefully*, the tubing through the hole. There *never* should be more than 3 cm of glass tubing between your hand and the stopper. Wash off any excess glycerol.

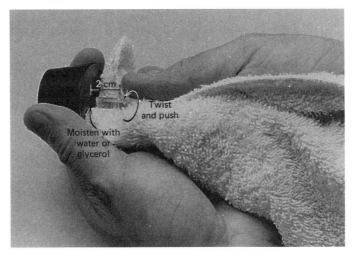

Figure T.1

Inserting glass tubing into a rubber stopper

2. Clean Glassware and Lab Bench

Cleanliness is very important to a chemist. Data can be misinterpreted if an analysis is conducted in contaminated glassware. Therefore, it is most important that clean glassware is maintained. The best time to clean glassware is *immediately* after its use (analogous to the best time to wash dirty dishes in the kitchen).

a.   Clean all glassware with soap and warm tap water. Rinse first with *tap* water and then once or twice with *small* amounts of distilled (or deionized) water. Never use distilled or deionized water for washing glassware; it is too expensive.

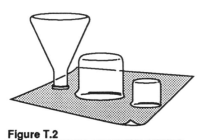

b.   Invert clean glassware such as beakers and flasks on a paper towel to dry (Figure T.2); do not wipe or air-blow dry because of possible contamination. Do not dry calibrated or heavy glassware (graduated cylinders, volumetric flasks, or bottles) in an oven or over a direct flame. If the glassware is cleaned as described, but is needed for immediate use, rinse the glassware with small amounts of the solvent or solution being used in the experiment and discard.

Figure T.2

Invert glassware on a paper towel

c.   At the end of the laboratory period, wipe clean the surface of the la bench near your work station clean with a damp towel. Also attend to the area where the laboratory chemicals are placed and the hood, balance, and sink areas where you worked.

3. Microscale Analyses

Microscale: a term used to identify relative (very small) amounts of chemicals used for an experiment.

Many of the Procedures in these experiments require a microscale technique for analysis. In this technique, small volumes of solutions, generally less than 2 mL are needed. A special apparatus called a **microcell plate** having either 24, 48, or 96 microcells (most often 24 in this manual) has been designed for these experiments. Each microcell has a volume of about 3.5 mL for the 24-cell plate, but only $1/2$ mL for the 96-cell plate. Additionally small quantities of test reagent are added to the microcells with a small plastic (throw-away) pipet, called a Beral pipet. The Beral pipet is designed to add small diameter drops of test reagent to the microcell (Figure T.3).

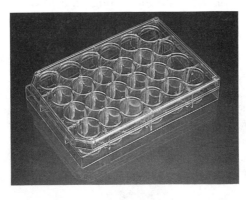

Figure T.3
24-cell tray and Beral pipet used for the microscale analyses

4. Handling Chemicals

Chemicals are very safe if handled and disposed of properly. When handling an unfamiliar chemical, ask your instructor for the precautions in handling the chemical and and how you are to properly dispose of it.

a.   **Contact with chemicals.** Avoid direct contact with all laboratory chemicals. Avoid breathing chemical vapors. Never taste or smell a chemical unless specifically directed to do so. Any interaction with a chemical may cause a skin, eye, or mucous irritation. In other words, **play it safe!** The international caution sign is used throughout this manual to cite where extra care should be taken in handling a chemical.

b.   **Transferring a chemical (solid or solution).** Read the label twice and transfer only the amount needed. The use of the wrong chemical (or wrong concentration of chemical) can lead to an "unexplainable" accident or result (and an argument with your laboratory instructor). *Never* use more reagent than the procedure calls for; *do not return the excess* to the reagent bottle—share it with a friend.

c.   **Mixing chemicals in solutions.** Add a reagent slowly; never dump it in. While stirring, slowly pour the *more* concentrated solution into water or into a less concentrated solution, never the reverse! This is especially true when diluting concentrated (conc) sulfuric acid, H_2SO_4, with water.

d.   **Disposal of chemicals.** Each experiment will give you guidelines for the disposal of test chemicals—**Disposal Information**—having the format as follows:

Disposal Information:	Dispose of the waste chemical(s) in the...(for example)
	•sink: nonflammable, nontoxic, water-soluble liquids followed by large amounts of water • waste container (properly labeled): water-insoluble liquids, solids, and toxic wastes • waste basket: paper products (such as litmus paper, filter paper, and matches) and broken glassware •covered containers (properly labeled): volatile liquids or very reactive chemicals

LIQUID REAGENTS AND SOLUTIONS

5. Transferring Liquids

When the reagent is being transferred from a reagent bottle, remove the glass stopper and hold it between the fingers of the hand used to grasp the reagent bottle (Figure T.5a). *Never* lay the stopper from a reagent bottle on a laboratory bench; impurities may be picked up and thus contaminate the solution.

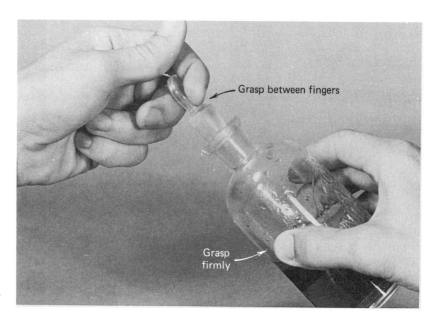

Grasp between fingers

Grasp firmly

Figure T.5a
Grasp the stopper

To transfer a liquid from one vessel to another, hold a stirring rod against the lip of the vessel containing the liquid and pour the liquid down the rod which is touching the inner wall of the receiving vessel (Figures T.5b and c). This avoids any splashing of the liquid in the vessel and any loss of reagent down the side of the reagent bottle. *Never* transfer more liquid than is required for the experiment. *Never* return unused chemicals to the reagent bottle.

Grasp stopper
Touch stirring rod to lip of bottle
Pour down stirring rod
Touch to side of beaker

Figure T.5b
Pour the liquid with the aid of a stirring rod

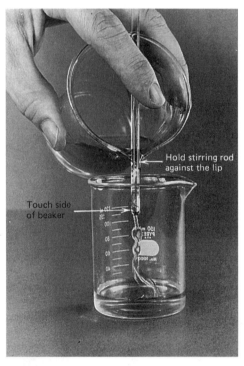
Hold stirring rod against the lip
Touch side of beaker

Figure T.5c
Transfer a liquid with the aid of a stirring rod

6. Heating Liquids

Bumping: the sudden formation of superheated vapor at the bottom of a test tube, flask, or beaker near the point at which the flame is applied.

a. **Test tube.** The test tube should be no more than one-third full. Place a gentle flame at the same level as the top of the liquid, not at the base (Figure T.6a). Move the test tube in and out of the flame while swirling the contents. Never point the test tube toward anyone; the contents may be ejected violently if the test tube is not properly heated.

b. **Erlenmeyer flask.** An Erlenmeyer flask may be heated directly over a flame. Hold it with a piece of tightly folded paper or tongs and gently swirl (Figure T.6b). *Do not* place the hot flask on the laboratory bench; allow to cool by setting the flask on a wire gauze.

c. **Beaker (or flask).** Support the beaker (or flask) on a wire gauze. Place a support ring around the upper part of the beaker. Place a glass stirring rod or boiling chip in the beaker; this minimizes the problem of bumping. Position the flame directly under the tip of the stirring rod(Figure T.6c).

d. **A hot water bath.** A small quantity of solution in a test tube that needs to be heated to a constant temperature can be placed in a hot water bath (Figure T.6d). If the solution is in a beaker or Erlenmeyer flask instead of a test tube, place it in a beaker (next available size), one-fourth filled with water, and heat to the desired temperature.

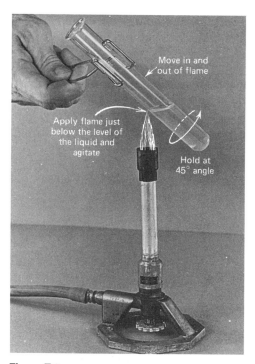

Figure T.6a
Heating a liquid in a test tube with a burner—apply the flame at the *top* of the liquid

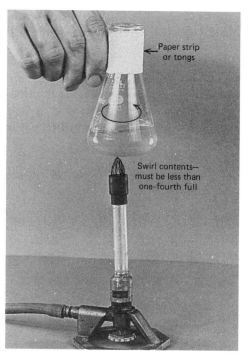

Figure T.6b
Heating a liquid in a flask directly over the flame while *swirling*

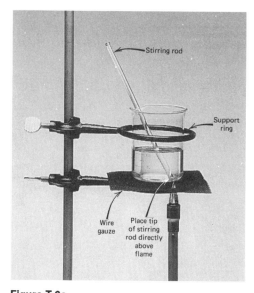

Figure T.6c
Place the flame directly beneath the *tip* of the stirring rod in the beaker

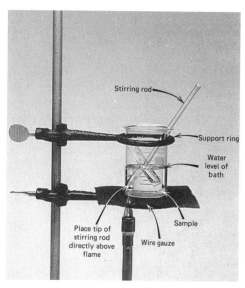

Figure T.6d
A hot water bath

7. Evaporation of Liquids

a. **Nonflammable liquids** may be evaporated from an evaporating dish with a *gentle* direct flame (Figure T.7a) or over a steam bath (Figure T.7b). Gentle boiling is more efficient that rapid boiling.

b. **Flammable liquids** may be similarly evaporated from an evaporating dish using a heating mantle (Figure T.7c) The use of a fume hood (Technique 17c) is suggested if large amounts are evaporated in a laboratory with inadequate ventilation; consult with your laboratory instructor.

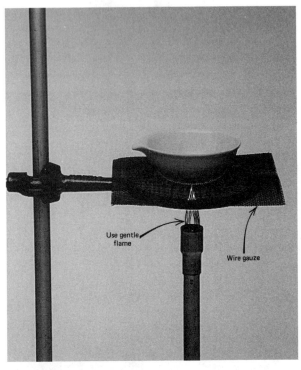

Figure T.7a
Evaporation of a *non*flammable liquid over a low, direct flame

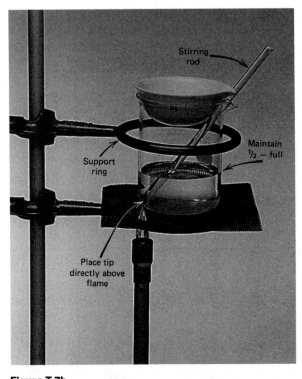

Figure T.7b
Evaporation of *non*flammable liquid over a steam bath

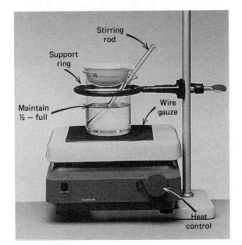

Figure T.7c
Evaporation of a *flammable* liquid over a steam bath

8. Reading a Meniscus

For exacting measurements of clear or transparent liquids in graduated cylinders, pipets, burets, and volumetric flasks, the volume of a solution is read at the *bottom* of the meniscus. Steady the eye horizontal to the surface of the liquid (Figure T.8a); position (horizontally) the top edge of a black mark (made on a white card) just below the level portion of the liquid. The reflection of the black background off the bottom of the meniscus better defines the lowest mark of the liquid (Figure T.8b). Substituting a finger for the black mark on the white card is not as effective, but it does help.

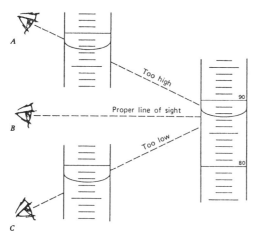

Figure T.8a
Reading the meniscus with the eye *horizontal* to the *bottom* of the meniscus

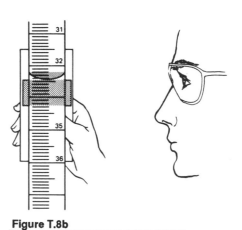

Figure T.8b
Reading the meniscus using a black mark on a white card

9. Pipetting a Liquid or Solution

a. **Pipet preparation.** Clean the pipet with a soap solution; rinse with several portions of tap water and then with distilled (or deionized) water. No water droplets should adhere to the inner wall of the pipet. Transfer the solution that you intend to pipet from the reagent bottle to the beaker. Do *not* insert the pipet tip directly into the reagent bottle. Dry the pipet tip with a clean, dust-free towel or tissue (e.g., a Kimwipe). Using suction from a collapsed rubber (pipet) bulb (**Caution:** *Never use your mouth*), draw several 2–3 mL portions into the pipet as a rinse. Roll each rinse around in the pipet to make certain that the solution washes the entire surface of the inner wall. Deliver each rinse to a waste beaker.

b. **Filling and operating the pipet.** Do not insert your own pipet, medicine dropper, or spatula into reagent bottles. Transfer the reagents as shown in Technique 5. To fill the pipet, place the tip well below the surface of the solution. Then using the collapsed rubber (pipet) bulb, draw the solution into the pipet until the level is 2–3 cm above the "mark" (Figure T.9a). Remove the bulb and *quickly* cover the top of the pipet with your *forefinger*, *not* your thumb. If you are right handed, operate the rubber bulb with your left hand and use the forefinger of your right hand to cover the pipet.

c. **Pipet delivery.** Remove the tip from the solution, dry the tip with a dust-free towel, and holding in a vertical position over a waste beaker, control the delivery of the excess solution from the pipet with the forefinger until the bottom of the meniscus is at the mark. Practice!! (Figure T.9b). See Technique 8 for reading the meniscus. Remove any drops suspended from the pipet tip by touching it to the wall of the waste beaker. Again wipe off the tip with a clean, dust-free towel or tissue, and deliver the solution to the receiving vessel (Figure T.9c); keep the tip above the level of the liquid and against the wall of the receiving vessel. Do *not* blow or shake out the last bit of solution that remains in the tip; this liquid has been included in the calibration of the pipet. See Experiment 1 for a discussion and procedure for the calibration of pipets.

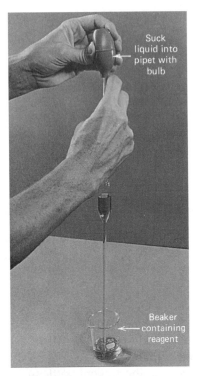

Figure T.9a
Drawing the solution into the pipet with a rubber bulb

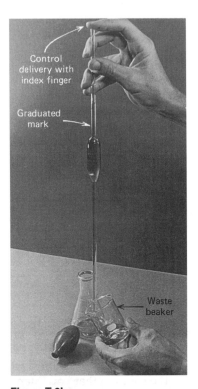

Figure T.9b
Controlling the delivery with the *forefinger*

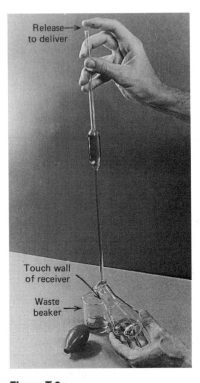

Figure T.9c
Delivering the solution with the pipet touching the side of the receiving flask

10. Titrating a Solution

a. **Cleaning the buret.** Clean the buret with a soap solution. If a buret brush is used, prevent the wire handle from scratching the wall. Rinse the soap solution from the buret several times *through the stopcock*, first with tap water and then with distilled (or deionized) water. *No drops should adhere to the inner wall of a clean buret.* Close the stopcock. Rinse the buret with several 3–5 mL portions of titrant. Tilt and roll so that the rinse comes into contact with the entire inner surface. Drain each rinse through the buret tip into a waste beaker.

b. **Filling and operating the buret.** Support the buret with a buret clamp (Figure T.10a). Close the stopcock and, with the aid of a funnel, fill the buret to just above the zero mark. The solution in the buret is called the **titrant.** Open the stopcock briefly to remove any air bubbles in the tip. Allow 30 seconds for the titrant to drain from the wall; record the volume (±0.02 mL). Note that the graduations increase in value from top to bottom on the buret. Operate the stopcock with your left hand (if right-handed, Figure T.10b) or right hand (if left-handed, Figure T.10c) during the titration to prevent the stopcock from sliding out of its barrel. Fill the buret after each titration.

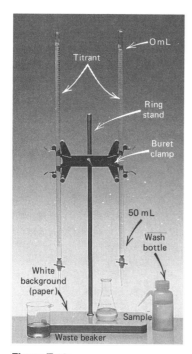

Figure T.10a
A titration apparatus

Titrant: the solution contained in the buret and added to the analyte.

Analyte: the substance being analyzed which is usually in the receiving flask in the titration procedure.

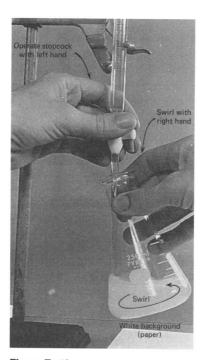

Figure T.10b
**Titration technique for *right–*
handed chemists**

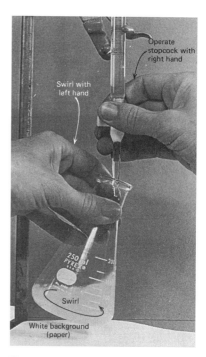

Figure T.10c
**Titration technique for *left–*
handed chemists**

c.   **Transferring the titrant to the receiving vessel.** Place a piece of white paper beneath the receiving vessel, generally an Erlenmeyer flask, so that the appearance of the endpoint (the point at which the indicator turns color) is more visual. If the endpoint results in the appearance of a light or white color, a black background is preferred.

During the titration, operate the stopcock with your left hand (if right-handed) and constantly swirl the flask with the right hand. The buret tip should extend 2–3 cm inside the mouth of the receiving vessel. Periodically stop adding the titrant to wash the wall of the flask (Figure T.10d). Near the endpoint (slower color fade, Figure T.10e), slow the rate of titrant addition until a single drop makes the color change persist for 30 seconds. **STOP**, allow 30 seconds for the titrant to drain from the wall, read and record the volume (±0.01 mL).

To add less than one drop of titrant, suspend the drop on the buret tip and wash it into the flask with distilled water.

d.   **Clean up.** After completing your titrations, drain the titrant from the buret, rinse with several portions of distilled water, and drain through the tip. Store the buret as directed by your instructor.

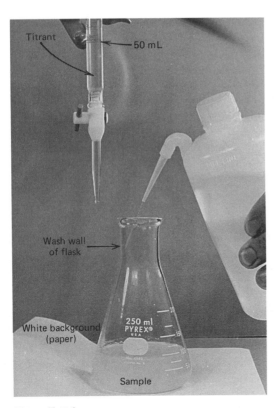

Figure T.10d
Wash the wall of the receiving flask *frequently* **during the titration**

Figure T.10e
Note the slow color fade near the endpoint in the titration

11. Testing with Litmus

To test the acidity/basicity of a solution with litmus paper, insert a stirring rod into the solution, withdraw it, and touch it to red or blue litmus (Figure T.11). Acidic solutions turn blue litmus red and basic solutions turn red litmus blue. *Never* place the litmus paper directly into the solution.

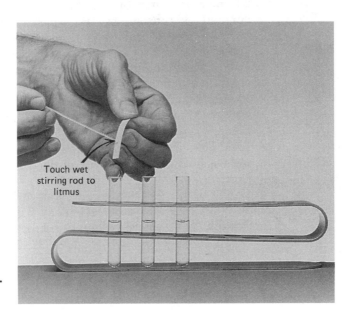

Figure T.11
Litmus test for pH. Touch wet stirring rod to litmus paper

SOLID REAGENTS

12. Transferring Solid Reagents

First, read the label *twice* on the bottle to ensure the use of the correct reagent. Place the lid (glass stopper or screw cap) of the reagent bottle top-side-down. Hold the bottle with the label against your hand, tilt, and roll back and forth (Figure T.12) until the desired amount has been dispensed; try not to dispense more reagent than is needed. If too much reagent is removed, do *not* return the excess, but rather, share it with a friend. Do *not* insert a spatula or other object into the bottle to remove or to return reagent unless you are specifically told to do so. When finished return the lid to the reagent bottle.

Figure T.12

Dispensing a solid from a reagent bottle

13. Using the Laboratory Balance

A balance is used to measure the mass of a chemical or a small piece of laboratory equipment. Some various types and models of laboratory balances are shown. Select the balance that provides the sensitivity listed in the Procedure of the experiment.

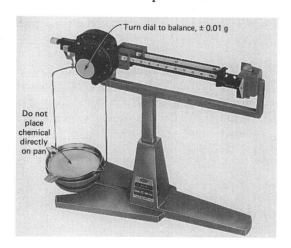

Figure T.13a

Triple-beam balance, sensitivity, ±0.01 g

Balance/Sensitivity
- triple-beam balance (Figure T.13a)/±0.01 g
- top-loader balance (Figures T.13b)/±0.01 g and/or ±0.001 g
- analytical balance (Figure T.13c)/±0.0001 g

Figure T.13b
Top-loader balance, sensitivity, ±0.01 g and/or ±0.001 g

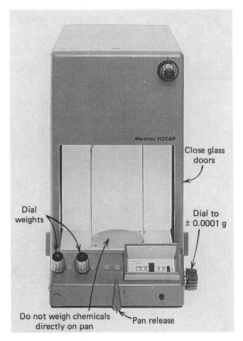

Figure T.13c
Analytical balance, sensitivity, ±0.0001 g

In using and caring for a balance, follow these guidelines:
- Handle with care; balances are expensive.
- Level the balance; see the instructor for assistance.
- Use a beaker, weighing paper, watch glass, or some other container to measure the mass of laboratory chemicals. Do *not* place laboratory chemicals directly on the pan.
- Do not drop anything on the pan.
- Do not attempt to be a handyman. If the balance is not operating properly, see your instructor.
- After the measurement, return the balance to the zero reading.
- Clean up any spillage of chemicals on the balance or in the balance area.

14. Separation of a Liquid from a Solid

a. **Decanting a liquid from a solid.** Allow the solid to settle in the beaker (Figure T.14a) or test tube. Then transfer the liquid, called the **supernatant**, with the aid of a stirring rod (Figure T.14b) as described in Technique 5. Do this slowly so as not to disturb the solid that has settled.

Figure T.14a
Settling of a precipitate

Spout

Stirring rod or object
of equal thickness

Precipitate
to settle
below spout
of beaker

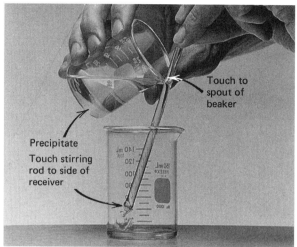

Figure T.14b
**Transferring the supernatant
liquid from a precipitate**

Touch to
spout of
beaker

Precipitate

Touch stirring
rod to side of
receiver

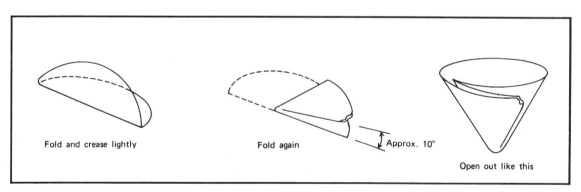

Fold and crease lightly

Fold again

Approx. 10°

Open out like this

Figure T.14c
**Sequence of folding filter paper
for a gravity—filtering funnel**

b. **Gravity filtration.** Fold the filter paper in half (Figure T.14c); refold to within about 10° of a 90° fold; tear off the corner unequally; open. Place the folded filter paper snugly into the funnel. Steady the funnel with a support clamp. Moisten the folded filter paper with the solvent being filtered and press the filter paper against the funnel's top wall to form a seal. Filter as shown in Figure T.14d. The funnel's tip should touch the beaker wall to reduce splashing. *Never* fill the funnel more than two-thirds full. Keep the funnel's stem filled with **filtrate** (the liquid passing through the funnel); the filtrate's weight creates a slight suction at the base of the barrel of the funnel which hastens the filtration process.

Apex: the point of the folded filter paper when placed into the funnel.

c. **Vacuum filtration.** Figure T.14e shows a typical setup for a vacuum filtration. Although a gravity funnel/filter apparatus, described in 14b, can be used, the apex of the filter paper is easily ruptured when the vacuum is applied. A Büchner funnel is normally used; a disc of filter paper fits over the flat, perforated bottom of this funnel. Applying a light suction to the filter paper, moistened with the solvent, creates a seal. When applying a vacuum with a water aspirator, open fully the faucet so that a maximum suction can be applied.

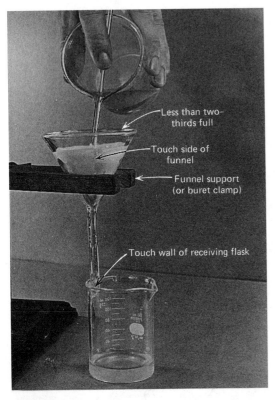

Figure T.14d
Gravity filtering apparatus

Figure T.14e
Vacuum filtering apparatus

d. **Centrifugation (use of a centrifuge).** Precipitates that form in solution can be made more compact at the bottom of a test tube (or centrifuge tube) by using a centrifuge (Figure T.14f). The supernatant is then easily decanted without any loss of the precipitate (Figure T.14g). This quick and efficient separation requires 20–40 seconds.

Observe these precautions while operating a centrifuge:

• Never fill the centrifuge tubes to a height more than 1 cm from the top.
• Label the test (centrifuge) tubes to avoid confusion.
• Always operate with an even number of test (centrifuge) tubes, containing equal volumes of liquid, placed opposite one another; this *balances* the centrifuge and eliminates excessive vibration and wear. If only one tube needs to be centrifuged, balance it with a tube containing the same volume of solvent (Figure T.14h)

Figure T.14f
Centrifuge

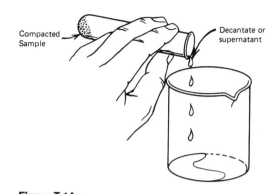

Figure T.14g
Decanting the supernatant liquid
from a compacted precipitate in
the bottom of a centrifuge tube

Figure T.14h
Balance the centrifuge. Place
tubes with equal volumes of
solution opposite each other
inside the rotor's metal sleeves.

Figure T.15
Flushing a precipitate from a
beaker

15. Flushing a Compacted Precipitate from a Beaker or Test Tube

Flush the precipitate with a wash bottle while holding the beaker or test tube over the receiving container (Figure T.15).

16. Ignition of a Crucible

Clay triangle: a porcelain covered wire triangle used to heat porcelain labware to high temperatures over a direct flame. The porcelain diffuses the heat to minimize hot spots.

a. Drying and/or firing the crucible. Support the crucible on a clay triangle (Figure T.16a) and heat the crucible in a hot flame until it glows red. Rotate the crucible with tongs to ensure complete ignition. Allow the crucible and lid to cool to room temperature. Use crucible tongs for any transfer of the crucible and lid. If the crucible still remains dirty, add 2 mL of 6 M HNO₃ (**Caution:** *avoid skin contact*) and evaporate to dryness.

b. Ignition of contents in the absence of air. Set the crucible upright in the clay triangle; adjust the lid slightly off the lip of the crucible (Figure T.16b). Use the tongs for adjustment.

c. Ignition of contents for complete combustion. Tilt the crucible and adjust the lid so that only about half of the crucible is covered (Figure T.16c).

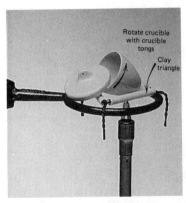

Figure T.16a
Drying and/or firing of a crucible and cover

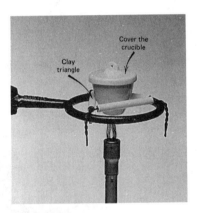

Figure T.16b
Ignition of a crucible's content *without* **air**

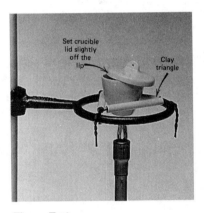

Figure T.16c
Ignition of a crucible's contents for *complete* **combustion**

GASES

17. Handling Gases

a. Testing for odor. An educated nose is an important and a very useful asset in the chemistry laboratory. Use it with caution, however, because some vapors are toxic. Fan some of the vapor toward your nose (Figure T.17a); never hold your nose directly over the vessel.

b. Collecting gases

i. By air displacement. Gases more dense that air may be collected by using the experimental setup shown in Figure T.17b. Gases less dense than air are collected using the apparatus in Figure T.17c.

ii. By water displacement. Gases that are relatively insoluble in water are collected using the apparatus shown in Figure T.17d.

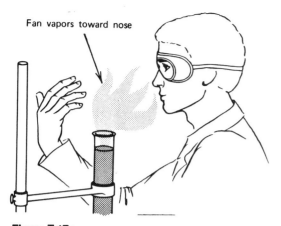

Figure T.17a
Testing odors. Fan the vapor gently toward the nose

Fan vapors toward nose

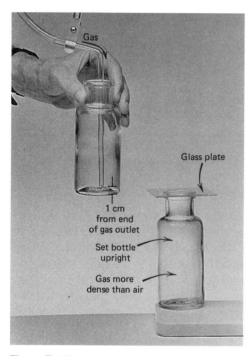

Gas

Glass plate

1 cm from end of gas outlet

Set bottle upright

Gas more dense than air

Figure T.17b
Collection of gases *more* dense than air

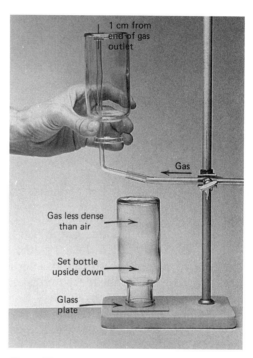

1 cm from end of gas outlet

Gas

Gas less dense than air

Set bottle upside down

Glass plate

Figure T.17c
Collection of gases *less* dense than air

c. Removal of irritating or toxic gases

i. *Laboratory fume hood.* When fume hood space is available, use it; do not substitute the improvised hood.

ii. *Improvised hood.* If space in the fume hood is inadequate to remove *small* quantities of the toxic or nauseating vapors, use the improvised hood (Figures T.17e). Use this method *only* with advice of the instructor.

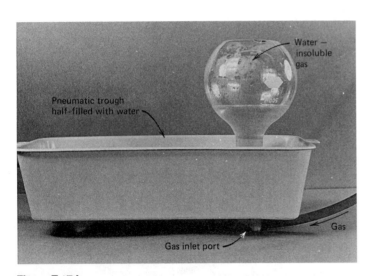

Figure T.17d
Collection of a *water-insoluble* gas by displacement

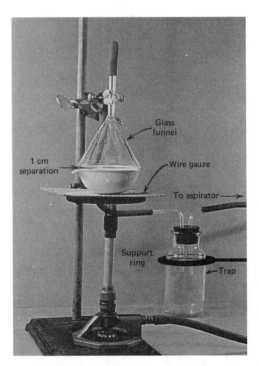

Figure T.17e
The use of an improvised hood over an evaporating dish

HANDLING DATA

18. Graphing Techniques

The most common graph used in the chemical laboratory is the line graph. The line graph has two lines drawn perpendicular to each other; these are called the axes: the x-axis (called the **abscissa**) is horizontal and the y-axis (the **ordinate**) is vertical. Each axis is scaled according to the units and range of the measurements; the scale on each axis need not be the same units or the same divisions—they only need to be consistent with the measurements. Each division or subdivision on the graph must have a constant integral value along each axis.

To illustrate the proper construction of a line graph, let's plot the following data for helium gas:

Volume, mL	20	24	30	35	37	41	46
Temperature, K	100	120	150	175	185	205	230

a. **Draw and label the axes.** Generally the dependent variable (that which you measure) is plotted along the y-axis and the independent variable (that which you control in the experiment) is plotted along the x-axis. For example, if we control the temperature of the system and observe the new volume, we study volume (V) as a function of temperature (T), or V *vs* T, and plot volume along the y-axis and temperature along the x-axis. The label for each axis should identify the parameter that is being plotted and its units (Figure T.18a).

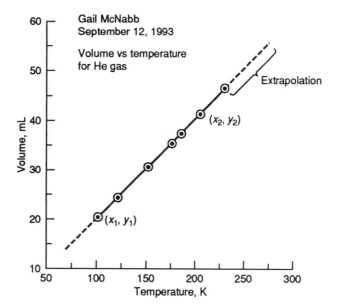

Figure T.18a
A line graph of volume (V) vs temperature (T) with labeled and scaled axes; a smooth curve is drawn through the data points; the data is also extrapolated for further interpretation

Extrapolation: to extend the experimental relationship beyond the collected data points of the graph.

b.   **Select scales so that the range of data points fill, or nearly fill, the entire space allotted for the graph.** The subdivisions of the scale should be easy to interpret. The intersection of the two axes does not have to be the zero point for each. In our example, the volume units measure from 20 to 46 mL; let's select an ordinate scale from 10 to 60 mL with 5 mL subdivisions. The temperature units range from 100 to 230 K; let's select an abscissa scale from 50 to 250 K with 25 K subdivisions. This selection of ranges allows us to not only fill the space allotted for the graph with data but also allows us to easily interpret the value of each subdivision along each axis. Also our scale selection allows for some extrapolation of units beyond our experimental range of data. (See Figure T.18a)

c.   **Place each data point at the appropriate place on the graph.** Draw a circle around each point. Ideally, the size of the circle should approximate the error in the measurement.

d.   **Draw the best smooth line through the data points.** The line does not have to pass through all of the points or, for that matter, any of the points—it must merely represent the best averaging of all the data points. Notice that the line passes "undrawn" through the circled data points. The extrapolation of data (the dashed lines) extends the data for additional interpretations.

e.   **Place a title on the graph well away from the plot.** If possible, the title, along with your name and date, should be placed in the upper portion of the graph.

f.   **For straight-line graphs,** the drawn line is represented algebraically by the equation

$$y = mx + b$$

m is the slope and b is the intersection of the line along the y-axis at $x = 0$. The slope is the ratio of the change in the ordinate (y-axis) values to that of the abscissa (x-axis) values, $\Delta y / \Delta x$ (Figure T.18b).

$$\text{slope, } m = \frac{(y_2 - y_1)}{(x_2 - x_1)} = \frac{\Delta y}{\Delta x}$$

In our example, the slope can be calculated from

$$x_2 = 200 \text{ K, } y_2 = 40 \text{ mL; } x_1 = 100 \text{ K, } y_1 = 20 \text{ mL}$$

$$m = \frac{(40 - 20)\text{mL}}{(200 - 100)\text{K}} = 0.20 \text{ mL/K}$$

We can conclude that the volume changes 0.20 mL for each 1 K.

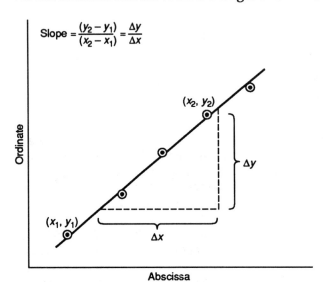

Figure T.18b
Determination of the slope of a straight line

EXPERIMENT 1

SAFETY AND TECHNIQUES

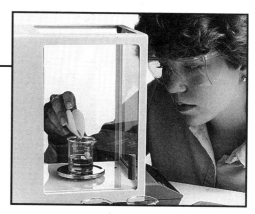

The laboratory is the source of all science. It is in the laboratory where new discoveries are sought; progress toward cures of diseases such as cancer, sickle cell anemia, and AIDS is slow and meticulous, but only through this kind of careful and reproducible experimentation or *research* can we ever expect to eliminate such devastating ailments in our society. The vaccine for polio, a very debilitating ailment prior to the mid-1950s, was discovered in the laboratory. Other plagues to modern society, such as anthrax and small pox, have been virtually eliminated through research.

High-purity silicon for semiconductors, synthetic fibers (such as nylon, acrilan, and polyester), and prescription drugs for heart disease, the common cold, and birth control were developed in the laboratory environment. These examples are only a small sampling of the effect that chemistry has had on modern society. You are about to begin a very exciting discipline where chemistry (and science in general) has produced materials that have changed the course of history more than once. New, yet to be discovered breakthroughs, will make it happen again. Will you be ready?

OBJECTIVES

- To check-in to the laboratory
- To learn various safety rules and procedures
- To identify and locate common laboratory and safety equipment
- To properly select and operate a balance
- To calibrate laboratory glassware

PRINCIPLES

The chemical principles that you learn in lecture are conclusions that are based upon careful laboratory experimentation. Years of observation, experimentation, interpretation, and prediction are needed before a "principle" finds its way into the textbook.

The techniques, procedures, and equipment used in this laboratory assist in providing a clearer understanding and interpretation of those principles. No textbook or lecturer can substitute for the observations that are made, the data that are collected and analyzed, and the interpretations that must be reached. The understanding of these chemical principles can help us to interpret the real world of science and the science of everyday living.

You will use laboratory equipment in designing experiments aimed at gathering reliable data. As you work, record the data and use your knowledge of chemical principles to explain what is seen. A good scientist is a thinking scientist. Cultivate self-reliance and confidence in your work, even it it doesn't look right; this is how scientific breakthroughs occur.

In the first few laboratory periods, you are introduced to basic rules, tools, and techniques of chemistry and situations requiring their use. The Techniques section (pages 1–20) illustrates many laboratory techniques that will make your laboratory experience more meaningful. Other techniques are introduced as the course advances.

Volumetric Glassware

Solute: the substance dissolved in the solvent.

The amount of chemical that is needed for a procedure in the laboratory is often measured and transferred as the solute of a solution. The volumes of the solutions are measured with glassware that has been previously calibrated by the supplier. Since the glassware is "mass produced" at the factory, occasionally the graduations on the glassware are not as exact as what the supplier has quoted. A scientist who strives to do very quantitative work in the laboratory always calibrates the glassware before it is used. In this experiment, you will calibrate test tubes and a pipet, but the procedure can be used to calibrate any glassware.

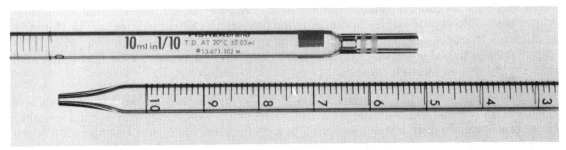

Figure 1.1
Pipet calibrated to deliver "TD"

Commercial pipets and some graduated cylinders are labeled as TD 20°C (Figure 1.1) or TC 20°C. TD 20°C (to deliver at 20°C) means that the volume of the pipet is calibrated according to the volume it delivers from gravity flow only at 20°C. The liquid that is retained in the pipet tip is considered in the calibration of the pipet and therefore is *not* to be blown out. Pipets labeled TD 20°C are used exclusively in this course. Glassware that is labeled TC 20°C (to contain, or rinse out, at 20°C) means that all of the liquid in the container is considered in the calibration and therefore must be removed, generally with a rinse. Therefore a TC 20°C pipet is designed to contain the exact volume of the liquid whereas a TD 20°C pipet contains the calibrated volume plus the volume of liquid in the tip.

Density

Intensive property: a property of a substance that is independent of sample size.

Density is an intensive property of all pure substances. Its measurement is expressed as a ratio of mass to volume; substances having a large density have a large mass occupying a small volume. Water has a density of 1.0 g/cm^3, or 62.4 lb/ft^3; osmium metal (atomic number 76) has the greatest mass of any of the elements, 22.6 g/cm^3. In this experiment the density of a liquid and/or solid is determined.

$Density = \frac{mass}{volume}$

The density of a liquid is determined by measuring the mass of a known volume of the liquid; the density of a water-insoluble solid is determined by measuring its mass and then the volume of water it displaces.

TECHNIQUES

- Technique 2, page 1   Clean Glassware and Lab Bench
- Technique 8, page 7   Reading a Meniscus
- Technique 9, page 7   Pipetting a Liquid or Solution
- Technique 13, page 11   Using the Laboratory Balance

PROCEDURE

A. Laboratory Check-In

In this first laboratory period you are assigned a drawer or locker containing laboratory equipment. Place the equipment on the laboratory bench and, with the check-in list (page ix); check off your equipment. If you are unsure of the names of the items, refer to the list of chemical "kitchenware" on page viii. See your instructor about any item that is not in the drawer or locker.

A good scientist is always neat and well-organized; keep your equipment clean and neatly arranged so that it is ready for instant use in an experiment. Always have a dishwashing detergent available for cleaning glassware and paper towels on which to set cleaned items for drying.

B. Laboratory Safety

You and your laboratory partners must *always* practice laboratory safety. It is your responsibility, not the instructor's, to *play it safe*. On the inside front cover of this book, there are spaces to list the location of important safety equipment, telephone numbers, and reference information. Fill this out now.

Study each of the following safety rules and laboratory guidelines to answer the questions on the Lab Preview.

1. Self-Protection

a. *Safety glasses, goggles, or eye shields must be worn* at all times to guard against the laboratory accidents of others as well as your own. Contact lenses should not be worn; if they are, additional eye protection is absolutely necessary.

b. Laboratory aprons or coats (with snap fasteners only) should be worn to protect clothing. Above all, wear old clothing. The wearing of slacks (or jeans) and skirts that extend below the knee is permitted. Shorts and short skirts are not permitted.

c. Only wear shoes that shed liquids. Sandals or canvas shoes are not permitted.

d. Confine long hair. Hair will ignite near an open flame!

e. Wash your hands and arms before leaving the laboratory. Toxic chemicals may be transferred to the mouth.

f. Whenever your skin (hands, arms, face,...) comes into contact with laboratory chemicals, wash it quickly and thoroughly with soap and warm water. Use the eye-wash fountain to flush chemicals from the eyes or face. Do *not* rub the affected area, especially the face or eyes, with your hands before washing.

g. If a chemical is spilled over a large part of the body, use the safety shower and flood the affected area for 5 minutes. Remove all contaminated clothing. Get medical attention; you are not a physician (not yet anyway!).

h. Report any accident or injury to your instructor, even if it seems apparently minor.

i. Learn how to use the fire extinguisher and other safety devices.

2. Laboratory Rules

a. Do *not* work alone in the laboratory. At least the instructor must be present.

b. Unauthorized experiments, including variations of those in the laboratory manual, are forbidden.

c. *Prepare* for the lab. Complete the Lab Preview and study the Objectives, Principles, Techniques, and Procedure for the experiment *before* lab. Always try to understand what you are doing and to *think* (whistle if you like) while working. Note beforehand the need for any extra equipment that is to be checked out from the stockroom.

d. No smoking, drinking, eating, or chewing is permitted. Chemicals may possibly enter through the mouth or lungs.

e. Horseplay or other careless acts are prohibited. Do not entertain guests. Do *not* waste time.

f. Maintain an orderly, clean lab bench, drawer, and work area, including the balance area. Clean up all chemical and water spills with a damp towel (and discard); discard paper scraps in marked containers; and return the extra glassware to the drawer (Be sure that it is clean!). Clean the sinks of all debris. Keep drawers closed while working and the aisles free of any obstructions, such as book bags and athletic equipment.

g. Be aware of your neighbors' activities; you may be the victim of their mistakes. Advise them of improper techniques or unsafe practices. If necessary, report your concern to the instructor.

h. Carefully complete your Data Sheet for each experiment. Believe in your data. A scientist's most priceless possession is integrity. If the results "do not look right", repeat the experiment; don't alter the data to make it look better—be a scientist!

Record data *in ink* as you perform the experiment. Data on scraps of paper will be confiscated. Where calculations are required, be orderly for the first set of data analysis. Do not clutter the Data Sheet with arithmetic details.

i. *Think* while you perform the experiment. Questions at the end of the Data Sheet are designed to test your understanding of the experiment and the principles involved. Discussions with other scientists (your lecturer, laboratory instructor, or other students (although the latter source may be less reliable)), your text, or various reference books (such as the Chemical Rubber Company's *Handbook of Chemistry and Physics*) are often reliable sources of information.

C. The Laboratory Balance

1. Several types of balances may be found in the chemistry laboratory. The triple-beam balance and the top-loading balance are the most common. Read **Technique 13** carefully. As suggested on the Data Sheet, determine the mass of several objects. Use the top-loader and the analytical balances only *after* the instructor explains their operation.

D. Volumetric Glassware

1. **Commercial calibration of graduated cylinders.** Look closely at the 10 mL and 50 mL graduated cylinders. How accurately can each be read? Record this data on the Data Sheet as, for example, ±0.5 mL, ±0.1 mL, ±0.05 mL, or ±0.01 mL.

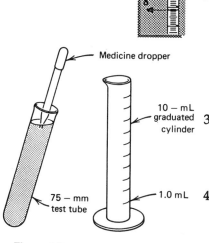

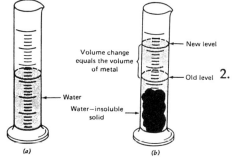

Figure 1.2
Determination of the number of drops in 1.0 mL of water

2. **Volumes of common test tubes.**
   a. Fill a clean, dry 150 mm test tube with water. Transfer the water to a 50 mL graduated cylinder, read and record the volume of water as accurately as the calibration allows (Technique 8).

   b. Repeat with a clean, dry 75 mm test tube. Determine its volume in the same way.

3. **Drops in 1.0 mL of water.** With a clean, dry medicine dropper (or Beral pipet) transfer 1.0 mL of water to a clean, dry 10 mL graduated cylinder (Figure 1.2). Count and record the number of drops. What is the average volume per drop?

4. **Calibration of a pipet.**
   a. Obtain a pipet from the stockroom with any volume between 5 and 25 mL. Look at its calibration. How accurately can its volume be recorded? If there is no indication of accuracy, it is called a Class B pipet having an accuracy range from ±0.012 to ±0.06 mL. If the pipet is labeled "A", for Class A pipet, the accuracy range is ±0.006 to ±0.03 mL.

   b. Determine the mass (±0.001 g) of a clean, dry 100 mL beaker. Fill the pipet to the etched mark with water (Technique 8). Wipe the tip dry with a lint-free tissue, and transfer the water (the tip should be touching the beaker wall) to the beaker (read Technique 9b carefully). Again determine the mass of the beaker and water on the same balance. Assuming the density of water to be 0.998203 g/mL at 20°C, calculate the delivered volume of the pipet. Calculate the percent error in the calibration of the pipet.

E. Density

Ask your instructor to which balance you are assigned for Part E. Write the balance number on the Data Sheet.

1. **Density of a liquid.**
   a. Determine the mass (±0.001 g) of the smallest beaker (with a volume ≥ 25 mL) in your drawer. Use the pipet that was calibrated in Part D.4. Rinse the pipet several times with your unknown sample and discard each rinse.

   b. Fill the pipet to the mark. Deliver the liquid to the beaker. Measure the combined mass of the liquid and beaker on the same balance. Calculate the density of the liquid.

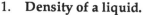

2. **Density of a solid.**
   a. Determine the mass of a sheet of weighing paper. Place the solid on the weighing paper and measure the combined masses.

   b. Half-fill a 50 mL graduated cylinder with water and record its volume to the nearest 0.02 mL. See Figure 1.3.

Figure 1.3
Apparatus for measuring the density of a solid

   c. Carefully transfer the solid to the graduated cylinder. Roll the solid around in the water to remove any trapped air bubbles that adhere to the solid. Record the new volume of water in the graduated cylinder. Calculate the density of the solid.

Disposal Information. Check with your instructor for the proper disposal of the liquid and solid samples.

Notes, Observations, and Calculations

SAFETY AND TECHNIQUES

Date _____ Name _____ Lab Sec. _____ Desk No. _____

1. True or False

    _____ a.   Paper towels are to be used for wiping dry your clean glassware.

    _____ b.   Sandals are *not* to be worn in the laboratory.

    _____ c.   Eye protection is worn only while you are conducting an experiment.

    _____ d.   Slacks, jeans, or skirts that extend below the knee are considered appropriate attire in the laboratory.

    _____ e.   You are encouraged to conduct experiments beyond those presented in the laboratory manual.

    _____ f.   Record data in pencil so that any errors in measurement can be corrected without presenting a messy Data Sheet.

    _____ g.   Do not discuss data, calculations, or the interpretations of the data with others.

    _____ h.   Eating is permitted in the laboratory *after* lunch.

    _____ i.   To determine the safety of a chemical, read the label.

    _____ j.   Do *not* invite friends into the laboratory to show them your work.

    _____ k.   A glass surface is considered clean when, after a rinse with deionized water, no water droplets cling to the surface.

    _____ l.   One safety rule that must always be followed in the laboratory is to wear flammable clothing.

2. Diagram the cross section of a graduated cylinder, illustrating how a meniscus is read (Technique 8).

3. a. Explain how to fill a pipet to the mark (Technique 9).

    b. How is the delivery rate of the liquid from the pipet controlled?

4. What does the label "TD 20°C" mean on a pipet?

5. A Beral pipet delivers 153 drops of alcohol for each milliliter. Calculate the volume, in mL, of each drop.

6. A 4 g sample of zinc is measured on a triple-beam balance. Which mass should recorded (Technique 13)?
    a. 4.0 g
    b. 3.99 g
    c. 3.9941 g
    d. 3.994 g

7. How should the mass in Question 6 be recorded if an analytical balance is used?

8. If 50.0 mL of an alcohol has a mass of 40.3 g, what is its density?

9. The density of diamond is 3.51 g/cm^3 and the density of lead is 11.3 g/cm^3. If, according to the Procedure, Part E.2, equal masses of diamond and lead were transferred to equal volumes of water in separate graduated cylinders, which graduated cylinder would have the highest volume reading? Explain.

SAFETY AND TECHNIQUES

Date _____Name _____ Lab Sec. _____Desk No. _____

A. Laboratory Check-In

Instructor's approval_____

B. Laboratory Safety

Complete the information on the inside front cover of the manual.

Instructor's approval_____

C. The Laboratory Balance

1. Measure the mass of the following objects on the triple-beam balance. Express your answer to the nearest 0.01 g.

    a. 250 mL beaker (g)   _____

    b. 150 mm test tube (g)   _____

    c. a piece of glass tubing (g) _____

    d. a quarter (g)   _____

    e. a key (g)   _____

2. Measure the mass of the following objects on the top-loading balance. Express your answer according to the sensitivity of the balance.

    a. a dime (g)   _____

    b. a dollar bill (g)   _____

    c. an empty 10 mL graduated cylinder(g)   _____

    d. a 10 mL graduated cylinder
       containing 6 mL of water (g) _____

    e. 150 mm test tube(g)   _____

D. Volumetric Glassware

1. Commercial calibration of graduated cylinders

    a. Accuracy of 10 mL graduated cylinder (mL)   _____

    b. Accuracy of 50 mL graduated cylinder (mL)   _____

2. Volumes of common test tubes

    a. Volume of 150 mm test tube (mL)   _____

    b. Volume of 75 mm test tube (mL)   _____

3. Drops in 1.0 mL of water

    a. Drops in 1.0 mL water   _____

    b. Volume of each drop (*mL*)   _____

4. Calibration of pipet

    a. Accuracy of pipet (*mL*)   _____

    b. Mass of 100 mL beaker (*g*)   _____

    c. Mass of beaker + water delivered from pipet (*g*)   _____

    d. Mass of water delivered (*g*)   _____

    e. Calculated volume of pipet (*mL*)   _____

    f. Percent error in calibration (%)*   _____

$$*\% \text{ error in calibration} = \left[\frac{\text{absolute difference between calculated and labeled volume}}{\text{calculated volume}} \right] \times 100$$

E. Density

Number of assigned balance   _____

1. Density of a liquid

    a. Corrected volume of pipet, Part D.4 (*mL*)   _____

    b. Mass of beaker (*g*)   _____

    c. Mass of beaker + liquid delivered from pipet (*g*)   _____

    d. Mass of liquid delivered (*g*)   _____

    e. Density of liquid (*g/mL*)   _____

2. Density of a solid

    a. Mass of weighing paper (*g*)   _____

    b. Mass of weighing paper + solid (*g*)   _____

    c. Volume of water in graduated cylinder (*mL*)   _____

    d. Volume of water after solid is transferred (*mL*)   _____

    e. Density of solid (*g/mL*)   _____

1. Water drops will adhere to the inner wall of a dirty pipet. Does a dirty pipet deliver more or less that its calibrated volume? Explain.

2. A TC 20°C pipet was assumed to be a TD 20°C pipet. Will a larger or smaller volume be delivered? Explain.

3. Suppose that in calibrating the volume of a pipet, sufficient time is not allowed for the liquid to drain from the pipet. Does this cause a higher or lower calculated volume for the pipet? Explain.

4. Suppose that in calibrating the volume of the pipet, the temperature of the water is not 20°C, but instead is 25°C. How does this affect the calibrated volume of the pipet? The density of water decreases with increasing temperature.

5. If, in the determination of the density of the solid, air bubbles adhered to its surface. Will the reported density of the solid be reported as high or low? Explain.

CHEMISTRY OF COMBUSTION

Fire is a fascinating phenomenon. The burning of wood has been known throughout recorded history, and its fascination has continued to intrigue each generation since then. Nearly everyone has experienced the consequences of "playing with fire" even though no disastrous event may have occurred as a result. Fire was recognized as being so important to the development of science that it was designated as one of the four basic "elements," the others being earth, air, and water.

The rapid burning of wood (or any fossil fuel) is called combustion, but slower types of combustion also occur, such as the "combustion" of iron, although we commonly call that corrosion or rusting, and the "combustion" of foods, such as sugars in our bodies, although we call that spoilage (or oxidation).

OBJECTIVES

- To determine the various temperature regions of a flame
- To determine the combustion products of a candle and a Bunsen flame

PRINCIPLES

Fuel: the substance that burns.

Oxidizer: the substance that supports combustion.

Luminesce: to give off a glowing light.

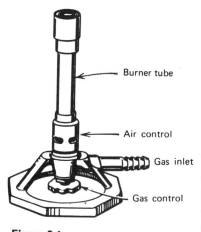

Burner tube

Air control

Gas inlet

Gas control

Figure 2.1
Bunsen-type burner

This experiment characterizes a flame. A flame requires a fuel, often the wood of a match, the wax of a candle, hydrogen, or natural gas, and an oxidizer, most commonly the oxygen in air although solid chemicals that function as oxidizers are common to the laboratory. We will study the flame of a candle and that of a laboratory Bunsen burner.

The candle provides a very inefficient, low temperature flame, with the candle wax serving as the fuel. When there is an insufficient supply of oxygen, small particles of carbon are produced which, when heated to incandescence, produce a yellow, luminous flame. The characteristics of the flame can be changed by the presence of air currents, the shape of the wick, and the candle wax.

Laboratory burners come in many shapes and sizes, but all accomplish one main purpose: to provide a combustible gas-air mixture that produces a high temperature flame. Because Robert Bunsen (1811-1899) was the first to construct this type of burner, his name is commonly given to all burners of this type (Figure 2.1).

The natural gas used in most laboratories has methane, CH_4, as its major component. With a proper mixture of methane and oxygen, a blue, *non*luminous flame yields carbon dioxide and water at temperatures approaching $1500°C$ in certain regions of the flame. Small concentrations of carbon monoxide are also produced. With an insufficient supply of oxygen, small particles of carbon form from the unburned fuel, which then luminesce resulting in a yellow flame.

TECHNIQUES

- Technique 17a, page 16   Handling Gases

PROCEDURE

A. The Candle

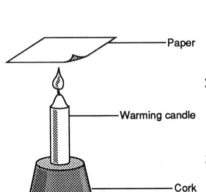

Figure 2.2
A candle flame tested for heat zones

Refer to the Data Sheet as you progress through the Procedure to record and/or account for your observations.

1. a. Mount a warming candle on a large cork. Light the candle. Study the flame closely: Why is the entire flame not the same color? Account for the color at the base of the flame being different from that at the top of the flame.

   b. Position a 3″ x 5″ white index card horizontally at the half-height position of the flame until the card is scorched (see Figure 2.2). **Caution**: *do not let the card catch fire!*

   c. Now hold the index card vertically in the flame until the card is scorched. Record and account for these observations. Characterize the horizontal and vertical profiles of the flame.

2. Place a slotted piece of aluminum foil around the wick just below the base of the flame, but above the melted wax. Leave the foil there for about 30 seconds. How does the shape and appearance of the flame change?

3. With a lighted match in hand, blow out the flame of the candle; hold the lighted match 2–3 cm from the wick in the column's "smoke". Account for your observation.

4. a. Place an inverted 800 mL beaker down over the candle flame. What substance collects on the wall of the beaker? Determine the time required to extinguish the flame.

   b. Repeat with a 250 mL beaker. Why is the flame extinguished?

5. a. Exhale for 30 s into a wide-mouth 1000 mL Erlenmeyer flask; invert the flask over the candle flame. Determine the time to extinguish the flame.

   b. Repeat the procedure with a wide-mouth 250 mL Erlenmeyer flask.

6. a. Set the 250 mL Erlenmeyer flask from Part 5b upright and add 25 mL of limewater; stopper the flask, and swirl. What happens?

   b. Place 25 mL of limewater in a clean 250 mL Erlenmeyer flask. Use a soda straw to exhale into the solution.

   c. Use the straw to *gently* exhale into the candle flame.

7. a. Half-fill a pie plate (or any flat dish) with limewater. Float the lighted candle (on the cork) on the limewater. Cover the candle with an 800 mL beaker.

   b. Note changes of the water level in the beaker and note the *appearance* of the water.

   c. Repeat substituting tap water for the limewater. Account for the differences in Parts 7b and 7c.

8. a. Place a spatula-full sample of baking soda into a 250 mL Erlenmeyer flask. Add 10 mL of vinegar. Smell the gas.

b. Hold a lighted match with tweezers and insert (do not drop) it into the mouth of the flask.

B. The Bunsen Burner

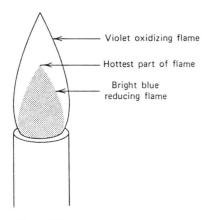

Violet oxidizing flame

Hottest part of flame

Bright blue reducing flame

Figure 2.3
Flame of a properly adjusted Bunsen burner

1. **Lighting the burner.**
a. Attach the burner's tubing to the gas outlet on the lab bench. Turn off the gas control at the base of the burner and fully turn on the gas valve at the outlet. Close the air holes at its base and open the gas control slightly.

b. Bring a lighted match or striker up the outside of the burner tube until the escaping gas at the top ignites. After it is lighted, adjust the gas control *at the base of the burner* until the flame is pale blue and has two or more distinct cones.

c. Opening the air control valve produces a slight hissing sound characteristic of a burner's hottest flame. The addition of too much air may blow out the flame. When the best adjustment is reached, three distinct cones are visible (Figure 2.3) and the flame should "sing". Obtain an instructor's approval of your well-adjusted flame.

2. **Flame temperatures using a wire gauze**
a. Test the temperature in different flame zones by holding with crucible tongs a wire gauze horizontally about 1 cm above the burner (Figure 2.4). Note the color and appearance of the gauze. Now move it up through the flame until it no longer glows.

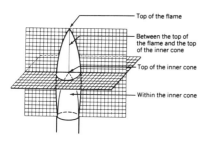

Top of the flame

Between the top of the flame and the top of the inner cone

Top of the inner cone

Within the inner cone

Figure 2.4
Regions of temperature measurement in a Bunsen flame

b. Now position the wire gauze vertically in the flame. This shows a vertical profile of the temperature regions of the flame.

c. Briefly place an 800 mL beaker over the flame.

d. Close the holes of the *air* control and repeat the Part B.2a,b,c with a luminous flame.

3. **Flame temperatures using the melting points of metals**
The melting points of iron, copper, and aluminum are 1535°C, 1083°C, and 660°C respectively. Hold the metals, with crucible tongs in the four zones of a nonluminous flame shown in Figure 2.5. Estimate the temperatures at each zone.

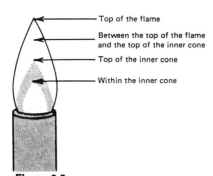

Top of the flame

Between the top of the flame and the top of the inner cone

Top of the inner cone

Within the inner cone

Figure 2.5
Regions of the flame for temperature measurements

Notes and Observations

CHEMISTRY OF COMBUSTION

Date _____ Name _____ Lab Sec. _____ Desk No. _____

1. What is the dominant color of the nonluminous flame from a Bunsen burner? Explain.

2. a. What is the fuel for the burning of a candle?

    b. Why are candle flames yellow?

3. Arrange the following steps in the proper sequence when lighting a Bunsen burner:
    a. Open the gas control valve slightly
    b. Turn on the gas outlet valve
    c. Close the air control valve
    d. Close the gas control valve
    e. Open the air control valve
    f. Connect the burner's tubing to the gas outlet
    g. Light the escaping gas
    h. Open the gas control valve

    _____ , _____, _____, _____, _____ , _____ , _____

4. Consider the fire in a fire place. Assume that the wood in the following situations is not a factor in discussing the observation.

    a. Why does the fire burn yellow rather than blue?

    b. The flame of burning embers is nearly colorless, but the embers are yellow. Can you explain why?

    c. Is the flame of a rapidly burning fire hotter or cooler than the flame of a slow burning fire?

5. A fire in the "pits" at the Indianapolis 500 motor speedway is especially dangerous because the flame is nearly colorless and nonluminous, unlike that of a gasoline fire. How might the fuel used in Indianapolis 500 race cars differ from that of gasoline?

CHEMISTRY OF COMBUSTION

Date _____ Name _____ Lab Sec. _____ Desk No. _____

A. The Candle

1. a. Briefly account for the various color regions of the candle flame.

    b. At right, sketch the burn pattern
    for the horizontal cross section of
    the flame and the vertical cross
    section of the flame.

    Horizontal   Vertical

    c. Write a short description of the temperature characteristics of the flame, based on the above tests.

2. Account for the observations resulting from the placement of the aluminum foil.

3. Explain the effect of the smoke in your observation.

4. a. time (800 mL beaker) _____

    b. time (250 mL beaker) _____
    Account for the time differential and explain why the flame was extinguished.

    What substance collects on the wall of the beaker? _____

5. a. time (1000 mL flask) _____

    b. time (250 mL flask) _____
    Account for the time differences from those in Part 4.

6. a. What does limewater test indicate about the combustion products?

    b. How does exhaled air relate to the combustion reaction for the candle?

    c. What analogy exists between the air exhaled through the straw and Part 5 of the Procedure?

7. a. Does the water level rise or lower? Why?

    b. What is the reaction occurring in the water? Where have you seen a similar reaction in this experiment?

    c. What is observed in this procedure that is different from the observations in Part 7b? Account for these observations. In your discussion consider the following facts:
    • the candle does *not* combine with *all* of the O_2 in the air;
    • the formation of CO_2 form O_2 does *not* produce a volume change;

• the change in the water level is *not* as great as in Part 7.

8. What happened when the match was immersed in the gas?

    Because of the observations in Part 8b, what can you conclude about the nature of the gas? Cite evidence from earlier in the experiment to substantiate your prediction.

B. The Bunsen Burner

1. Instructor's approval of a well-adjusted Bunsen flame. _____

2. a and b. Use your observations from the horizontal and vertical wire gauze tests to sketch the profile of a nonluminous Bunsen flame; label the "cool" and "hot" regions.

                                                              Flame Regions

    c. What appears on the inner surface of the beaker? Is this consistent with the observations in Part A? Explain.

    d. What differences in the flame temperatures are observed between the nonluminous and luminous flames. Use supporting data to substantiate your conclusions.

3. At right, sketch Figure 2.5 for a nonluminous flame and indicate the approximate temperature at each of the four zones.

1. a. What are the products of combustion?

    b. Describe the physical properties of the product(s).

    c. Describe the chemical properties of the product(s).

2. Why is it possible to extinguish the flame of a candle or match by *very gently* blowing on it?

3. What substance is "burning" and what substance is supporting the combustion for the candle and for the Bunsen burner?

4. The candle flame is typically yellow whereas the properly adjusted Bunsen flame is blue. Account for these differences.

EXPERIMENT 3

KITCHEN CHEMISTRY

The very thought that chemistry is only done in the laboratory or in that huge chemical plant down the street is absurd. Chemistry and chemical reactions occur in and around us every minute of every day. . . we breathe, we work, we write, we run, we study; all of these processes require the use of chemicals and the completion of many, and often very complex, chemical reactions. In our everyday living environment the kitchen is a source of many, relatively simple, yet common, chemicals; often more "laboratory" chemicals exist in the kitchen than what exists in many chemistry laboratories.

Many of the cleaning agents, pesticides, and fertilizers that we buy for the home can be very hazardous if not handled or stored properly according to the directions on the label. Daily reports in the newspapers or on television and admissions to emergency rooms at hospitals are constant reminders that not only children, but adults as well, can have accidents that result from the careless use or storage of these chemicals. The poison control center in any state or municipality is inundated with daily phone calls questioning the danger to a child or adult that has been afflicted. For examples, children have permanent scars as a result of investigating the can of Drano stored under the kitchen sink and homes have been lost from fire due to the storage of fertilizers near gasoline. The message here is "chemicals in the home or the laboratory are safe, if they are used and stored properly, i.e., according to the labels on the product."

TECHNIQUES

- Technique 3, page 2   Microscale Analyses
- Technique 5, page 3   Transferring Liquids
- Technique 6c, page 4   Heating Liquids
- Technique 11, page 10   Testing with Litmus

PRINCIPLES

The chemical principles that are used to account for many of the observations in this experiment are explained in more detail later in the general chemistry laboratory. Chemical reactions that are called acid-base, oxidation-reduction, precipitation, and complex formation are encountered in this experiment; but, for now, consider yourself a scientist and account for your observations as best possible. Later in the course you may want to again review this experiment to realize a more complete description of your observations.

In this experiment we will observe a number of chemical reactions that can be done in the laboratory of your own kitchen; for convenience and for safety reasons, we have brought many of these chemicals to the laboratory for you to observe their chemical properties.

Qualitative experiment: the detection of a particular substance in a mixture and the determination of its characteristic properties.

Quantitative experiment: the measurement of "how much" of a particular substance is present in a mixture.

Additional qualitative experiments with household chemicals are completed in Experiments 17 and 27. The quantitative analysis of vitamin C (Exp't 20), household bleach (Exp't 21), vinegar (Exp't 29), common antacids (Exp't 30), and the chemistry of hard water (Exp't 18) and an aluminum can (Exp't 22) are also included in this laboratory manual.

PROCEDURE

While many of the chemicals in this experiment are from the kitchen, they should be handled with care and the cautions should be closely followed. The same care should be taken when handling these chemicals in the home.

A. Reactions of Bleach

24-cell plate: chemical apparatus used to contain about 3.5 mL of solution per cell.

1. a. Place several drops of red food coloring into a microcell of a 24-cell plate. Add drops of liquid bleach to the colored water until the color disappears. Account for the reaction that you observe.

   b. Add several additional drops of liquid bleach to the water. Now add 1 or 2 drops of red food coloring to the water and swirl. Account for your observation.

   c. Repeat the experiment, substituting dry bleach for the liquid bleach.

B. Reactions of Iodine

1. a. Place two thumb tacks in a 50 mL beaker. Just cover the tacks with a tincture of iodine solution. Set the mixture aside and continue with the rest of the experiment, but occasionally observe any changes that are occurring.

   b. After the color of the iodine has nearly disappeared add drops of bleach. Account for the chemical change.

2. Place a small drop of tincture of iodine solution on a white paper towel and spray the wetted portion with spray starch. What has happened? How might you account for the observation?

3. Place a small drop of tincture of iodine solution on a piece of white bread, on a cracker, and on a cube of sugar. What conclusions can you make? How might you determine whether the salt on the cracker or on a salted peanut is regular or iodized salt?

Pinch: a pinch approximates the volume of a grain of rice.

Fresh Fruit™: a commercially available, concentrated powder of ascorbic acid, used to keep canned fruit looking fresh.

4. a. Place a small drop of tincture of iodine solution in a microcell of a 24-cell plate. Add several drops of concentrated lemon juice or a pinch of Fresh Fruit. Record your observation.

   b. Place a small drop of tincture of iodine solution on a piece of white bread in a microcell. Again add several drops of concentrated lemon juice or a pinch of Fresh Fruit™. What happened? Observe the mixture during the remaining time in the laboratory. Account for your observation.

5. a. Half-fill two microcells in a 24-cell plate with iodized table salt. To only one microcell add one drop of starch (use the liquid from the can of spray starch) to the salt. At the wetted edge of the starch in the salt add 2–3 drops of 3% hydrogen peroxide (household antiseptic). Account for the color change.[1]

   b. To the other microcell add only 2–3 drops of 3% hydrogen peroxide (no starch). The color change is faint, a repeat of the test may be necessary. After the appearance of the iodine is evident, add several

---

[1]A similar reaction of iodized salt occurs with liquid bleach although it is not as evident.

drops of concentrated lemon juice or Fresh Fruit. How does this reaction relate to previous observations?

C. Reactions of Baking Soda

1. Set up a 1 x 5 array of microcells in a 24-cell plate. Half-fill each microcell successively with water, vinegar, orange juice, (reconstituted) lemon juice, and household ammonia. Place a pinch of baking soda into each microcell and observe closely. Account for your observation.

2. Number a 1 x 4 array of microcells. Place a pinch of baking powder in microcells 1 and 2 and baking soda in microcells 3 and 4. Using a Beral pipet slowly half-fill microcells 1 and 3 with water and microcells 2 and 4 with vinegar. Closely observe what happens in each microcell and account for the chemical change.

D. Preparation and Testing of Acid/Base Indicators

A number of acid-base indicators are prepared from common household products. The acidity/basicity of a number of household products will be tested in Part E with these indicators. Consult with your instructor for determining which indicators you are to prepare and use in the experiment.

1. Prepare a solution of red cabbage and/or tannin.

   a. **Red Cabbage**. Place a leaf of red cabbage in 150 mL of hot, distilled (or deionized) water in a 250 mL beaker. Allow the juice to leach from the cabbage until the water is cool. Decant about 50 mL of the solution into a small Erlenmeyer flask.

   b. **Tannin**. Place 4–6 teabags in a 250 mL beaker containing 100 mL of hot, distilled (or deionized) water until a "strong" tea solution results. Discard the tea bags.

2. Prepare a solution of turmeric and/or grape juice.

   a. **Turmeric**. Place a full spatula of tumeric in 50 mL of isopropyl (rubbing) alcohol. The solution becomes a dark yellow.

   b. **Grape Juice**. Use as is.

3. **Ex-Lax** (or any laxative that contains "yellow phenolphthalein"). Crush an Ex-Lax tablet and add 50 mL of isopropyl (rubbing) alcohol. Swirl and dilute to one cup with water.

4. **Litmus**. Use the litmus paper.

5. Set up three 1 x 6 arrays of microcells in the 24-cell plate. Half-fill the first 1 x 6 array with water, followed by 1 x 6 arrays of 0.1 M HCl and 0.1 M NaOH. Place a Beral pipet in each prepared indicator solution. Add several drops of the respective indicator to the water, 0.1 M HCl and the 0.1 M NaOH. Record your observation on the Data Sheet.

E. Testing the pH of Various Household "Chemicals"

1. Set up six 1 x 4 arrays of microcells in the 24-cell plate. Select 4 indicators and 6 test solutions as listed on the Data Sheet. Solid test chemicals need to be dissolved in a small portion of water (for tablets you will have to crush and grind the sample).

2. Half-fill the first 1 x 4 array of microcells with water and then successive 1 x 4 arrays with the remaining five test solutions.

3. Test each array with several drops of the four selected indicators. Determine whether the test solution is acidic or basic and arrange, if possible, the test solutions in order of increasing acidity.

4. Discard and rinse the 24-cell plate and test same six solutions other indicators or six additional solutions with the same indicators. Ask your instructor for advice.

F. Reactions of Acids and Bases

1. Prepare *four* sets of the following solutions, each in a 1 x 6 array of microcells in a 24-cell plate.

Microcell No.	1	2	3	4	5	6
Drops of vinegar	10	8	6	4	2	0
Drops of sudsy ammonia	0	2	4	6	8	10

2. Select four different indicators. Add 2 drops of a selected indicator to one set of solutions. Repeat with the remaining three indicators. Discuss your observations for each indicator.

EXPERIMENT 3

KITCHEN CHEMISTRY

Date _____ Name _____ Lab Sec. _____ Desk No. _____

1. Prior to the beginning of this experiment, read the label to determine a (major) component in five of the following kitchen chemicals:

    a. liquid laundry bleach   _____

    b. sudsy ammonia   _____

    c. an antacid (e.g., milk of magnesia)   _____

    d. laundry soap   _____

    e. shampoo   _____

    f. a carbonated soft drink   _____

    e. a tub and tile (bathroom) cleaner   _____

2. From your home list five items, exclusive of those listed in Question 1, that you consider "chemicals".

    _____

    _____

    _____

    _____

    _____

3. List five common substances used in today's experiment that can serve as acid-base indicators.

4. The chemistry of the following "kitchen chemicals" is reviewed in this experiment. Write the formula for the chemical that is being tested in each formulation.

    a. bleach _____

    b. tincture of iodine _____

    c. baking soda _____

    d. vinegar_____

    e. lemon juice _____

5. Red cabbage juice undergoes a number of color changes over a wide range of pH, similar to what is observed for "universal indicator." Use your text to determine the colors at various pH's for red cabbage or for universal indicator.

EXPERIMENT 3

KITCHEN CHEMISTRY

Date _____ Name _____ Lab Sec. _____ Desk No. _____

A. Reactions of Bleach

	Observation	Explanation
1. a. food coloring/ liquid bleach		
b. liquid bleach/food coloring		
c. food coloring/dry bleach		

What happens when bleach is applied to a stain on clothes?

B. Reactions of Iodine

	Observation	Explanation
1. a. tacks/tincture of iodine		
b. addition of bleach		
2. tincture of iodine/starch		
3. tincture of iodine/bread		
tincture of iodine/cracker		
tincture of iodine/sugar		
4. a. tincture of iodine/lemon juice		
b. tincture of iodine/bread/lemon juice (initial)		
tincture of iodine/bread/lemon juice (final)		
5. a. iodized salt/starch/hydrogen peroxide		
b. iodized salt/hydrogen peroxide		
iodized salt/hydrogen peroxide/lemon juice		

Procedure for determining the kind of salt on a peanut/cracker.

C. Reactions of Baking Soda

	water	vinegar	orange juice	lemon juice	ammonia
1. baking soda					

	#1 baking powder	#2 baking powder	#3 baking soda	#4 baking soda
2. water		xxxx		xxxx
vinegar	xxxx		xxxx	

D. Preparation and Testing of Acid/Base Indicators

Fill in the table with the observed color from the test, using only those indicators that you have prepared.

	Water	Acid (0.1 M HCl)	Base (0.1 M NaOH)
Red cabbage			
Tannin			
Tumeric			
Grape juice			
Ex-Lax			
Litmus			

E. Testing the pH of Various Household "Chemicals"

Complete the following table with only those chemicals that you are testing, labeling each observation as acid or base. A degree of acidity or basicity should also be indicated.

Indicator (Names)			
Water			
Vinegar			
Sudsy ammonia			
Tums*			
Rolaids*			

Soapy water			
Aspirin			
Lemon juice			
Vanish*			
Glass cleaner			
Vitamin C			
Cream of tartar			
Oven cleaner			
Baking powder			
Baking soda			
Washing soda			
Fresh Fruit*			
Drain-O*			
Alum			
Orange juice			
Other 1			
Other 2			
Other 3			

*Product Names

F. Reactions of Acids and Bases

Account for the general change in appearance in the four sets of solutions.

Specifically describe the change in appearance for

a. Indicator name (_____) _____

b. Indicator name (_____) _____

c. Indicator name (_____) _____

d. Indicator name (_____) _____

EXPERIMENT 4

WATER ANALYSIS: SOLIDS

When you are thirsty and need a drink of water, do you take for granted that the tap water is safe to drink? Do you think about what might be dissolved or suspended in the water? Is your criterion for safe water, "if it tastes good, then it must be safe?" The tap water of public drinking water supplies must meet certain federal health standards; for example, the water at the tap must have a chlorine concentration of one part per million to protect against harmful bacteria. The raw water from surface reservoirs and underground aquifers, that are used for public drinking water supplies, contain dissolved salts, mostly soluble carbonate and bicarbonate salts.

Surface water is used as the primary drinking water source for most large municipalities. The water is piped into a water treatment facility where impurities are removed and bacteria are killed before the water is placed into the distribution lines. The contents of the surface water must be known and predictable so that the treatment facility can properly and adequately remove these impurities. Tests are used to determine the contents of the surface water.

The U.S. Public Health Service recommends that drinking water not exceed 500 mg total solids/kg water. However, in some localities, the total solids content may range up to 1000 mg/kg of potable water; that's 1 g/L!! An amount over 500 mg/kg water does not mean the water is unfit for drinking; an excess of 500 mg/kg is merely not recommended.

OBJECTIVE

• To determine the total, dissolved, and suspended solids in a water sample.

PRINCIPLES

Suspended solids: solids that exhibit colloidal properties or solids that remain in the water because of turbulence.

Watershed: the land area over which the natural drainage of water occurs.

The total, dissolved, and suspended solids in a water sample are determined in this experiment. The water sample may be from the ocean, a lake, a stream, or from an underground aquifer.

Water in the environment has a large number of impurities with an extensive range of concentrations. **Dissolved solids** are water soluble substances, usually salts. Naturally-occurring dissolved solids generally result from the movement of water over or through mineral deposits, such as limestone. These dissolved solids, characteristic of the watershed, generally consist of the sodium, calcium, magnesium, and potassium cations and the chloride, sulfate, bicarbonate, carbonate, bromide, and fluoride anions. The dissolved solids are responsible for the "hard" water that exists in some locales. Anthropogenic (human-related) dissolved solids include nitrates from fertilizer runoff and human wastes, phosphates from detergents and fertilizers, and organic compounds from pesticides and industrial wastes. **Salinity**, a measure of the dissolved solids in a water sample, is defined as the grams of dissolved solids per kilogram of water.

Suspended solids are very finely divided particles that are water *insoluble* but are filterable. These particles are kept in suspension by the turbulent action of the moving water. Examples of suspended solids include decayed organic matter, sand, salt, and clay.

Total solids are the sum of the dissolved and suspended solids in the water sample. In this experiment the total solids and the dissolved solids are determined directly; the suspended solids are assumed to be the difference, since

$$total\ solids = dissolved\ solids + suspended\ solids$$

TECHNIQUES

- Technique 2, page 1   Clean Glassware and Lab Bench
- Technique 7a, page 6   Evaporation of Liquids
- Technique 8, page 7   Reading a Meniscus
- Technique 9, page 7   Pipetting a Liquid or Solution
- Technique 13, page 11   Using the Laboratory Balance
- Technique 14b, page 12   Separation of a Liquid from a Solid

PROCEDURE

Obtain 100 mL of a water sample from your instructor. If permitted, bring your own water sample to the laboratory for analysis. Check out one additional evaporating dish from the stockroom.

Clean, dry, and determine the mass (±0.001 g) of two evaporating dishes. Be certain that you can identify each. Use the same balance for the remainder of the experiment.

A. Dissolved Solids

Cool flame: a Bunsen flame of low intensity—a slow rate of natural gas is flowing through the burner barrel.

1. Gravity filter about 50 mL of a thoroughly stirred or shaken water sample into a clean, dry 100 mL beaker. While waiting for the filtration to be completed, proceed to Part B.

2. Pipet a 25 mL aliquot (portion) of the filtrate into one of the evaporating dishes. Redetermine the mass of the combined evaporating dish and sample. Place the dish/sample on the wire gauze (Figure T.7a) and *slowly* heat—do not boil—the mixture to dryness. As the mixture nears dryness, cover with a watch glass, and reduce the intensity of the flame.[1] If spattering does occur, allow the dish to cool to room temperature, rinse the adhered solids from the watch glass (Figure 4.1) and return the rinse to the dish.

3. Again heat slowly, being careful to avoid further spattering. After all of the water has evaporated, maintain a "cool" flame beneath the dish for 3 minutes. Allow the dish to cool to room temperature and determine its final mass.

B. Total Solids and Suspended Solids

1. a. Thoroughly stir or agitate 100 mL of sample; pipet a 25 mL aliquot of this sample into the second evaporating dish. Evaporate the sample to dryness as described in Part A.2.

    b. Calculate the mass of total and suspended solids in the original water sample, using data from Part A.

---

[1]This reduces the spattering of the remaining solid and its subsequent loss in the analysis

Figure 4.1
Wash the spattered material from the convex side of the watch glass

C. Analysis of Data

1. Compare your data with three other chemists in your laboratory who analyzed the *same* water sample. Record their results on the Data Sheet.

2. The standard deviation, σ, for n sets of data is calculated using the equation

$$\sigma = \sqrt{\frac{d_1^2 + d_2^2 + d_3^2 + \ldots + d_n^2}{(n-1)}}$$

d_1 is the difference between your suspended solids value (Data set #1) and the average suspended solids value, x, obtained from the four other chemists. The meaning of the standard deviation value is that 68% of all subsequent solids determinations for that particular water sample will fall within the range of $x + \sigma$ to $x - \sigma$. Calculate the standard deviation of the suspended solids from the four analyses on the water sample.

Portable water quality test kits can be used to perform many "quick" tests on a water sample. The following portable water quality test kits are commercially available from the Hach Company. The parameters for testing, along with their range of sensitivity, follow.

Tests	Concentration Range	Tests	Concentration Range
aluminum	0-0.7 mg/L	manganese (EPA)	0-10 mg/L
arsenic (EPA)	0-0.2 mg/L	mercury	0-3 µg/L
barium	0-250 mg/L	molybdenum	0-25 mg/L
benzotriazole	0-15 mg/L	nickel (EPA)	0-1.5 mg/L
boron	0-15 mg/L	nitrogen, ammonia (EPA)	0-2 mg/L
cadmium	0-70 µg/L	nitrogen, nitrate	0-30 mg/L
chloride	0-20 mg/L	nitrogen, nitrite (EPA)	0-200 mg/L
chlorine, free (EPA)	0-1.7 mg/L	nitrogen, total	0-200 mg/L
chlorine, total (EPA)	0-1.7 mg/L	oil in water	0-60 ppm
chlorine dioxide	0-0.055 mg/L	oxygen, dissolved	0-10 mg/L
chromium (EPA)	0-0.5 mg/L	oxygen demand, chemical (EPA)	0-800 mg/L
cobalt	0-1.2 mg/L	ozone	0-1.3 mg/L
color, true	0-70 units	pH	pH 4-10
copper (EPA)	0-150 µg/L	phenols (EPA)	0-0.2 mg/L
cyanide(EPA)	0-0.12 mg/L	phosphonates	0-20 mg/L
cyanuric acid	0-50 mg/L	phosphorus, reactive (EPA)	0-2 mg/L
N,N-diethylhydroxylamine	0-300 µg/L	polyacrylic acid	0-20 mg/L
detergents, anionic	0-0.2 mg/L	potassium	0-6 mg/L
erythorbic acid	0-1000 µg/L	residue, suspended solids	0-500 mg/L
fluoride (EPA)	0-2 mg/L	selenium	0-1.0 mg/L
formaldehyde	0-650 µg/L	silica	0-15 mg/L
hardness	0-2.5 mg/L	silver	0-0.5 mg/L
hydrazine	0-0.015 mg/L	sodium chromate	0-1000 mg/L
iodine	0-7 mg/L	sulfate (EPA)	0-100 mg/L
iron, ferrous	0-2 mg/L	sulfide (EPA)	0-0.5 mg/L
iron, total (EPA)	0-2 mg/L	tannin and lignin	0-5 mg/L
lead (EPA)	0-150 µg/L	tolyltriazole	0-15 mg/L
		turbidity	0-400 FTU
		volatile acids	0-2500 mg/L
		zinc (EPA)	0-1 mg/L

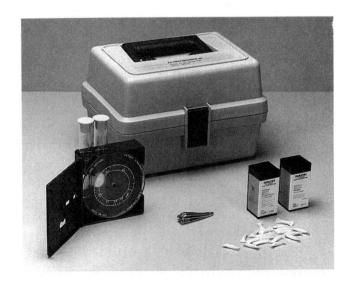

WATER ANALYSIS: SOLIDS

Date _____ Name _____ Lab Sec. _____ Desk No. _____

1. a. Characterize a suspended solid.

    b. Identify three kinds of suspended solids.

    c. How do suspended solids stay suspended?

2. In evaporating a solution to dryness in an evaporating dish, why must the heating rate be decreased as the mixture nears dryness?

3. A 25 mL aliquot of a well-shaken sample of river water is pipetted into a 27.211 g evaporating dish. After the mixture is evaporated to dryness, the dish and remaining sample has a mass of 43.617 g. Determine the total solids in the sample; express in units of g/kg sample. Assume the density of the sample to be 1.01 g/mL.

4. a. What is an aliquot of a sample?

    b. What is the filtrate in a filtration procedure?

    c. How full (the maximum level) should a funnel be filled with solution in a filtration procedure?

    d. What finger should be used in controlling the volume of solution in a pipet?

    e. How is a suspended drop removed from the tip of a pipet after delivery of the filtrate to the evaporating dish?

    f. What device or procedure is used to draw the filtrate into the pipet in Part A.2 of the procedure?

5. List several cations and anions that contribute to the salinity of a water sample.

6. The reported concentration of suspended solids in a water sample is 460 g/kg. The standard deviation for six trials is 12 g/kg.

    a. List the range in g/kg of precision for the suspended solid determinations.

    b. What percent of all subsequent determinations will lie within this range.

WATER ANALYSIS: SOLIDS

Date _____ Name _____ Lab Sec. _____ Desk No. _____

Sample Number: _____

Describe the nature of your water sample, i.e., its color, turbidity, etc.

A. Dissolved Solids

Data Set #1

1. Mass of evaporating dish (g) ....

2. Mass of evaporating dish + water sample (g) ....

3. Mass of water sample (g) ....

4. Mass of evaporating dish + *dried* sample (g) ....

5. Mass of dissolved solids in 25 mL aliquot (g) ....

6. Mass of dissolved solids per total mass of sample (g solids/g sample) ....

7. Dissolved solids or salinity (g solids/kg sample) ....

B. Total Solids and Suspended Solids

1. Mass of evaporating dish (g) ....

2. Mass of evaporating dish + water sample (g) ....

3. Mass of water sample (g) ....

4. Mass of evaporating dish + *dried* sample (g) ....

5. Mass of total solids in 25 mL aliquot (g) ....

6. Mass of total solids per total mass of sample (g solids/g sample) ....

7. Suspended solids or salinity (g solids/kg sample), ss_1 ....

C. Analysis of Data

	Chemist #1 (you)	Chemist #2	Chemist #3	Chemist #4
Dissolved Solids (g/kg)				
Total Solids (g/kg)				
Suspended Solids (g/kg)	$ss_1=$	$ss_2=$	$ss_3=$	$ss_4=$

1. Average value of suspended solids from four chemists (x) = _____

	$d_n = x - ss_n$	$d_n^2 = (x - ss_n)^2$
Chemist #1		
Chemist #2		
Chemist #3		
Chemist #4		

2. Standard deviation, σ = _____
   Show work here.

QUESTIONS

1. If the sample in the evaporating dish in Part B had not been heated to total dryness, how would this have affected the reported value for

   a. total solids? Explain.

   b. suspended solids? Explain.

2. Suppose that the water sample has a relatively high percent of volatile solid material. How would this affect the reported mass of

   a. dissolved solids? Explain.

   b. total solids? Explain.

   c. suspended solids? Explain.

3. Suppose that in Part A of the Procedure, the evaporating dish had not been properly cleaned of a volatile material before the mass of the sample was determined. How would this error in technique affect the reported mass of dissolved solids in the water sample? Explain.

EXPERIMENT 5

CHEMICALS AND
THEIR REACTIONS

All matter consists of chemicals. Foods consist solely of chemicals, automobiles are built with and powered by chemicals, and our bodies consist of chemicals. Each chemical interacts with its environment in its own unique way; foods spoil, automobiles rust, and bodies age. A passing knowledge of chemicals and their properties is important in forming the basis for an understanding of the chemical world.

The chemist communicates in a technical language that is unlike that of any other profession. Often the layman does not understand a lawyer's legal jargon, a physician's medical terms, or an announcer's descriptions of actions in a baseball game. However, if one is familiar with any of these professions, the terms are understood. We will attempt to make the language of the chemist a bit more understandable in this experiment.

OBJECTIVES

- To recognize and to write
  - symbols for elements
  - formulas for compounds
  - balanced equations for chemical reactions
- To characterize some physical properties of some compounds

PRINCIPLES

A chemist's basic language consists of *symbols* for elements, *formulas* for compounds, and *equations* to describe chemical *reactions*. In this experiment we will observe a number of chemicals and chemical reactions, identify chemical change, and write balanced equations.

Symbol: an abbreviation that represents the name and also one atom of an element.

Symbols for elements. The elements discovered by early chemists are generally given Latin names that describe their appearance or chemical properties. The more recently discovered elements have English names. For example, **Au** represents aurum (gold), **Cu** is the symbol for cuprum (copper), but **Ni** represents nickel and **As** is arsenic. The periodic table (see back cover) lists the symbols for the elements, their atomic numbers (the whole number), and their relative atomic masses.

Formula: combination of the symbols of elements that represents the name and also one molecule (or formula-unit) of a compound.

Formulas for compounds. Subscripts in the formula indicate the number of atoms of each element in the compound. For example, H_2O represents water, a chemical combination of 2 H atoms and 1 O atom; $C_{12}H_{22}O_{11}$ is the formula for sucrose (table sugar), a chemical combination of 12 C atoms, 22 H atoms, and 11 O atoms. A change in subscript represents a change in the chemical composition of the compound.

Chemical equation: a brief description of a chemical reaction in which products form from reactants.

Equations for chemical reactions. A chemical equation has the general form,

$$\text{reactants} \rightarrow \text{products}$$

For example, the equation,

$$HCl(g) + NH_3(g) \rightarrow NH_4Cl(s)$$

indicates that the reactants, gaseous HCl and gaseous NH_3, react to produce the product, solid NH_4Cl. More specifically, the chemical equation states that one molecule of gaseous hydrogen chloride reacts with one molecule of gaseous ammonia to produce one formula-unit of ammonium chloride—that's a pretty wordy statement! Notice how the equation simplifies the presentation of the chemical reaction.

A chemical equation not only simplifies a statement, it also indicates the number of atoms involved in the reaction and how they rearrange to form products. The above equation also states that 4 H atoms, 1 Cl atom, and 1 N atom rearrange to form a product that also contains 4 H atoms, 1 Cl atom, and 1 N atom—the atoms are conserved; since each atom has its own mass, then mass is also conserved. A correct equation, then, must have the same number of atoms of each element appearing on both sides of the "→".

Chemical reaction: the production of a substance having properties different from the substances from which it was formed.

Chemical reactions. The product of a chemical reaction has chemical and physical properties unlike those of the reactants. For example, sodium is a very reactive, silvery-white metal and chlorine is a very greenish-yellow, toxic gas; and yet, when they react, sodium chloride, a white, crystalline solid that is a part of our diet, is produced. Clearly, sodium chloride is unlike the elements from which it forms.

How does a chemist recognize chemical change? Generally, a change of color, the formation of a precipitate or gas, the detection of a different odor, and/or the absorption or evolution of heat accompanies a chemical reaction. Other evidence of chemical change may be more subtle.

Simple reactions can be categorized into several groups.

- A single displacement reaction occurs when an element, A, (usually a metal) displaces another element, B, from a compound.

$$A + BY \rightarrow AY + B$$

- A double displacement reaction occurs when two elements substitute for each other in their respective compounds.

$$AX + BY \rightarrow AY + BX$$

- A direct combination reaction occurs when two elements combine to form a single compound.

$$X + Y \rightarrow XY$$

- A decomposition reaction occurs when heat or some other influence causes a compound to dissociate or decompose.

$$AQ \rightarrow A + Q$$

TECHNIQUES

- Technique 3, page 2   Microscale Analyses
- Technique 6a, page 4   Heating Liquids
- Technique 11, page 10   Testing with Litmus
- Technique 17a, page 16   Handling Gases

PROCEDURE

A. Formulas and Compounds

A large display of compounds in their most stable state at room temperature is arranged in test tubes. Complete the table as suggested on the Data Sheet. Based upon your general observations, develop some generalizations about compounds as suggested on the Data Sheet.

B. Single Displacement Reactions.

24-cell plate: chemical apparatus used to contain about 3.5 mL of solution per cell.

Set up the following experiments in a 24-cell plate.

1. *Polish* a 2 cm Zn strip with steel wool and insert it into 6 M HCl (**Caution:** *avoid skin contact; flush with water*). Record your observations on the Data Sheet.

2. Dissolve several crystals of $CuSO_4 \cdot 5H_2O$ in 1 mL of water and insert a polished Zn strip. Allow the solution to set for the duration of the laboratory period. Wipe off and examine the reddish-black deposit and the zinc surface for pits or pock marks. Record.

3. Dissolve several crystals of $ZnSO_4$ in 1 mL of water and insert a polished Cu strip. Allow the solution to set for the duration of the laboratory period. Do you see evidence of a chemical change?

Disposal Information:	Dispose of the test solutions in the "Waste Salts" containers.

C. Double Displacement Reactions

1. Set up and label nine microcells in a 24-cell plate and transfer the following quantities of solutions.

   #1 1 mL of 0.1 M Na_2CO_3
   #2 1 mL of 0.1 M $BaCl_2$
   #3 1 mL of 0.1 M NH_4Cl
   #4 1 mL of 0.1 M $CuSO_4$
   #5 1 mL of 0.1 M $FeSO_4$
   #6 1 mL of 0.1 M Na_3PO_4
   #7 1 mL of 6 M H_2SO_4 (**Caution:** *handle with care and avoid skin contact*)
   #8 1 mL of 6 M NaOH (**Caution:** *handle with care and avoid skin contact*)
   #9 1 mL of 6 M NH_3 (**Caution:** *handle with care and avoid skin contact*)

Beral pipet: a small plastic apparatus used for extracting small volumes of solutions.

2. Using clean Beral pipets (one per solution) combine the contents of the microcells as directed. Record your observations on the Data Sheet.

   a. Add $^1/_2$ of #8 to #3. Warm the bottom of the microcell with your finger. Smell *cautiously*. Test the vapor with moistened red litmus paper.[1]
   b. Add drops of #5 to #6.
   c. In another microcell add $^1/_2$ of #1 to $^1/_2$ of #7. Smell *cautiously*. Test the vapors with blue litmus paper.
   d. Add #4 to #9.
   e. Add #2 to the remainder of #1.
   f. Insert a thermometer into the remainder of #7 and add the remainder of #8. This, an acid-base reaction, is called a **neutralization** reaction.[2]

---

[1]To test the vapors with litmus paper, moisten it with water place it across the diameter of the microcell. If the blue litmus turns red, then the vapors are acidic; if the red litmus turns blue, the vapors are basic.

[2]A more quantitative investigation of the heat evolved in a neutralization reaction is performed in Experiment 11.

Pinch: a volume about the size of a grain of rice.

3. Place a pinch of FeS in a 75 mm test tube and add $^1/_2$ mL of 3 M HCl. Cautiously and with proper technique test the odor, the odor of rotten eggs. (**Caution:** *the odor is due to H_2S, a very poisonous gas and should be tested only long enough to detect the odor. Place the test tube in the fume hood*). Test the $H_2S(g)$ with moistened blue litmus paper.

| **Disposal Information:** | Dispose of the test solutions in the "Waste Salts" containers. |

D. Combination Reactions

1. Grip a 2 cm piece of Mg ribbon with a pair of crucible tongs. Heat it directly in a Bunsen flame until it ignites. (**Caution:** *do not look directly at the burning Mg ribbon!*)

2. Place a pinch of sulfur on the end of a spatula and heat directly with a Bunsen flame. Test the gas (SO_2) with moistened blue litmus paper.

E. Decomposition Reactions

1. a. Place several crystals of $NaHSO_3$ in a 200 mm test tube and heat cautiously with a Bunsen flame. Test and describe the odor (Use the proper technique for testing!). Test the vapors with blue litmus paper. Where, during this experiment, have you previously detected this odor?

    b. Now heat the contents more strongly until a red-brown color appears. Allow the contents to cool to room temperature. Add distilled (or deionized) water (the test tube should be $^1/_2$ to $^1/_3$ full) and agitate the test tube until the contents dissolve. Divide the solution into 2 equal volumes.

    #1. Add several drops of 0.1 M $BaCl_2$.
    #2. Add several drops of 3 M HCl. Test for odor—does it smell familiar?

    c. Dissolve several crystals of $NaHSO_3$ in water and repeat the test #1 and #2 in Part E.1b. How do the tests differ? Record.

| **Disposal Information:** | Dispose of the test solutions in the "Waste Salts" containers. |

2. **Demonstration only.** Place about 3 g of $C_{12}H_{22}O_{11}$ into a porcelain evaporating dish. Place the dish in a fume hood. Add 3 mL of conc H_2SO_4 (**Caution:** *conc H_2SO_4 causes severe skin burns and clothing to disappear! Immediately flush the skin with water*). Conc H_2SO_4 is a strong dehydrating agent, extracting H_2O molecules from the $C_{12}H_{22}O_{11}$ molecule. Record your observations and write a balanced equation.

CHEMICALS AND THEIR REACTIONS

Date _____ Name _____ Lab Sec. _____ Desk No. _____

1. Write the English and Latin names for elements with the following symbols.

Symbol	Latin Name	English Name	Matching Latin Names
K			natrium
Pb			ferrum
Ag			aurum
Fe			stibium
Hg			kalium
Sb			plumbum
Sn			hydragyrum
Au			argentum
Na			stannum

2. At least 109 elements are known; however, only a few make a significant contribution to the Earth's crust. Complete this table of elements that are of greatest abundance in the Earth's crust.

Name of Element	% of total atoms	Atomic Number	Symbol	Atomic Mass	Metal or Nonmetal
oxygen	46.6				
silicon	27.2				
aluminum	8.13				
iron	5.00				
calcium	3.63				
sodium	2.83				
potassium	2.59				
magnesium	2.09				

3. All of the naturally occurring gases (except the noble gases), bromine, and iodine are diatomic molecules. Write formulas for the elements that occur as diatomic molecules in their natural state.

    _____, _____, _____, _____, _____, _____, _____

4. Write the formula for the following compounds.

    a. ammonia   1 N-atom, 3 H-atoms   _____

    b. lye   1 Na-atom, 1 O-atom, 1 H-atom   _____

    c. calcium carbonate   1 Ca-atom, 1 C-atom, 3 O-atoms   _____

    d. glucose   6 C-atoms, 12 H-atoms, 6 O-atoms _____

    e. hydrogen peroxide   2 H-atoms, 2 O-atoms   _____

    f. alcohol (ethanol)   2 C-atoms, 6 H-atoms, 1 O-atom   _____

    g. octane   8 C-atoms, 18 H-atoms   _____

5. a. Describe the technique for testing a vapor with litmus paper.

    b. What is indicated when a vapor turns blue litmus red?

6. Describe the technique for testing the odor of a vapor.

CHEMICALS AND THEIR REACTIONS

Date_____Name_____ Lab Sec. _____Desk No._____

A. Formulas and Compounds

1. On a separate sheet of paper, construct a table with the headings shown below and complete it as best possible.

	Formula	Name	State (g, l, s)	Color	Physical Appearance
a.					
b.					
•					
•					

2. Generalizations based on observations.

    a. Blue-colored salts often contain the element _____.

    b. Green-colored salts often contain the element _____.

    c. MnO_4^- salts exhibit a _____ color.

    d. $Cr_2O_7^{2-}$ salts exhibit a _____ color.

    e. Most CO_3^{2-} salts exhibit a _____ color. Exceptions contain the elements

    _____ and _____.

B. Single Displacement Reactions

Part	Chemical Reactants	Evidence of Reaction	Chemical Products	Balanced Equation
1.			$ZnCl_2 + H_2$	
2.			$ZnSO_4 + Cu$	
3.			no reaction	

C. Double Displacement Reactions

Part	Chemical Reactants	Evidence of Reaction	Chemical Products	Balanced Equation
2a.	$NaOH + NH_4NO_3$		$NH_3 + H_2O + NaCl$	

Result of litmus test

2b.			$Fe_3(PO_4)_2 + NaCl$	
2c.			$Na_2SO_4 + CO_2 + H_2O$	

Result of litmus test

2d.			$[Cu(NH_3)_4]^{2+} + SO_4{}^{2-}$	
2e.			$BaCO_3 + NaCl$	
2f.			$Na_2SO_4 + H_2O$	
3.			$FeCl_2 + H_2S$	

Result of litmus test

D. Combination Reactions

Part	Chemical Reactants	Evidence of Reaction	Chemical Products	Balanced Equation
1.			MgO	
2.			SO_2	

E. Decomposition Reactions

1.   a. Odor _____. Conclusion from litmus test_____.

   Formula of gas _____.

   b. Observation from $BaCl_2$ test _____. $BaSO_4$ is an insoluble salt; therefore what is
   one of the decomposition products of $NaHSO_3$?_____

   Observation from 3 M HCl test _____. From a detection of the odor, what is a
   second decomposition product of $NaHSO_3$?_____

   Write a balanced equation for the thermal decomposition of $NaHSO_3$.

   c. Observation from $BaCl_2$ test _____. How does the appearance of the system differ
   from that in Part E.1b (#1)?

   Observation from 3 M HCl test _____. How does the odor differ from that in Part
   E.1b (#2)?

   Write a balanced equation for the the reaction of HCl(*aq*) with $NaHSO_3$.

2. What evidence of a chemical reaction is indicated?

The products of the reaction are carbon and water. Write a balanced equation for the decomposition of sugar.

QUESTIONS

1. In Part B.2, what happens with time to the color of the solution?_____ Explain.

2. In Parts B.2 and B.3, the relative chemical reactivities of Zn and Cu are determined. Which metal is more reactive? _____ Explain.

3. What is the gas evolved in Part C.2a?_____ In Part C.2c? _____ In Part C.3?_____

4. When an antacid neutralizes excess stomach acid, is heat evolved or absorbed?_____ What is *always* a product of the reaction?_____

5. What is the color of a cotton shirt or jeans after having spilled conc H_2SO_4 on them? (Note: cotton is chemically similar to sugar.)

EXPERIMENT 6

IDENTIFICATION OF A SALT

Isn't salt those white crystals that come from a salt shaker, you know, the stuff we use to enrich the flavor of foods? Yes, but it is more than that to a chemist! Do we mean to say that salt is not just "salt"?

A salt to a chemist consists of a mixture of cations and anions in a ratio that results in a neutral, solid compound. The salt may be colored, toxic, insoluble in water, and/or chemically reactive. Each physical and chemical property may be dependent upon the cation, the anion, or both. In this experiment, we will focus on the chemical properties of several anions by looking at the formation of their precipitates.

OBJECTIVES

- To observe the chemical properties of SO_4^{2-}, CO_3^{2-}, PO_4^{3-}, Cl^-, and S^{2-}
- To identify the SO_4^{2-}, CO_3^{2-}, PO_4^{3-}, Cl^-, and S^{2-} anions in various salts

PRINCIPLES

Cation: an atom or group of atoms with a positive charge.

Anion: an atom or group of atoms with a negative charge.

One property of an anion is the solubility of its salts in neutral and acidic solutions. The solubility of a salt, and therefore the amount of anion dissolved in solution, is dependent upon temperature and the cation present; for example, sodium salts of the carbonate and phosphate ions are considered soluble, whereas their calcium salts are considered to be only very slightly soluble.

The solubility of a salt is also dependent upon the acidity of the solution. A sulfate salt, such as $BaSO_4$, that has a relatively low solubility in water generally has a low solubility in an acidic solution, whereas most carbonate salts of low solubility, such as $BaCO_3$, dissolve in an acidic solution because of the reaction of the carbonate ion with the hydrogen ion of the acid:

$$CO_3^{2-}(aq) + 2\,H^+(aq) \rightarrow H_2CO_3\,(aq) \;\rightarrow\; H_2O(g) + CO_{2(g)}$$

Therefore, the use of an acid quickly distinguishes SO_4^{2-} from CO_3^{2-} salts with low solubilities. Many phosphate salts, such as calcium phosphate, $Ca_3(PO_4)_2$, also have limited solubility and, like the carbonate salts, dissolve in an acidic solution. How then could we distinguish between the CO_3^{2-} ion the PO_4^{3-} ion if they were present in a mixed precipitate? One distinguishing chemical property is that the PO_4^{3-} ion does not produce a gas in the presence of hydrogen ion, but merely forms the dihydrogen phosphate ion:

$$PO_4^{3-}(aq) + 2\,H^+(aq) \;\rightarrow\; H_2PO_4^-(aq)$$

Occasionally we find that an additional test is necessary to confirm our predictions. Ammonium molybdate, $(NH_4)_2MoO_4$, is used to confirm the presence of the PO_4^{3-} ion; its addition to a solution containing the $H_2PO_4^-$ ion causes the formation of a yellow precipitate:

Precipitate: an insoluble substance that forms in water.

$$H_2PO_4^-(aq) + 12\ (NH_4)_2MoO_4(aq) + 22\ H^+(aq) \rightarrow$$
$$(NH_4)_3PO_4(MoO_3)_{12}(s) + 21\ NH_4^+(aq) + 12\ H_2O(l)$$
$$\text{(yellow)}$$

Neither the SO_4^{2-} or the CO_3^{2-} salts behave in a similar manner. Therefore, we can distinguish between the SO_4^{2-}, CO_3^{2-}, or PO_4^{3-} anions in salts that are very slightly soluble.

Obviously there are many other anions that form precipitates. One of the most common anions is the chloride ion, Cl^-. Most chloride salts are considered soluble, but the silver salt, $AgCl$, is an exception.

$$Ag^+(aq) + Cl^-(aq) \rightarrow AgCl(s)$$

Another anion of salts is the sulfide ion, S^{2-}. A quick test for presence of the S^{2-} anion is to acidify a concentrated solution of its salt and note the odor of hydrogen sulfide gas and (with nitric acid) the formation of yellow sulfur; its smell is characteristic—you'll know!!

In today's experiment you will determine which of the SO_4^{2-}, CO_3^{2-}, PO_4^{3-}, Cl^-, and S^{2-} anions are present in an unknown salt mixture by conducting the tests that have been described.

TECHNIQUES

- Technique 3, page 2 — Microscale Analyses
- Technique 17a, page 16 — Handling Gases

PROCEDURE

Your instructor will issue two unknown samples. Sample A consists of a single salt and Sample B consists of two salts. In each sample you are to identify the anion(s) present.

1. Set up *two* separated 1 x 4 arrays of microcells in a 24-cell plate. Transfer a small amount (20–40 crystals) of the salts Na_2SO_4, Na_2CO_3, Na_3PO_4, and $NaCl$ in the first 1 x 4 array and Na_2S and your two unknowns into the second 1 x 4 array.

2. Half-fill each cell with distilled (or deionized) water and agitate the solution for about 30 s. Note how quickly the salts dissolve? Record on the Data Sheet.

3. Add 1 drop of 6 M NH_3 (**Caution:** *do not inhale*) to each solution; then add several drops of a 0.2 M $Ba(NO_3)_2$ solution and stir or swirl. Does a precipitate appear? Record.

4. Add 5 drops of 6 M HNO_3 (**Caution:** *do not allow HNO_3 to contact the skin. If it does, wash immediately with plenty of water.*) to each solution. Look at the solution; is there any evidence of a reaction? Test for any odor with your nose (Technique 17a). Record all of your observations.

5. Carefully divide each solution into two equal volumes in adjacent microcells using a Beral pipet; avoid transferring any precipitate.
   a. To one, add 5 drops of 0.1 M $AgNO_3$.
   b. To a second, add 5 drops of 0.1 M $(NH_4)_2MoO_4$, agitate. Record evidence of any precipitate (the precipitate may be slow in forming).

Disposal Information: Dispose of all test solutions in the "Waste Salts" container.

IDENTIFICATION OF A SALT

Date _____ Name _____ Lab Sec. _____ Desk No. _____

1. Describe the procedure for testing the odor of a chemical.

2. What chemical test(s) can you use to distinguish between calcium chloride, $CaCl_2$, and calcium carbonate, $CaCO_3$, both of which are white solids?

3. Write the formulas for the compounds that confirm the presence of the five cations tested in this experiment.

   a. SO_4^{2-} _____

   b. CO_3^{2-} _____

   c. PO_4^{3-} _____

   d. Cl^- _____

   e. S^{2-} _____

IDENTIFICATION OF A SALT

Date _____ Name _____ Lab Sec. _____ Desk No. _____

Analyzing your unknown. The test salts and the unknown samples are *pure* samples. You must realize that these tests could not be as well defined if impurities were present in the salt. Sample A will contain one salt, a salt that contains only one of the anions. However it may be a salt that is nearly insoluble. If so, you should begin your testing with Part 3 and continue through the procedure. Sample B will be more difficult. For example, if a precipitate in Part 2 dissolves in Part 3, you may not know immediately whether PO_4^{3-} or CO_3^{2-} is present; you will need to complete Part 4 to determine the presence of PO_4^{3-}. Also the $CO_2(g)$ formation may be difficult to observe—you must watch closely. It would be advisable for you to check Sample B two or three times to ensure confidence in your data.

Sample A: Identification Number _____

Sample B: Identification Number _____

	Na_2SO_4	Na_2CO_3	Na_3PO_4	NaCl	Na_2S	Sample A	B
Soluble in Water?							
Precipitate with $Ba(NO_3)_2$?							
Reaction with HNO_3?							
Odor Test							
Precipitate with $AgNO_3$?							
Precipitate with $(NH_4)_2MoO_4$?							

Sample A: Anion present _____

Sample B: Anions present _____

QUESTIONS

1. What single test reagent will distinguish a soluble CO_3^{2-} salt from a soluble PO_4^{3-} salt. Explain. Assume an acidic solution.

2. What single test reagent will distinguish a soluble Cl⁻ salt from a soluble SO_4^{2-} salt? Explain.

3. What single test reagent will distinguish $BaSO_4$ from $BaCO_3$? Explain.

4. Describe how you could identify the presence of *both* Na_2CO_3 and $BaCO_3$ in a salt mixture?

5. Describe how you could identify the presence of *both* Na_3PO_4 and $Ba_3(PO_4)_2$ in a salt mixture.

EXPERIMENT 7

INORGANIC NOMENCLATURE

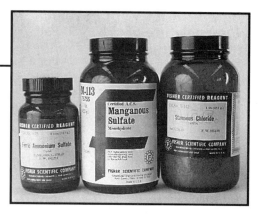

Every profession seems to have its own language; physicians can talk to other physicians, but can have a difficult time communicating to patients; lawyers write legal jargon for insurance policies or deeds so that it is understood by other lawyers, but not necessarily by the person buying the policy. Even chemists have a language that is used to more explicitly characterize a reaction or identify a compound.

For chemists to communicate research data and results internationally, a standardization of the technical language has been necessary. Historically, common names for many compounds were used and are still universally understood—for examples, water, sugar, and ammonia—but with new compounds being prepared daily, a random or "convenient" system for naming compounds is no longer viable. In this assignment you will learn a few systematic rules, established by the International Union of Pure and Applied Chemistry (IUPAC), for naming and writing the formulas for some inorganic compounds.

OBJECTIVES

- To name common inorganic compounds
- To write formulas for common inorganic compounds

PRINCIPLES

You are undoubtedly already familiar with some symbols of the elements and the names of several common compounds (see Experiments 5 and 6); for example, NaCl is sodium chloride. Continued practice and work in writing formulas and naming compounds will make you even more knowledgeable of the chemist's vocabulary.

The naming of chemical compounds can be categorized into the nomenclature for inorganic compounds and for organic compounds. This assignment focuses on the inorganic compounds; Experiment 39 introduces organic compounds and its nomenclature.

Inorganic compounds can be classified into several groups:

Binary Compounds. These compounds consist of only two elements. Three major subclassifications of binary compounds are:
- **Salts** which consist of a metal cation and nonmetal anion
- Compounds which consist of **two nonmetals**
- **Binary acids** which consist of hydrogen and a more electronegative nonmetal dissolved in water.

Polyatomic ion: a group of covalently bonded atoms that carry a charge.

Compounds with Polyatomic Ions. These compounds *generally* consist of a metal cation and a polyatomic anion. One element in the polyatomic anion is almost always oxygen and the other is *usually* a nonmetal.

Nomenclature of Binary Compounds

Salts, a Metal Cation and a Nonmetal Anion

For a binary salt, the more metallic (more electropositive) element is named first. The root of the second element is then named with the suffix -*ide* added to it. NaBr is sodium brom*ide*; Al_2O_3 is aluminum ox*ide*. The names of some binary salts have the -*ide* ending, but are not binary compounds; these compounds contain the NH_4^+, OH^-, and CN^- ions. NH_4Cl is ammonium chlor*ide*; KOH is potassium hydrox*ide,* and NaCN is sodium cyan*ide.*

Some metals, especially transition and post-transition metals, form more than one cation resulting in the formation of more than one compound with a given anion. For example, copper forms Cu^+ and Cu^{2+} ions and therefore forms two salts with chlorine: CuCl and $CuCl_2$. Two systems are used to distinguish the names of these salts. The Stock system uses the English name for the metal cation followed immediately by the charge on the ion with a Roman numeral in parentheses. The older system uses the Latin root name for the metal cation and applies a suffix that is dependent upon the charge: an -*ous* ending designates the ion with the lower charge and an -*ic* ending designates the ion with the higher charge. The CuCl and $CuCl_2$ salts, as well as examples for iron and tin, are:

- CuCl is copper(I) chloride and cupr*ous* chloride
- $CuCl_2$ is copper(II) chloride and cupr*ic* chloride
- $FeBr_2$ is iron(II) bromide and ferr*ous* bromide
- $FeBr_3$ is iron(III) bromide and ferr*ic* bromide
- SnO is tin(II) oxide and stann*ous* oxide
- SnO_2 is tin(IV) oxide and stann*ic* oxide

In the naming of salts, the Stock system is preferred.

Two Nonmetals

Two nonmetals may combine to form more than one compound; for examples carbon and oxygen combine to form carbon monoxide, CO, and carbon dioxide, CO_2, and nitrogen and oxygen combine to form N_2O, NO, NO_2, N_2O_3, N_2O_4, and N_2O_5. To distinguish between the nitrogen oxides, we use the same system as that for the carbon oxides—Greek *prefixes* are used to indicate the number of atoms of each element in the compound. The common prefixes are in Table 7.1.

Table 7.1   Greek Prefixes

mono-*	one	hexa-	six
di-	two	hepta-	seven
tri-	three	octa-	eight
tetra-	four	nona-	nine
penta-	five	deca-	ten

*mono- is seldom used, since "one" is generally implied

Binary Acids

To name a binary acid, the prefix *hydro-* and the suffix *-ic* are added to the root name for the nonmetal and the word "acid" is added. Remember that compounds are not named as acids unless they are dissolved in water.

- HCl is *hydrochloric acid* when hydrogen chloride is dissolved in water
- HBr is *hydrobromic acid* when hydrogen bromide is dissolved in water
- H_2S is *hydrosulfuric acid* when hydrogen sulfide is dissolved in water

Nomenclature of Compounds Containing Polyatomic Ions

Salts, a Metal Cation and a Polyatomic Anion Containing Oxygen

A metal cation and a polyatomic anion form a salt. The metal ion is named as it was for the binary salts (either using the Stock system or the "old" system); the polyatomic anion is named using the root of the "other" element, *not* the oxygen, and the suffix *-ate* or *-ite*, depending on the number of oxygens—the polyatomic anion with the greater number of oxygens receives the *-ate* suffix. The sulf*ate* ion is SO_4^{2-}, the nitr*ate* ion is NO_3^-, and arsen*ate* ion is AsO_4^{3-}; the sulf*ite* ion is SO_3^{2-}, the nitr*ite* ion is NO_2^-, and the arsen*ite* ion is AsO_3^{3-}.

- Na_2SO_4 is sodium sulf*ate*; Na_2SO_3 is sodium sulf*ite*
- KNO_3 is potassium nitr*ate*; KNO_2 is potassium nitr*ite*

Acids of Polyatomic Anions (Oxoacids)

Hydrogen substitutes for the metal ion in the above salts to form oxoacids. To name these acids we do *not* name the hydrogen, but rather look at the polyatomic anion and change its *-ate* name to an "*-ic acid*" or its *-ite* name to an "*-ous acid*".

- Na_2SO_4 is sodium sulf*ate*; H_2SO_4 is sulfur*ic acid*
- Na_2SO_3 is sodium sulf*ite*; H_2SO_3 is sulfur*ous acid*
- KNO_3 is potassium nitr*ate*; HNO_3 is nitr*ic acid*
- KNO_2 is potassium nitr*ite*; HNO_2 is nitr*ous acid*

Notice that the *-ate* suffix for the salt becomes the *-ic* suffix for the acid, while the *-ite* suffix becomes the *-ous* suffix for an acid.

Some polyatomic anions have more than just a higher or lower number of oxygens, especially those having a halogen as the "other" element. For these polyatomic ions other prefixes are added. For example, the chloro oxoacids and an appropriate salt are

Per-: a prefix meaning completely; a radical or polyatomic ion containing an element with its maximum oxidation number.

- $HClO_4$ is *per*chlor*ic* acid; $KClO_4$ is potassium *per*chlor*ate*
- $HClO_3$ is chlor*ic* acid; $KClO_3$ is potassium chlor*ate*
- $HClO_2$ is chlor*ous* acid; $KClO_2$ is potassium chlor*ite*
- $HClO$ is *hypo*chlor*ous* acid; $KClO$ is potassium *hypo*chlor*ite*

Hypo-: a prefix meaning below; a radical or polyatomic ion containing an element with a very low oxidation number.

Bromine and iodine form similar oxoacids and salts. Notice again the *-ic, -ate* and the *-ous, -ite* relationships between the acids and the salts; the prefixes remain unchanged.

Acid Salts

When both a metal and hydrogen serve as cations to the anion in a salt, it is called an **acid salt**. The number of hydrogens present is indicated by the Greek prefix. The suffix for the anion is that used for salts (binary or those with a polyatomic anion).

- NaHS is sodium hydrogen sulfide
- NaH_2PO_4 is sodium dihydrogen phosphate
- $CaHPO_4$ is calcium hydrogen phosphate
- $Ca(H_2PO_4)_2$ is calcium dihydrogen phosphate

An older system of naming acid salts substitutes the prefix *bi-* for a single hydrogen before naming the polyatomic anion; for example $NaHSO_4$ is sodium *bi*sulfate and $NaHCO_3$ is sodium *bi*carbonate.

Writing Formulas

In writing formulas the combined charges for the cations and anions in the salts/acids must equal zero. Table 7.2 lists a number of common cations and anions. You should learn most, if not all, of those listed—others you will learn with experience. Lets try writing formulas for a few compounds (we'll need to use Table 7.2):

- Barium fluoride. Barium is Ba^{2+} and fluoride is F^-. For the sum of the charges to be zero, one Ba^{2+} ion must combine with two F^- ions; the formula must be BaF_2.
- Calcium nitride. Calcium is Ca^{2+} and nitride is N^{3-}. Three Ca^{2+} provides a 6^+ charge; this balances two N^{3-} (a 6^- charge). The formula is Ca_3N_2.
- Potassium oxalate. Potassium is K^+ and oxalate is $C_2O_4^{2-}$. Two K^+ balances one $C_2O_4^{2-}$; the formula is $K_2C_2O_4$.
- Ferric chromate. Ferric is Fe^{3+} and chromate is CrO_4^{2-}. Two Fe^{3+} (a 6^+ charge) balances three CrO_4^{2-} (a 6^- charge); the formula is $Fe_2(CrO_4)_3$.

Table 7.2. Name and Charge of Common Ions for Elements and Polyatomic Ions

Metallic and Polyatomic Cations

Charge of 1$^+$

NH_4^+	ammonium	H^+	hydrogen	K^+	potassium
Cu^+	copper(I), cuprous	Li^+	lithium	Ag^+	silver(I)
				Na^+	sodium

Charge of 2$^+$

Ba^{2+}	barium	Fe^{2+}	iron(II), ferrous	Ni^{2+}	nickel(II), nickelous
Cd^{2+}	cadmium	Pb^{2+}	lead(II), plumbous	Sr^{2+}	strontium
Ca^{2+}	calcium	Mg^{2+}	magnesium	Sn^{2+}	tin(II), stannous
Cr^{2+}	chromium(II), chromous	Mn^{2+}	manganese(II), manganous	UO_2^{2+}	uranyl
Co^{2+}	cobalt(II), cobaltous	Hg^{2+}	mercury(II), mercuric	VO^{2+}	vanadyl
Cu^{2+}	copper(II), cupric	Hg_2^{2+}	mercury(I), mercurous	Zn^{2+}	zinc(II)

Charge of 3$^+$

Al^{3+}	aluminum	Cr^{3+}	chromium(III), chromic	Fe^{3+}	iron(III), ferric
As^{3+}	arsenic(III), arsenious	Co^{3+}	cobalt(III), cobaltic	Mn^{3+}	manganese(III), manganic

Charge of 4$^+$

Pb^{4+}	lead(IV), plumbic	Sn^{4+}	tin(IV), stannic

Charge of 5$^+$

V^{5+}	vanadium(V)	As^{5+}	arsenic(V), arsenic

Nonmetallic and Polyatomic Anions

Charge of 1$^-$

$CH_3CO_2^-$	acetate	F^-	fluoride	NO_2^-	nitrite
Br^-	bromide	OH^-	hydroxide	ClO_4^-	perchlorate
ClO_3^-	chlorate	H^-	hydride	IO_4^-	periodate
Cl^-	chloride	ClO^-	hypochlorite	MnO_4^-	permanganate
ClO_2^-	chlorite	I^-	iodide		
CN^-	cyanide	NO_3^-	nitrate		

Charge of 2$^-$

CO_3^{2-}	carbonate	O^{2-}	oxide	SO_3^{2-}	sulfite
CrO_4^{2-}	chromate	$C_2O_4^{2-}$	oxalate	S^{2-}	sulfide
$Cr_2O_7^{2-}$	dichromate	O_2^{2-}	peroxide	$S_2O_3^{2-}$	thiosulfate
SiO_3^{2-}	silicate	SO_4^{2-}	sulfate		

Charge of 3$^-$

N^{3-}	nitride	PO_3^{3-}	phosphite	BO_3^{3-}	borate
PO_4^{3-}	phosphate	P^{3-}	phosphide	AsO_3^{3-}	arsenite
				AsO_4^{3-}	arsenate

PROCEDURE

A. A Survey of the Ions

1. List and name (Stock system and the "old"-ic, -ous system) all of the cations having two common charges.

2. List and name all anions that have a chlorine atom.

3. List and name all anions that have a sulfur atom.

4. a. What is the common charge of the Group I metals of the Periodic Table?
   b. What is the common charge of the Group II metals of the Periodic Table?

B. Nomenclature

Your instructor will indicate which of the following (parts of) exercises to complete.

1. Name the binary compounds.

a. $NaCl$	e. Li_2S	i. SrS	m. Li_3N	q. IF_7
b. $CsOH$	f. Al_2O_3	j. V_2O_5	n. Ag_2O	r. $TiCl_4$
c. $CaBr_2$	g. ZnH_2	k. CaC_2	o. KCN	s. N_2O
d. MgO	h. AgI	l. BaI_2	p. NH_4Cl	t. NI_3

2. Name the following salts.

a. Na_2SO_4	g. Ag_2CrO_4	m. K_3PO_4	s. $VOCl_2$
b. KNO_3	h. KCH_3CO_2	n. $(NH_4)_3PO_4$	t. UO_2Cl_2
c. Li_2CO_3	i. $NaMnO_4$	o. KIO_4	u. $NaClO$
d. $Cd(OH)_2$	j. $Li_2S_2O_3$	p. $KClO_3$	v. Na_2O_2
e. $Ca_3(PO_4)_2$	k. $Ba(NO_2)_2$	q. $Ca(CH_3CO_2)_2$	w. $(NH_4)_2S_2O_3$
f. $K_2Cr_2O_7$	l. $AgClO_4$	r. $MgSO_4$	x. $K_2C_2O_4$

3. Name the following binary salts using the Stock system and the "old"-ic, -ous system.

a. CrS	g. FeS	l. PbO_2	q. $FeCl_3$
b. Fe_2O_3	h. As_2O_5	m. As_2S_3	r. SnO_2
c. CrI_3	i. $CuBr_2$	n. SnF_4	s. SnO
d. $CuCl$	j. $CoBr_3$	o. CoO	t. $PbCl_2$
e. PbO	k. $SnCl_2$	p. HgO	u. Mn_2O_3
f. Hg_2Cl_2			v. NiS

4. Name the following salts using the Stock system and the "old"-ic, -ous system.

a. $FeSO_4$	f. $CoSO_4$	k. $Pb(CH_3CO_2)_4$	p. $MnSO_4$
b. $As(NO_3)_3$	g. $Fe(OH)_3$	l. $As_2(SO_3)_3$	q. $Hg_2(ClO_2)_2$
c. $CuCN$	h. $Cr(CN)_2$	m. $FePO_4$	r. $Hg(ClO_2)_2$
d. $Hg(NO_3)_2$	i. $Sn(NO_2)_4$	n. $Ni(NO_3)_2$	s. $Co_3(PO_4)_2$
e. $CuCO_3$	j. $Co_2(CO_3)_3$	o. $CuClO_4$	t. $PbSO_4$

5. Name the following acids.

a. HBr	e. H_3PO_4	i. H_3BO_3	m. H_2SO_4	q. CH_3COOH
b. $HClO$	f. H_3AsO_3	j. HCN	n. HNO_3	r. H_3PO_3
c. H_2S	g. $HMnO_4$	k. HNO_2	o. $H_2S_2O_3$	s. $H_2C_2O_4$
d. H_2SO_3	h. H_2CrO_4	l. HIO_4	p. $HClO_4$	t. H_2CO_3

6. Name the following acid salts. In addition, use the "bi-" method wherever appropriate.
   a. $KHCO_3$ d. $Ca(HCO_3)_2$ g. NaHS   i. $NH_4H_2PO_4$
   b. $NaHSO_4$ e. $KHSO_3$ h. $NaHCO_3$ j. KH_2AsO_4
   c. KHC_2O_4 f Li_2HPO_4

7. Write formulas for each of the following compounds.
a.	ferrous sulfate	r.	barium acetate
b.	potassium permanganate	s.	silver thiosulfate
c.	calcium carbonate	t.	nitrogen triiodide
d.	iron(III) oxide	u.	potassium iodide
e.	cupric hydroxide	v.	sodium silicate
f.	aluminum sulfide	w.	calcium hypochlorite
g.	mercury(II) chloride	x.	potassium chlorate
h.	cadmium sulfide	y.	cuprous iodate
i.	copper(I) chloride	z.	sodium dihydrogen phosphite
j.	ammonium cyanide	aa.	ammonium oxalate
k.	sodium chromate	bb.	potassium dichromate
l.	nickel(II) nitrate	cc.	copper(I) sulfate
m.	manganese(II) oxide	dd.	cobalt(II) phosphate
n.	manganese (IV) oxide	ee.	chromium(III) phosphate
o.	lead(II) carbonate	ff.	iron(II) oxalate
p.	stannous chloride	gg.	copper(I) sulfide
q.	sodium nitrite	hh.	potassium periodate

8. Write formulas for each of the following compounds.
a.	hydrogen fluoride	r.	gold(III) nitrate
b.	sulfuric acid	s.	nitric acid
c.	iodine pentafluoride	t.	hydrocyanic acid
d.	hydrobromic acid	u.	carbon monoxide
e.	hypobromous acid	v.	chlorous acid
f.	hydrogen sulfide	w.	uranyl sulfate
g.	sulfur trioxide	x.	potassium phosphide
h.	phosphorus pentafluoride	y.	oxalic acid
i.	potassium borate	z.	thiosulfuric acid
j.	perchloric acid	aa.	sodium bicarbonate
k.	nitrous acid	bb.	calcium hydride
l.	silicon dioxide	cc.	boric acid
m.	acetic acid	dd.	carbonic acid
n.	vanadyl acetate	ee.	zinc hydroxide
o.	tetraphosphorus decaoxide	ff.	dinitrogen pentaoxide
p.	dichlorine heptaoxide	gg.	hypochlorous acid
q.	mercurous cyanide	hh.	xenon hexafluoride

Notes

INORGANIC NOMENCLATURE

Date _____ Name_____ Lab Sec. _____ Desk No._____

Problems assigned by laboratory instructor: _____, _____, _____, _____. Use the following table to answer the problems that you were assigned.

Formula	Name	Formula	Name

Instructor's signature. _____

EXPERIMENT 8

FORMULA OF A HYDRATE

Water seems to be a part of everything that we see or touch! Water is a component of our atmosphere, water is the media by which many nutrients and salts are transported within our bodies, and water adsorbs to solid surfaces or binds to the internal network of a solid's structure.

It is this latter phenomenon that we study in this experiment. Many salts that crystallize in nature do so with the accompaniment of water molecules. In fact, many salt crystals, for example those that are "grown" in physical science laboratories, have the shiny appearance and the many brilliant facets because of the (internal) presence of water molecules. When these *hydrated* salt crystals are heated, the water molecules escape, the salt lattice crumbles and the brilliance is lost.

OBJECTIVES

- To determine the percent by mass of water in a hydrate
- To establish the mole ratio of salt to water

PRINCIPLES

Adsorb: to physically bind to the surface of a substance.

Hydrate: water molecules are chemically bound to the salt as part of the internal structure of the compound.

"$\xrightarrow{\Delta}$": a symbol indicating that heat is required for the reaction.

Many naturally occurring salts or salts purchased from chemical suppliers are hydrated; that is, water molecules are bound to the ions in the crystalline structure of the salt. The number of moles of water per mole of salt is usually constant. For example, iron(III) chloride is normally purchased as $FeCl_3 \cdot 6H_2O$, not as $FeCl_3$; copper(II) sulfate as $CuSO_4 \cdot 5H_2O$ or $CuSO_4 \cdot H_2O$, not as $CuSO_4$. Epsom salt has the formula $MgSO_4 \cdot 7H_2O$. For some salts, such as washing soda, $Na_2CO_3 \cdot 10H_2O$, heat can remove the hydrated water molecules:

$$Na_2CO_3 \cdot 10H_2O(s) \quad \xrightarrow{\Delta} \quad Na_2CO_3(s) + 10\,H_2O(g)$$

In sodium carbonate decahydrate, $Na_2CO_3 \cdot 10H_2O$, 10 moles of H_2O are bound to each mole of Na_2CO_3 or 180 g of H_2O are bound to 106 g of Na_2CO_3. The percent H_2O by mass in the hydrated salt is

$$\frac{180 \text{ g } H_2O}{(180 \text{ g } H_2O + 106 \text{ g } Na_2CO_3)} \times 100 = 62.9 \% \ H_2O$$

In other salts the water molecules cannot be easily removed, no matter how the intense the heat, e.g., $FeCl_3 \cdot 6H_2O$.

TECHNIQUES

- Technique 13, page 11   Using the Laboratory Balance
- Techniques 16a, c, page 16   Ignition of a Crucible

PROCEDURE

Complete 3 trials in this experiment. Obtain 3 crucibles and lids, identify each crucible and lid as a matched pair, and simultaneously perform the experiment in triplicate. While 1 crucible is cooling, another sample can be heated.

1. Support a clean crucible and lid on a clay triangle and heat with an intense flame for 5 minutes. Cool. If the crucible remains dirty, add a few milliliters of 6 M HNO_3 (**Caution:** *avoid skin contact, flush immediately with water*) and evaporate to dryness. Handle the crucible and lid with the crucible tongs for the rest of the experiment; do not use your fingers. Determine the mass (±0.001 g) of the cooled crucible and lid.[1]

2. Add *at most* 3 g of an unknown hydrate salt to the crucible and again determine the mass (±0.001 g) of the salt, the lid, and sample.

3. Return the crucible with the sample to the clay triangle and set the lid off the crucible's edge to allow evolved gases to escape.

Anhydrous: without water.

4. At first, heat the sample slowly and then gradually intensify the heat. Do not allow the crucible to become red hot. This can cause the anhydrous salt to decompose. Heat the sample for 15 minutes. Cover the crucible with the lid, cool to room temperature, and determine the mass of the salt, the lid, and crucible.

Desiccant: a substance that absorbs water.

5. Reheat the sample for 5 more minutes. Again measure the mass. If the second mass measurement disagrees (>± 1%) with first, repeat the heating until a mass of <1% is achieved.

Disposal Information: Discard the anhydrous salt in the "Waste Salts" container.

---

[1]Place the crucible and lid in a desiccator (if available) for cooling. Cool the crucible and lid in the same manner for the remainder of the experiment.

FORMULA OF A HYDRATE

Date _____ Name _____ Lab Sec. _____ Desk No. _____

1. Naturally occurring gypsum is a hydrate of $CaSO_4$. A 4.335 g sample of gypsum is heated in a crucible until a constant mass is obtained. The mass of the anhydrous (without water) salt, $CaSO_4$, is 3.428 g.

   a. Calculate the percent by mass of water in the gypsum sample.

   b. Calculate the moles of H_2O removed and moles of $CaSO_4$ remaining in the crucible.

   c. What is the formula of gypsum?

2. Anhydrous $CaCl_2$ is used as a desiccant in desiccators—it removes water from the atmosphere within the desiccator to form a hydrate. A 16.43 g $CaCl_2$ sample has a mass of 21.75 g after being left in a desiccator for several weeks. What is the formula of the hydrated $CaCl_2$ salt?

3. In today's experiment what error in the data is likely to occur if the hydrated salt is heated too strongly? Read the Procedure.

4. How long should the hydrated salt be heated in removing the water?

5. After the mass of the crucible is determined for the first time in Part 1 of the Procedure, you are advised to handle the crucible and lid with the crucible tongs only, *not* your fingers. Explain why this is good technique.

FORMULA OF A HYDRATE

Date _____ Name _____ Lab Sec. _____ Desk No. _____

Name of salt	Trial 1	Trial 2	Trial 3
1. Mass of crucible and lid (g)			
2. Mass of crucible, lid, and hydrated salt. (g)			
3. Mass of hydrated salt (g)			
4. Mass of crucible, lid, and anhydrous salt 1st mass measurement (g)			
2nd mass measurement (g)			
3rd mass measurement (g)			
5. Mass of anhydrous salt (g)			
6. Moles of anhydrous salt (mol)			
7. Mass of water lost (g)			
8. Moles of water lost (mol)			
9. Percent by mass of H_2O lost from hydrated salt (%)			
10. Average % H_2O in hydrated salt (%)			
11. Mole ratio of anhydrous salt to water	*		
12. Formula of hydrate			

*Calculation for Trial 1. Show your work.

1. If some volatile impurities are not burned off in Part 1, but are removed in Part 4, is the mass of anhydrous salt too high or too low? Explain.

2. If the hydrated salt is not heated to a high enough temperature for a long enough period of time in Part 4, will the reported moles of water in the hydrated salt be too high or too low? Explain.

3. What happens to the sample's reported percent water if the salt decomposes, yielding a volatile product?

4. Anhydrous $CaCl_2$ removes water vapor from the atmosphere in a desiccator. Explain how $CaCl_2$ removes the water vapor.

EMPIRICAL FORMULA OF A COMPOUND

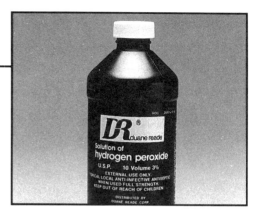

Did you ever stop to think where all of these chemical formulas that you have been reading about in the textbook and heard about in lecture come from? How do we know they are what we say they are? All chemicals have formulas! How can you possibly learn them all? Can you just look at a compound and say it necessarily has a particular formula? Is it like naming a child, once you give it a name, it will be that forever? Who says that the formula for water is H_2O and that for table sugar is $C_{12}H_{22}O_{11}$? How do we know that?

Formulas are determined from a chemical analysis, but the analysis must be quantitative. Formulas are to chemists as the keyboard is to typists; in each case, letters are placed in a certain sequence by a set of rules that is understood by everyone to create a representation of an image. For example, the word "ocean" represents an image, just as the formula H_2O represents and image, but one that is quite different from that of H_2O_2.

To determine the formula for water we need to know the amount (the number of moles) of hydrogen that reacts with the available oxygen to produce the measured mass of water; or in the decomposition of a measured mass of table sugar, the number of moles of carbon, hydrogen, and oxygen present. Thereafter, we can apply facts that we have already learned about the relative masses of the elements to determine a formula of the compound.

OBJECTIVE

• To determine the chemical formula of a compound of magnesium and oxygen or magnesium and chlorine.

PRINCIPLES

Quantitative: a measurement made with calibrated equipment, such as a balance or volumetric glassware.

An empirical formula specifies the simplest, whole-number, mole ratio of atoms in a compound. This formula can be determined in the laboratory either by its synthesis from the elements or by a decomposition into its respective elements.

An example for determining the empirical formula from a decomposition reaction is given for mercuric oxide on the Lab Preview. Sodium chloride can be synthesized from its elements: 2.75 g of sodium react with 4.25 g of chlorine. The amounts (in moles) of the two elements are:

$$2.75 \text{ g Na } \times \frac{1 \text{ mol Na}}{23.0 \text{ g Na}} = 0.120 \text{ mol Na}$$

$$4.25 \text{ g Cl } \times \frac{1 \text{ mol Cl}}{35.45 \text{ g Cl}} = 0.120 \text{ mol Cl}$$

Therefore, the mole ratio of sodium to chlorine is 0.120 to 0.120. Since the empirical formula is a whole-number mole ratio, the mole ratio of sodium to chlorine is 1 to 1 and the empirical formula is Na_1Cl_1 or $NaCl$.

The empirical formula also provides the mass ratio of the elements in the compound. The formula $NaCl$ implies that 23.00 g (1 mole) of sodium combines with 35.45 g (1 mole) of chlorine. The percents by mass of sodium and chlorine in sodium chloride are

$$\frac{23.00 \text{ g}}{(23.00 \text{ g} + 35.45 \text{ g})} \times 100 = 39.35\% \text{ Na}$$

$$\frac{35.45 \text{ g}}{(23.00 \text{ g} + 35.45 \text{ g})} \times 100 = 60.65 \% \text{ Cl}$$

This experiment describes the syntheses of compounds between magnesium and oxygen and magnesium and chlorine. For each compound the empirical formula is determined using a series of techniques, measurements, and calculations.

TECHNIQUES

- Technique 13, page 11   Using the Laboratory Balance
- Techniques 16a,b, page 16   Ignition of a Crucible

PROCEDURE

Three trials for at least one of the synthesis reactions are to be completed. Check out and identify additional crucibles and lids from the stockroom and follow the procedure in triplicate—while one synthesis reaction is occurring, prepare the next.

A. Magnesium-Oxygen Synthesis

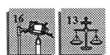

1. Fire a crucible and lid. Allow them to reach room temperature.[1] Determine the mass (± 0.001 g) of the crucible only. Repeat the firing of the crucible until a constant mass is achieved. For the rest of the experiment handle the crucible and lid with crucible tongs only.

2. Cut 0.2–0.3 g of polished (with steel wool) Mg ribbon into short lengths. Place them into the crucible, repeat the mass measurement, and record.

3. Return the crucible, lid, and Mg to the clay triangle. Heat slowly, occasionally lifting the lid to allow air to the Mg (Figure 9.1). If too much air comes in contact with the Mg, rapid oxidation of the Mg occurs and it burns brightly (Experiment 5, Part D.1) with some probable loss of product. If this happens, immediately return the lid to the crucible.

4. Continue heating until no change in appearance of the Mg ash is apparent. Remove the lid and heat the open crucible with a hot flame for several minutes. Remove the heat; add a few drops of water to decompose any magnesium nitride[2] formed during combustion. Dry the ash with a low flame, cool, determine the mass of the open crucible, and record.

5. Repeat Part A.4 until repeated mass measurements are less than $\pm 2\%$.

| Disposal Information: | Dispose of the magnesium oxide in the "Waste Salts" Container |

---

[1] Allow the crucible to cool in a desiccator if available. Continue to use the desiccator for the remainder of the experiment.

[2] $3 \text{ Mg}(s) + \text{N}_2(g) \xrightarrow{\Delta} \text{Mg}_3\text{N}_2(s)$

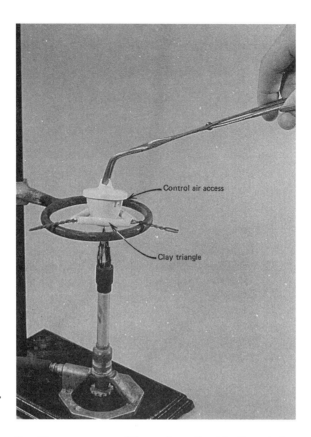

Control air access

Clay triangle

Figure 9.1
Controlling the access of air to the Mg ribbon

B. Magnesium-Chloride Synthesis

1. Prepare a crucible as described in Part A.1.

2. Cut 0.2–0.3 g of polished (with steel wool) Mg ribbon into short lengths. Place them into the crucible, measure their combined mass, and record.

3. Place the crucible containing the Mg on a clay triangle. Slowly add drops of 6 M HCl (**Caution:** *avoid skin contact)* as the Mg reacts. After no further reaction is apparent with the addition of the HCl, *gently* heat the reaction mixture to evaporate to dryness—do not boil.

4. When the sample appears dry, continue heating for an additional minute; avoid excessive heating. Cover, cool (in a desiccator if available), measure its mass, and record.

5. Repeat Part B.4 until repeated mass measurements are less than <±3%

Disposal Information: **Dispose of the magnesium chloride in the "Waste Salts" Container**

Notes, Observations, and Calculations

EMPIRICAL FORMULA OF A COMPOUND

Date _____ Name _____ Lab Sec. _____ Desk No. _____

1. Elemental oxygen was first discovered when mercuric oxide was decomposed with heat, forming mercury metal and oxygen gas. When a 1.048 g sample of mercuric oxide is heated, 0.971 g of mercury remains.

   a. Calculate the moles of mercury and oxygen in the sample.

   b. What is the empirical formula of mercuric oxide?

2. A 2.60 g sample of titanium metal chemically combines with 7.71 g of chlorine gas. Determine the empirical formula for the titanium chloride salt.

3. A 0.497 g sample of chromium metal forms an oxide compound with a mass of 0.726 g.
   a. Determine the empirical formula of the chromium oxide.

   b. Calculate the percent by mass of chromium and oxygen in the compound.

4. How many grams of oxygen gas combine with 0.843 g vanadium to form V_2O_5?

5. Calculate the percent by mass of sulfur and bismuth in Bi_2S_3.

6. Explain how the mass of chlorine is determined in the synthesis of the magnesium-chlorine compound in Part B of the Procedure.

EMPIRICAL FORMULA OF A COMPOUND

Date _____ Name _____ Lab Sec. _____ Desk No. _____

Synthesis	Trial 1	Trial 2	Trial 3
1. Mass of crucible and lid (*g*)			
2. Mass of crucible, lid, and Mg (*g*)			
3. Mass of Mg (*g*)			
4. Moles of Mg (*mol*)			
5. Mass of crucible, lid, and sample after reaction 1st mass measurement (*g*)			
2nd mass measurement (*g*)			
3rd mass measurement (*g*)			
6. Mass of compound (*g*)			
7. Mass of oxygen/chlorine (*g*)			
8. Moles of oxygen/chlorine (*mol*)			
9. Mole ratio of Mg to oxygen/chlorine			
10. Empirical formula of compound			
11. Percent Mg by mass in compound (%)			
12. Percent oxygen/chlorine by mass in compound (%)			

* Show sample calculation for Trial 1

1. If in Part A.3 the magnesium oxidation is uncontrolled (burns brightly), how will this error in experimental technique affect the reported mass of

   a. magnesium in the sample? Explain.

   b. oxygen in the sample? Explain.

2. In this experiment the mass of the magnesium is determined to the nearest milligram but not the oxygen or chlorine. Explain why an exact mass measurement of oxygen or chlorine is unnecessary.

3. Suppose that in the synthesis reactions, the magnesium metal is not polished.

   a. Would the moles of magnesium that actually react be greater or less than the amount measured?

   b. How would this same error affect the reported moles of oxygen/chlorine that combine with the magnesium?

4. In Part B, suppose all of the magnesium does not react with the HCl.

   a. How will this affect the reported mass of chlorine in the magnesium chloride compound?

   b. Does this affect the reported number of moles magnesium in the compound? Explain.

EXPERIMENT 10
LIMITING REACTANT

Every process, that requires two or more "things" to complete, is faced with the prospect that one of the two things will be used up first. Think about the situation where there are 10 hotdogs in a package but there are 8 hotdog buns per package. Two hotdogs will be left over unless another package of buns is purchased, but then 6 buns will be left without a hotdog, and the dilemma continues.

So it is with chemical reactions requiring two or more reactants. One of the chemicals is necessarily depleted before the other, in which case the reaction stops and there will be an excess of the other chemical. Seldom is a chemical system created in which there is the *exact* number of molecules for both reactants present in the system; if that situation ever did happen to exist it might take "forever" for the last two reactant molecules to find each other. This experiment reflects the dilemma of a limiting reactant that all scientists encounter in the "real life" of the laboratory.

OBJECTIVES

- To determine the limiting reactant in the formation of a precipitate
- To determine the percent composition of a salt mixture

PRINCIPLES

Stoichiometrically: according to a balanced equation.

Two factors that limit the yield of products in a chemical reaction are (1) the amounts of starting materials (reactants) and (2) the percent yield of the reaction. Many experimental conditions (for example, temperature and pressure) can be adjusted to increase the yield for a reaction, but because chemicals react according to fixed mole ratios (stoichiometrically), only a limited amount of product can form from given amounts of starting material. The reactant restricting the amount produced is the **limiting reactant** in that chemical system.

To better understand the limiting reactant concept, let's look at the reaction studied in this experiment. The molecular form of the equation is

$$BaCl_2(aq) + Na_2SO_4(aq) \rightarrow BaSO_4(s) + 2\,NaCl(aq)$$

Since $BaCl_2$ (in the form of the dihydrate, $BaCl_2 \cdot 2H_2O$) and Na_2SO_4 are soluble salts and since $BaSO_4$ is insoluble, the ionic equation is

$$Ba^{2+}(aq) + 2\,Cl^-(aq) + 2\,Na^+(aq) + SO_4^{2-}(aq) \rightarrow$$
$$BaSO_4(s) + 2\,Na^+(aq) + 2\,Cl^-(aq)$$

Spectator ions: ions that do not participate in a chemical reaction.

Cancelling spectator ions common to both sides of the equation, the net ionic equation is

$$Ba^{2+}(aq) + SO_4^{2-}(aq) \rightarrow BaSO_4(s)$$

Precipitate: the formation on an insoluble ionic compound, generally in water.

One mole of Ba^{2+} [from 1 mole of $BaCl_2 \cdot 2H_2O$ (244.2 g/mol)] reacts with 1 mole of SO_4^{2-} [from 1 mole of Na_2SO_4 (142.1 g/mol)] to produce 1 mole of $BaSO_4$ precipitate (233.4 g/mol), if the reaction proceeds to completion.

Suppose, however, only 2.00 g of $BaCl_2 \cdot 2H_2O$ and 1.20 g of Na_2SO_4 are in the reaction vessel. How many moles and grams of $BaSO_4$ are then produced? Since the equation reads in terms of moles, not grams, the number of moles (or millimoles, mmol) of each reactant must be determined. Therefore,

$$2.00 \text{ g } BaCl_2 \cdot 2H_2O \times \frac{1 \text{ mol } BaCl_2 \cdot 2H_2O}{244.2 \text{ g } BaCl_2 \cdot 2H_2O} = 0.00819 \text{ mol } BaCl_2 \cdot 2H_2O$$
$$= 8.19 \text{ mmol } Ba^{2+}$$

$$1.20 \text{ g } Na_2SO_4 \times \frac{1 \text{ mol } Na_2SO_4}{142.1 \text{ g } Na_2SO_4} = 0.00844 \text{ mol } Na_2SO_4 = 8.44 \text{ mmol } SO_4^{2-}$$

Since 1.0 mol Ba^{2+} reacts with only 1.0 mol SO_4^{2-}, then the 8.19 mmol Ba^{2+} in the reaction vessel can only react with 8.19 mmol SO_4^{2-} (of the 8.44 mmol present) producing a maximum of 8.19 mmol $BaSO_4$. This consumes all of the Ba^{2+} in the vessel (Ba^{2+} is the *limiting reactant*) and leaves an excess of (8.44 – 8.19 =) 0.25 mmol of SO_4^{2-} (SO_4^{2-} is called the *excess reactant*).

Since the limiting reactant (Ba^{2+}) is now known, the theoretical yield of product ($BaSO_4$) is calculated from the balanced equation. Ba^{2+}, the limiting reactant, controls the moles and mass of $BaSO_4$ produced.

From the balanced equation, 1.0 mol Ba^{2+} produces 1.0 mol $BaSO_4$

or 8.19 mmol Ba^{2+} produces 8.19 mmol $BaSO_4$ *or* 0.00819 mol $BaSO_4$

and $0.00819 \text{ mol } BaSO_4 \times \frac{233.4 \text{ g } BaSO_4}{1 \text{ mol } BaSO_4} = 1.91 \text{ g } BaSO_4$

Thus, 1.91 g $BaSO_4$ forms if the reaction is 100% complete.

In this experiment a solid, unknown mixture of Na_2SO_4 and $BaCl_2 \cdot 2H_2O$ is added to water and $BaSO_4$ precipitates from the solution. We will measure the mass and calculate the moles of $BaSO_4$ that precipitate; from the balanced equation we can calculate the moles and masses of $BaCl_2 \cdot 2H_2O$ and Na_2SO_4 in the original unknown mixture.

From a series of tests, the limiting and excess reactants are determined. The difference between the mass of the original salt mixture, m_{sm}, and the mass of the limiting reactant, m_{lr} allows us to calculate the mass of excess reactant, m_{xr}, in the salt mixture

$$m_{xr} = m_{sm} - m_{lr}$$

The percent composition of the salt mixture is also determined.

TECHNIQUES

- Technique 5, page 3 — Transferring Liquid Reagents
- Technique 6c, page 4 — Heating Liquids
- Technique 13, page 11 — Using the Laboratory Balance
- Technique 14a, b, c, page 12 — Separation of a Liquid from a Solid
- Technique 15, page 15 — Flushing a Precipitate from a Beaker

PROCEDURE

Two trials are recommended for this experiment. To hasten the analyses, determine the mass of duplicate unknown salt mixtures, and simultaneously follow the procedure for each. Label the beakers accordingly for Trial 1 and Trial 2 to avoid intermixing samples and solutions.

A. Precipitation of BaSO₄

1. Measure about 1.0 g (±0.001 g) of the unknown salt mixture on weighing paper. Transfer it to a 400 mL beaker and add 200 mL (±0.2 mL) of distilled (or deionized) water. Add, using a stirring rod, 1 mL conc HCl (**Caution:** conc HCl *is a severe skin irritant. Flush the affected area with a large amount of water*). Stir the aqueous mixture with the stirring rod for about 1 minute and then allow the precipitate to settle.

2. Cover the beaker with a watchglass and maintain the solution at a temperature between 80°C and 90°C on a steam bath or with a low flame for 40-50 minutes (Figure 10.1).[1] Remove the heat and allow the precipitate to settle.

3. Decant two 50 mL (±0.2 mL) volumes of the supernatant liquid into separate 100 mL beakers, labeled Beaker I and Beaker II; save for Part B.

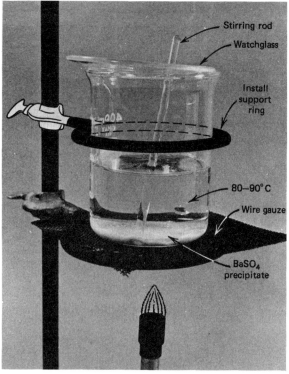

Figure 10.1
Warm the precipitate to digest the BaSO₄ precipitate

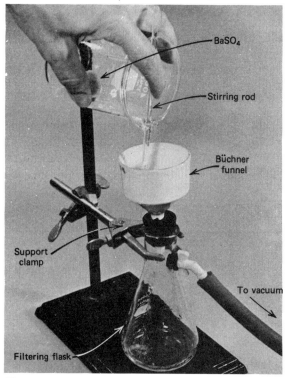

Figure 10.2
Quantitative Transfer of the Precipitate

---

[1]The purpose of this step is to digest the precipitate; that is, by allowing the precipitate to maintain an equilibrium with its ions in solution for an extended period of time, larger aggregates of particles will form, making the filtration of the precipitate more efficient.

4. The $BaSO_4$ precipitate may either be gravity filtered or vacuum filtered. In either case, use fine porosity filter paper, such as Whatman No. 42 or Fisher*brand* Q2. Premeasure the mass (±0.001 g) of the dry filter paper and seal it into the funnel with a few milliliters of distilled (or deionized) water. Discard this water from the receiving flask. Have your instructor approve your filtering apparatus.

5. While the solution is still warm, quantitatively transfer the precipitate to the funnel (Figure 10.2). Remove any precipitate from the beaker wall with a rubber policeman; use hot water to swirl and transfer the precipitate onto the filter. Wash the precipitate on the filter paper with two 5 mL portions of hot water.

6. First, dry the precipitate on the filter paper; then dry overnight in a constant-temperature drying oven set at 110°C, or until the next laboratory period. Measure and record the mass of the filter paper and precipitate.

B. Determination of the Excess (and the Limiting) Reactant

The limiting reactant in the salt mixture is determined in the following tests. Use the two 50 mL volumes collected in Part A.3. See Figure 10.3.

1. **Testing for excess Ba^{2+}.** Add 2 drops of 0.5 M SO_4^{2-} ion (from 0.5 M Na_2SO_4) to the 50 mL of solution in Beaker I. If a precipitate forms, Ba^{2+} is in excess and SO_4^{2-} is the limiting reactant in the original salt mixture.

2. **Testing for excess SO_4^{2-}.** Add 2 drops of 0.5 M Ba^{2+} ion (from 0.5 M $BaCl_2$) to the other 50 mL of solution in Beaker II. If a precipitate forms, SO_4^{2-} is in excess and Ba^{2+} is the limiting reactant in the original salt mixture.

Disposal Information: Dispose of the $BaSO_4$ precipitate in the "Solid Salts" container and the test solutions in the "Waste Salt Solution" container.

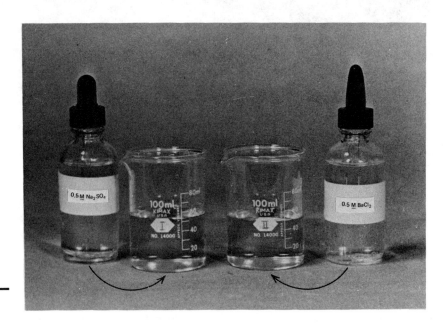

Figure 10.3
Testing for the excess (and the limiting) reactant

LIMITING REACTANT

Date _____ Name _____ Lab Sec. _____ Desk No. _____

1. If 1.668 g of $BaCl_2 \cdot 2H_2O$ is mixed with an excess of Na_2SO_4 to form 200 mL of solution, what mass of $BaSO_4$ can form?

2. A 1.668 g sample of $BaCl_2 \cdot 2H_2O$ is mixed with 1.492 g of Na_2SO_4 to form 200 mL of solution.

    a. Calculate the moles of $BaCl_2 \cdot 2H_2O$ and the moles of Na_2SO_4 in the mixture.

    b. What is the limiting reactant for the precipitation of $BaSO_4$?

    c. What mass of $BaSO_4$ can form?

3. How is the test for the presence of excess $BaCl_2 \cdot 2H_2O$ in your unknown salt mixture conducted in today's experiment?

4. Describe the technique for transferring a precipitate from a beaker to a funnel.

5. A 1.582 g sample of a $BaCl_2 \cdot 2H_2O/Na_2SO_4$ salt mixture, when mixed with water, filtered, and dried, produced 0.713 g of $BaSO_4$. The Na_2SO_4 salt was determined to be the limiting reactant.

   a. Calculate the mass of Na_2SO_4 in the mixture.

   b. Calculate the mass of $BaCl_2 \cdot 2H_2O$ in the mixture.

   c. What is the percentage of Na_2SO_4 and $BaCl_2 \cdot 2H_2O$ in the salt mixture?

EXPERIMENT 10

LIMITING REACTANT

Date _____ Name _____ Lab Sec. _____ Desk No. _____

A. Precipitation of BaSO₄

Unknown Number _____

	Trial 1	Trial 2
1. Mass of salt mixture (*g*)		
2. Mass of filter paper (*g*)		
3. Instructor's approval of filtering apparatus		
4. Mass of filter paper and precipitate (*g*)		
5. Mass of BaSO₄ precipitate (*g*)		

B. Determination of the Excess (and the Limiting) Reactant

1. Limiting reactant in original salt mixture _____

2. Excess reactant in original salt mixture _____

CALCULATIONS	Trial 1	Trial 2
1. Moles of BaSO₄ precipitate (*mol*)		
2. Moles of BaCl₂•2H₂O reacted (*mol*)		
3. Mass of BaCl₂•2H₂O reacted (*g*)		
4. Moles of Na₂SO₄ reacted (*mol*)		
5. Mass of Na₂SO₄ reacted (*g*)		
6. Mass of salt mixture (from A.1), (*g*)		
7. Mass of excess reactant (*g*)		
8. Percent BaCl₂•2H₂O in mixture		
9. Percent Na₂SO₄ in mixture		

1. Because $BaSO_4$ is a very finely divided precipitate, some is lost in the filtering process. If coarse filter paper is used instead of one with fine porosity, will the reported percent of limiting reactant in the original salt mixture be high or low? Explain.

2. The solubility of $BaSO_4$ at 25°C is 9.04 mg/L. How many grams and moles of $BaSO_4$ dissolve in the 200 mL of solution?

3. How do excessive quantities of wash water affect the amount of $BaSO_4$ collected on the filter (see Part A.5)? Explain.

4. a. Sodium carbonate, Na_2CO_3, is an inherent contaminant of Na_2SO_4. How does its presence affect the expected mass of $BaSO_4$ reported? $BaCO_3$ is also insoluble.

   b. Will this cause the reported percent of the limiting reactant to be high or low? Explain.

EXPERIMENT 11

CALORIMETRY

Nearly all chemical and physical changes that occur in the laboratory or in nature are accompanied by a change in energy, most often in the form of thermal energy or heat. Energy changes accompany the changing colors of the leaves in the Fall, the growing of a child to an adult, the blinking of an eye, and the corrosion of a metal; these examples are all energy changes accompanying chemical reactions. The melting of ice, the release of a ball toward the Earth, and the cooling of magma are energy changes accompanying physical changes. Thermal energy, then, is a very important part of nature; early scientists considered energy to be one of the four basic elements of nature, the other three being Earth, air, and water; the energy was needed to make changes from one basic element to another.

OBJECTIVES

- To determine the specific heat capacity of a metal
- To measure the heat of neutralization for a strong acid–strong base reaction
- To measure the heat evolved or absorbed for the dissolution of a salt

PRINCIPLES

A calorimeter is the apparatus used to measure the quantity of heat evolved or absorbed in a chemical or physical change. Changes which evolve heat are exothermic; their ΔH values (enthalpy of change) are negative. Changes that absorb heat are endothermic; their ΔH values are positive.

Enthalpy of change: the energy change accompanying a chemical or physical change at constant pressure.

Three calorimetric measurements are made in this experiment. Each requires the measurement of temperature changes, masses of materials, and the plotting of the data.

Specific Heat Capacity of a Metal

The energy (or heat in joules) required to change the temperature of 1 g of a substance by 1 °C is the specific heat capacity (commonly, specific heat) of that substance.

$$\text{specific heat capacity (sp. ht.)} = \frac{\text{heat (joules)}}{\text{mass(grams)} \times \Delta T(^\circ C)} \text{ or,}$$
$$\text{heat} = \text{sp. ht.} \times \text{mass} \times \Delta T$$

The temperature change of the substance is ΔT. The specific heat capacity for most substances changes only slightly with temperature; we will assume that it remains constant over the temperature range used in this experiment. The specific heat capacity of H_2O is 4.18 J/(g•°C).

The specific heat capacity of a metal (that is nonreactive with water) is measured by first heating a known mass of the metal to a measured temperature. The metal is then placed into a calorimeter containing a known mass of (cooler) water also at a measured temperature. The heat from the metal is transferred to the water until the metal and water reach the same temperature. This final equilibrium temperature is recorded. Expressed in equation form

$$\text{heat(J) lost by metal(M)} = \text{heat(J) gained by water(H}_2\text{O)}$$

A substitution of the specific heat capacity equation given above for heat in this equation gives

$$\text{sp. ht. (M)} \times \text{mass (M)} \times \Delta T(°C) = \text{sp. ht. (H}_2\text{O)} \times \text{mass (H}_2\text{O)} \times \Delta T(°C)$$

The specific heat capacity of the metal is calculated from a rearrangement of the equation,

$$\text{sp. ht. (M)} = \frac{\text{sp. ht. (H}_2\text{O)} \times \text{mass (H}_2\text{O)} \times \Delta T \text{ (H}_2\text{O)}}{\text{mass (M)} \times \Delta T \text{ (M)}}$$

Each equation assumes no heat loss to the calorimeter.

Enthalpy of Neutralization for an Acid-Base Reaction

The reaction of a strong acid with a strong base produces water and heat as products:

$$H^+(aq) + OH^-(aq) \rightarrow H_2O(l) + \text{heat} \quad (\Delta H_n \text{ is negative})$$

The enthalpy of neutralization, ΔH_n, is calculated by (a) assuming the densities and specific heat capacities of the acidic and basic solutions are the same as that for water and (b) measuring the temperature change when the two solutions are mixed.

$$\Delta H_n = \text{sp. ht. (H}_2\text{O)} \times \text{mass (acid + base)} \times \Delta T \text{ (solution)}$$

ΔH_n is generally expressed in units of kJ/mol of acid (or base) reacted.

Enthalpy of Solution for a Salt

The dissolving of a salt may be an endothermic or exothermic process depending upon two factors often considered in the dissolution of the salt: the lattice energy of the salt and the hydration energy of the ions. The lattice energy is the energy required (endothermic process, a $+\Delta H_{LE}$) to vaporize 1 mole of the solid salt into its gaseous ions; the hydration energy is the energy released (exothermic quantity, a $-\Delta H_{hyd}$) when the gaseous ions are attracted to and surrounded by water molecules. The enthalpy of solution of a salt, ΔH_s, is the sum of these two factors (Figure 11.1 for NaCl).

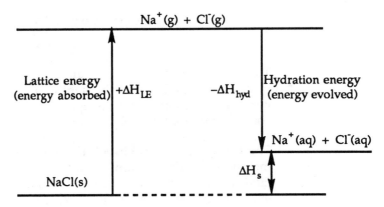

Figure 11.1
Energy changes in the dissolving of NaCl in water

The enthalpy of solution, ΔH_s, is determined by adding the heat changes for the salt and water.

$$\Delta H_S = \text{heat change } (H_2O) + \text{heat change (salt)}$$

$$\Delta H_S = \frac{\text{sp. ht.}(H_2O) \times \text{mass}(H_2O) \times \Delta T(H_2O) + \text{sp. ht.(salt)} \times \text{mass(salt)} \times \Delta T(\text{salt})}{\text{mass(salt)}}$$

ΔH_S is generally expressed in units of kJ/g salt. The specific heat capacities of some common salts are listed in Table 11.1.

Table 11.1. Specific Heat of Some Salts

Salt	Formula	Specific Heat Capacities(J/g•°C)
ammonium chloride	NH_4Cl	1.57
ammonium nitrate	NH_4NO_3	1.74
ammonium sulfate	$(NH_4)_2SO_4$	1.409
sodium hydroxide	$NaOH$	1.49
sodium sulfate	Na_2SO_4	0.903
sodium thiosulfate pentahydrate	$Na_2S_2O_3 \cdot 5H_2O$	1.45
potassium bromide	KBr	0.435
potassium chloride	KCl	0.688
potassium hydroxide	KOH	1.16
potassium nitrate	KNO_3	0.95

TECHNIQUES

- Technique 6c, page 4 — Heating Liquids
- Technique 8, page 7 — Reading a Meniscus
- Technique 13, page 11 — Using the Laboratory Balance
- Techniques 18, page 18 — Graphing Techniques

PROCEDURE

Ask your instructor which parts of the experiment you are to complete. You and a partner are to complete two trials for each part assigned. Two styrofoam cups, a lid, and a 110°C thermometer must be obtained from the stockroom—this will serve as your calorimeter. In this experiment do *not* use the thermometer as a stirring rod!!

A. Specific Heat of a Metal

Thermal equilibrium: the temperature of two objects in contact is the same.

1. a. Ask the instructor for a ≈10 g sample of an unknown metal. Measure its mass (±0.01 g) in a dry, previously determined (±0.01 g) 200 mm test tube.

   b. Place the 200 mm test tube in a 400 mL beaker filled with water well above the level of the metal sample in the test tube (Figure 11.2).

   c. Heat the water to boiling and maintain this temperature for at least 5 minutes so that the metal reaches thermal equilibrium with the boiling water. Measure the temperature of the water (±0.1 °C).

2. a. Set up the calorimeter (Figure 11.3). Using a graduated cylinder, add approximately 25.0 mL (±0.1 mL) of water to the calorimeter. Secure the thermometer with a small three-pronged clamp. Be certain the thermometer bulb is below the water level.

   b. Read the temperature of the water several times. Record the temperatures (±0.1°C) at 1 min intervals on the Data Sheet.

3. a. Once thermal equilibrium has been reached in Parts A.1 and A.2, remove the test tube from the boiling water; quickly transfer the metal to the calorimeter. Be careful not to break the thermometer and not to splash water from the calorimeter. Replace the lid and swirl gently.

   b. Record the temperature (±0.1 °C) as a function of time (10–20 s intervals) on the table on the Data Sheet.

Disposal Information:	Return your metal to the appropriately marked container for the unknown metals.

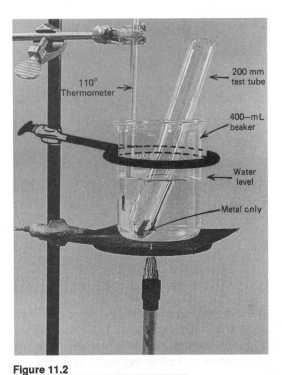

Figure 11.2
The placement of the metal below the level of the water in the beaker

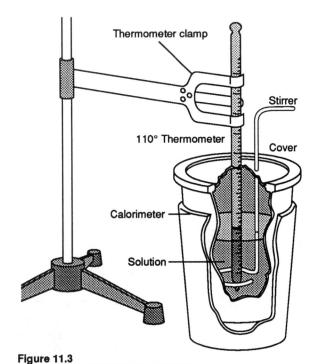

Figure 11.3
The setup of a calorimeter

 4. Plot on linear graph paper, temperature (ordinate) *vs* time (abscissa) (Figure 11.4). The maximum temperature reached by the mixture occurs at the intersection of two lines: a straight line drawn perpendicular to the initial temperature/time line at the time when the metal is added to the calorimeter and the best straight line drawn through the points after the recorded maximum temperature is reached. The maximum temperature is never actually measured because heat is always being transferred to or from the wall of the calorimeter. Have your instructor approve your graph.

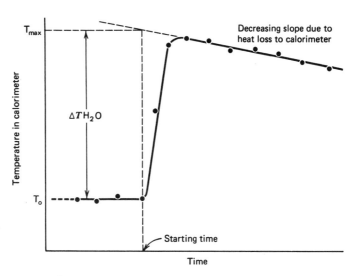

Figure 11.4

The temperature of the water in the calorimeter as a function of time

B. Enthalpy of Neutralization for an Acid-Base Reaction

1. a. Clean and dry the calorimeter. Using a graduated cylinder, pour 50.0 mL (±0.1 mL) of standard[1] 1.0 M NaOH solution into the calorimeter, secure and insert the thermometer through the lid into the solution.

    b. Record the temperature (±0.1 °C) of the NaOH solution at 1 min intervals. Record the exact concentration of the NaOH solution on the Data Sheet.

2. a. Measure 50.0 mL (±0.1 mL) of 1.1 M HCl in a clean, graduated cylinder. Rinse the thermometer to remove any NaOH solution and then measure the temperature of the HCl solution. The NaOH and the HCl solutions should be at the same temperature.

    b. Carefully but quickly, add the acid to the base, replace the lid, and swirl gently.

3. Record the solution temperature (±0.1 °C) as a function of time on the Data Sheet. Plot on linear graph paper and interpret the maximum temperature, as obtained in Part A.4. Have your instructor approve your graph.

C. Enthalpy of Solution for a Salt

1. Clean, rinse, and dry your calorimeter. Add approximately 25.0 mL (±0.1 mL) of distilled (or deionized) water to the calorimeter and record its temperature (±0.1 °C).

2. Measure about 5.0 g (±0.01 g) of your assigned salt on weighing paper. Add the salt to the calorimeter, replace the lid, and swirl gently. Measure and record the temperature as a function of time on the Data Sheet.

3. Plot on linear graph paper and interpret the maximum temperature *change* (Part A.4). Have your instructor approve your graph.

---

[1]A standard solution is one whose concentration has been very accurately determined.

This is a plain jacket bomb calorimeter that can be used for measuring the heat of combustion (at constant volume), ΔH_{comb}, for any solid or liquid fuel, such as the various foodstuffs found in diet books.

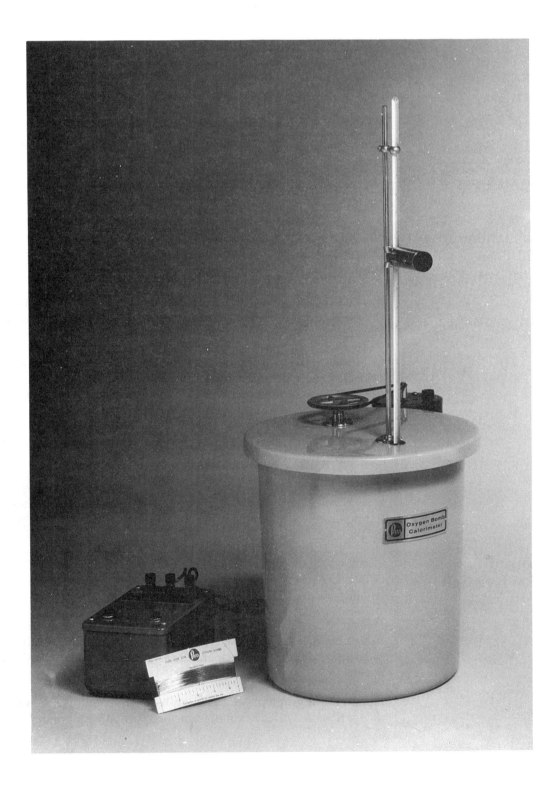

CALORIMETRY

Date _____ Name_____ Lab Sec. _____ Desk No. _____

1. A metal of mass 13.11 g and at 81.0 °C is placed in a calorimeter containing 25.0 mL of water at 25.0 °C. The final equilibrium temperature is 30.0 °C. What is the specific heat capacity of the metal? The specific heat capacity of H_2O is 4.18 J/(g•°C). Assume the density of water to 1.00 g/mL.

2. A 4.50 g sample of a salt dissolves in 30.0 mL of water initially at 25.0 °C. The final equilibrium temperature is 18.0 °C. What is the enthalpy of solution per gram of salt? The specific heat of the salt is 0.692 J/(g•°C).

3. In Part B, excess moles of HCl are added to the NaOH solution for the neutralization reaction. Why is this procedure preferred rather that adding a 1:1 mole ratio according to the balanced equation?

4. Will the recorded temperature change for an exothermic reaction inside a glass calorimeter be greater or less than that for today's styrofoam "coffee cup" calorimeter? Assume glass to be a better conductor of heat than styrofoam.

5. (a) How can "bumping" be avoided when water is being heated in a beaker?

(b) When determining the volume of solution in a graduated cylinder, you should always read the

_____of the meniscus.

(c) A balance with _____ g sensitivity is used in today's experiment.

(d) The margin of error in reading the thermometer in today's experiment is ± _____ °C.

(e) What are you *not* to do with your thermometer in today's experiment?

CALORIMETRY

Date _____ Name _____ Lab Sec. _____ Desk No. _____

A. Specific Heat Capacity of a Metal

Unknown number _____	Trial 1	Trial 2
1. a. Mass of test tube + metal (g)		
b. Mass of test tube (g)		
c. Mass of metal (g)		
d. Temperature of metal ($°C$)		
2. a. Volume of water in calorimeter (mL)		
b. Mass of water (g) Assume density of water is 1.0 g/mL		
c. Temperature of water ($°C$)		
3. a. Instructor's approval of graph		
b. Maximum temperature of metal and water from graph ($°C$)		

Calculations

	Trial 1	Trial 2
1. Temperature change of water, ΔT ($°C$)		
2. Heat gained by water (J)		
3. Temperature change of metal, ΔT ($°C$)		
4. Specific heat capacity of metal ($J/g\,°C$)		
5. Average specific heat capacity of metal ($J/g\,°C$)		

B. Enthalpy of Neutralization for an Acid-Base Reaction

	Trial 1	Trial 2
1. a. Concentration of NaOH solution (mol/L)		
b. Initial temperature of NaOH solution ($°C$)		
c. Volume of NaOH solution (mL)		
2. a. Initial temperature of HCl solution ($°C$)		
b. Volume of HCl solution (mL)		
3. a. Instructor's approval of graph		

b. Maximum final temperature of mixture from graph (°C) _____ | _____

Calculations

1. Average initial temperatures (NaOH + HCl) _____ | _____

2. Temperature change, ΔT (°C) _____ | _____

3. Mass of final mixture (g)
   Assume density of mixture is 1.0 g/mL _____ | _____

4. Specific heat of solution   4.18 J/g•°C _____ | _____

5. Heat evolved (J) _____ | _____

6. Moles of OH⁻ reacted (limit. reactant) (mol) _____ | _____

7. Moles of H_2O formed (mol) _____ | _____

8. Heat evolved per mole of water (J/mol H_2O) _____ | _____

9. Average ΔH_n (kJ/mol H_2O) _____

C. Enthalpy of Solution for a Salt

Salt_____

	Trial 1	Trial 2
1. a. Volume of water (mL)		
b. Mass of water (g). Assume density of H_2O is 1.0 g/mL		
c. Initial temperature of water (°C)		
2. Mass of salt (g)		
3. a. Instructor's approval of graph		
b. Maximum (or minimum) temperature of solution from graph (°C)		

Calculations

1. Change in temperature of solution, ΔT (°C) _____ | _____

2. Enthalpy change of water (J) _____ | _____

3. Enthalpy change of salt (J). See Table 11.1 _____ | _____

4. Enthalpy of solution, ΔH_s (J) _____ | _____

5. ΔH_s per gram of salt (J/g) _____ | _____

6. Average ΔH_s per gram salt (J/g) _____

Specific Heat Capacity of a Metal				Enthalpy of Neutralization for an Acid-Base Reaction				Enthalpy of Solution for a Salt			
Trial 1		Trial 2		Trial 1		Trial 2		Trial 1		Trial 2	
Temp	Time	Temp	Time	Temp	Time	Temp	Time	Temp	Time	Temp	Time

QUESTIONS

1. The coffee cup calorimeter, although a good insulator, absorbs some heat when the system is above room temperature.

   a. How does this affect the specific heat capacity of the metal?

   b. How does this affect the reported ΔH_n value in the acid-base reaction?

2. The specific heat capacity of the styrofoam calorimeter is 1.34 J/g°C. If we assume that the entire inner cup reaches thermal equilibrium with the solution, how much heat is lost to the calorimeter in Part B? Assume the inner cup has a mass of 2.35 g.

3. If the maximum recorded temperature in Part B is used to determine ΔT rather than the maximum extrapolated temperature, will the reported ΔH_n be greater or less than the actual ΔH_n for this reaction? Explain.

4. Suppose that when the metal was transferred to the calorimeter in Part A, some of the water splashed from the calorimeter. How does this affect the reported specific heat capacity of the metal? Explain.

EXPERIMENT 12

SPECTROSCOPY

The interaction of light with matter is a phenomenon that is intriguing to most everyone. Think about the effects of public lighting on the colors of skin tones, clothes, and automobiles. The mercury vapor lights tend to make objects appear blue, whereas the sodium vapor lights emit a yellow-orange hue. What this does to affect other visible colors is considered "strange"! For example, a red car appears gray under a low pressure sodium vapor light.

Sunlight is considered "white light" in that all colors of the visible spectrum are emitted. When sunlight strikes an object that absorbs none of the visible light, the object appears white, but if the object absorbs all of the visible light, then the object appears black. But if the object absorbs the blue light from the sunlight, it then appears yellow, the complimentary color of blue in the color wheel. How and why atoms or molecules of an object absorb or do not absorb sunlight is the basis of this experiment.

OBJECTIVES

- To account for a line spectrum
- To observe the flame emission colors from excited metal atoms
- To study the hydrogen spectrum
- To identify an element from its spectrum

PRINCIPLES

The current model of the atom includes a nucleus, comprised of protons and neutrons, surrounded by electrons. From experimental observations of atomic absorption and emission spectra, and from a quantum mechanical interpretation of spectra, the energy of an electron in an atom is assumed to be *quantized*; that is, an electron in an atom can have only discrete energies.

When an atom absorbs energy from a flame or electric discharge, it absorbs the energy necessary to excite one of its electrons to a higher energy state—the atom is now in an excited state. When the electron returns to its original energy state, it emits the energy previously absorbed in the form of one or more photons. The excited atom decreases in energy by ΔE_{atom} when the electron returns from the higher, E_h, to a lower, E_l, energy state. This energy change of the atom equals the energy of one photon.

Photon: a quantity of energy having mass-like properties.

E_h ___ e^- ___ higher energy state

$\sim\sim\sim \rightarrow E_{photon}$

E_l ___ lower energy state

$$\Delta E_{atom} = E_h - E_l = E_{photon}$$

The photon energy, E_{photon}, equal to the energy difference between the two energy states for the electron, is inversely proportional to its wavelength, λ, by the equation,

$$E_{photon} = h\frac{c}{\lambda} = h\nu$$

c is the speed of light, 3.00×10^8 m/s, h is Planck's constant, 6.63×10^{-34} J s/photon, ν is the frequency of the photon in s^{-1} (or Hertz, Hz), and λ is the wavelength of the photon in meters.

Many possible energy states exist for an electron in an atom. For example, suppose that the same electron in a large number of atoms of an element is excited to the same (higher) energy state. As the electrons of the various atoms return to the ground state (the lowest energy state), each according to its own pathway, photons of different energies are emitted; the pathway for the deexcitation of an electron depends on the stability of the intermittent energy states of the atom. When these photons pass through a prism, they produce a **line spectrum** (Figure 12.1); each line results from a particular electron transition in a large number of atoms.

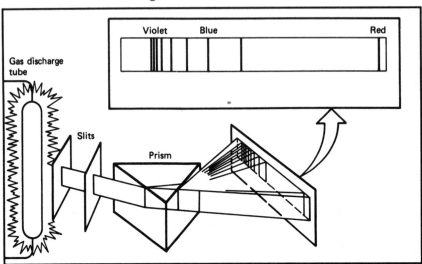

Figure 12.1
A line spectrum

Each element has its own line spectrum because the energy states for electrons in its atoms are unique. Sodium having 11 electrons has a different set of electronic energy states, and thus exhibits a different line spectrum, than does calcium (with 20 electrons) or mercury (with 80 electrons). The most prevalent mode of electron deexcitation produces the most intense line in the spectrum, producing the color characteristic of that element. For example, many of the electrons in excited sodium atoms commonly deexcite to produce a photon with a wavelength in the orange part of the visible spectrum. We, therefore, associate excited sodium atoms as being orange which explains the yellow-orange hue emitted by sodium vapor lights.[1] By the same account, we associate blue light with mercury vapor lamps (a deexcitation of electrons in excited mercury atoms emit a photon with a wavelength in the blue region of the spectrum) and green with excited barium atoms. The characteristic colors associated with a number of excited atoms are observed in their flame tests of Part A of this experiment.

Hydrogen Spectrum

The value or designation of each electronic energy state in a hydrogen atom is calculated in this experiment, using the equation,

$$\frac{1}{\lambda} = R\left[\frac{1}{n_l^2} - \frac{1}{n_h^2}\right]$$

The wavelength, λ, of photons emitted from excited hydrogen atoms is related to whole number integers, n, which identify the higher energy states for the electron before deexcitation, n_h, and the lower energy states after its deexcitation, n_l. For all electron deexcitations, $n_h > n_l$; for the visible part of

---

[1]High and low pressure sodium vapor lamps are now common for lighting streets and highways in public areas.

the hydrogen spectrum, n_l equals 2. R, the Rydberg constant, is approximately 1.1×10^{-2} nm^{-1}.

In Part B of this experiment, we will observe the hydrogen spectrum, estimate the wavelengths of the emitted photons, and calculate the energy change of the atom associated with each photon. Next we will calculate an n_h value for each line in the spectrum, which will then be rounded off to the *nearest* integer. A plot of the data will enable us to determine a more precise value of the Rydberg constant, R. This is accomplished by rearranging the equation $\dfrac{1}{\lambda} = R\left[\dfrac{1}{n_l{}^2} - \dfrac{1}{n_h{}^2}\right]$ into an equation for a straight line, y = mx + b.

$$y = m\ x\ +\ b$$
$$\frac{1}{\lambda} = -R\left[\frac{1}{n_h{}^2}\right] + \left[\frac{R}{n_l{}^2}\right]$$

Since $n_l = 2$ for the *visible* spectrum of hydrogen, a plot of $1/\lambda$ *versus* $1/n_h{}^2$ has a y-intercept of R/4 and a slope of -R. Remember n_h must be a whole number.

In Part C, an element is identified by determining the position of its "lines" in the line spectrum relative to the known wavelengths from the mercury line spectrum, Table 12.1. Then a match of the unknown's spectrum with the spectra for several elements in Table 12.2 is made for its identification.

PROCEDURE

A. Flame Tests

1. Dip the end of a platinum or nichrome wire into conc HCl (**Caution:** *Avoid skin contact. Flush affected area with large amounts of water*). Heat the wire in the hottest region of the flame (Figure 2.3) until there is no visible color (Figure 12.2). Repeat this cleaning procedure as necessary.

2. In a watch glass, add a few crystals of $CaCl_2$ to 2–3 drops of conc HCl. Stir. Dip the clean wire into the solution and return it to the flame. Note the color of the flame (view it through a spectroscope, if available). Correlate the color of the flame with its dominate wavelength, using the chart on the Data Sheet. Check the appropriate section.

3. Repeat Steps A.1 and 2 with $CuCl_2$, NaCl, $BaCl_2$, LiCl, $SrCl_2$, and KCl salts. For KCl view the flame through a cobalt glass plate.[2]

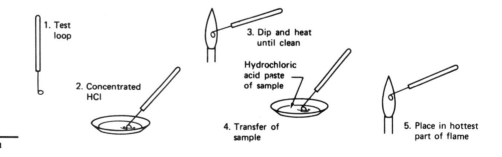

Figure 12.2
The procedure for performing a flame test

---

[2]A cobalt glass plate filters all wavelengths other than those emitted by excited K^+ ions.

B. Hydrogen Spectrum 1. A photographic slide is used to view the hydrogen spectrum. Insert the H/Hg slide into the projector. Notice the three spectra on the slide. A continuous spectrum is at the bottom of the slide, next the H-spectrum, and at the top the Hg-spectrum. Align the visible lines of the Hg-spectrum with the wavelengths in Table 12.1, which are also marked on the scale below.

Table 12.1. The Wavelengths of the Visible Lines in the Mercury Spectrum

Violet	404.7 nm
Violet	407.8 nm
Blue	435.8 nm
Green	546.1 nm
Yellow	577.0 nm
Yellow	579.1 nm

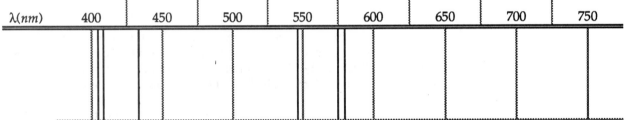

2. Mark the position of the mercury lines on a full size (22 x 28 cm) piece of graph paper. Display the spectra from the hydrogen slide on the graph paper, align the known mercury line spectrum on the graph paper, and then, as a result of the alignment, record the wavelengths of the hydrogen spectrum. Calculate $\Delta E_{atom} = E_{photon}$, $1/\lambda$, n_h, $1/n_h^2$. Plot $1/\lambda$ vs $1/n_h^2$ to determine the Rydberg constant. Have your instructor approve your graph.

C. Unknown Spectrum Using an unknown/Hg slide (see instructor), identify the wavelengths for the lines in the spectrum of the unknown. Using Table 12.2, identify the element having the unknown spectrum.

Table 12.2 Wavelengths and Relative Intensities of the Emission Spectra for Several Elements

Element	Wavelength (nm)	Relative Intensity	Element	Wavelength (nm)	Relative Intensity	Element	Wavelength (nm)	Relative Intensity
Argon	451.1	100	Helium	388.9	500	Rubidium	420.2	1000
	560.7	35		396.5	20		421.6	500
	591.2	50		402.6	50		536.3	40
	603.2	70		412.1	12		543.2	75
	604.3	35		438.8	10		572.4	60
	641.6	70		447.1	200		607.1	75
	667.8	100		468.6	30		620.6	75
	675.2	150		471.3	30		630.0	120
	696.5	10000		492.2	20			
	703.0	150		501.5	100	Sodium	466.5	120
	706.7	10000		587.5	500		466.9	200
	706.9	100		587.6	100		497.9	200
				667.8	100		498.3	400
Barium	435.0	80					568.2	280
	553.5	1000	Neon	585.2	500		568.8	560
	580.0	100		587.2	100		589.0	80000
	582.6	150		588.2	100		589.6	40000
	601.9	100		594.5	100		616.1	240
	606.3	200		596.5	100			
	611.1	300		597.4	100	Thallium	377.6	12000
	648.3	150		597.6	120		436.0	2
	649.9	300		603.0	100		535.0	18000
	652.7	150		607.4	100		655.0	16
	659.5	3000		614.3	100		671.4	6
	665.4	150		616.4	120			
				618.2	250	Zinc	468.0	300
Cadmium	467.8	200		621.7	150		472.2	400
	478.0	300		626.6	150		481.1	400
	508.6	1000		633.4	100		507.0	15
	610.0	300		638.3	120		518.2	200
	643.8	2000		640.2	200		577.7	10
				650.7	150		623.8	8
Cesium	455.5	1000		660.0	150		636.2	1000
	459.3	460					647.9	10
	546.6	60	Potassium	404.4	18		692.8	15
	566.4	210		404.7	17			
	584.5	300		536.0	14			
	601.0	640		578.2	16			
	621.3	1000		580.1	17			
	635.5	320		580.2	15			
	658.7	490		583.2	17			
	672.3	3300		691.1	19			

Hg Reference

375 400 425 450 475 500 525 550 575 600 625 650 675

Wavelength (nanometers)

SPECTROSCOPY

Date _____ Name _____ Lab Sec. _____ Desk No. _____

1. Distinguish between an absorption spectrum and an emission spectrum.

2. The radiation emitted from the deexcitation of an SrCl* molecule, formed in a starburst of a fireworks display, has a wavelength range of 620–750 nm.
   a. Calculate the energy range of these photons.

   b. What color range do these photons have?

3. a. Using the equation, $\frac{1}{\lambda} = R\left[\frac{1}{n_l^2} - \frac{1}{n_h^2}\right]$, calculate the wavelength for an electron transition from the

   $n_h = 4$ to the $n_l = 1$ energy level in the hydrogen atom. Assume $R = 1.1 \times 10^{-2}$ nm^{-1}.

b. Where in the electromagnetic spectrum (visible, ultraviolet, or infrared) does this line appear?

c. Calculate the energy of the photon resulting from this electron transition.

4. A large number of hydrogen atoms have electrons excited to the $n_h = 4$ energy state. How many possible spectral lines can appear in the emission spectrum as a result of the electron reaching the ground state ($n_l = 1$)? Remember a spectral line appears when an electron deexcites from a higher to a lower energy state. Diagram all possible pathways for deexcitation from $n_h = 4$ to $n_l = 1$.

n = 4_____

n = 3_____

n = 2_____

n = 1_____

5. What dominant color appears for the mercury street bulb? the sodium street lamp? an incandescent bulb?

SPECTROSCOPY

Date _____ Name _____ Lab Sec. _____ Desk No. _____

A. Flame Tests

$\lambda(nm)$	400	450	500	550	600	650	700	750
$CaCl_2$								
$CuCl_2$								
NaCl								
$BaCl_2$								
LiCl								
$SrCl_2$								
KCl								

| ultraviolet | violet | blue | green | yellow | orange | red | infrared |

B. Hydrogen Spectrum

$\lambda(nm)$	color	$\Delta E_{atom} = E_{photon}$	$1/\lambda$	n_h(calc)	n_h(integer)	$1/n_h^2$

Show sample calculation of E_{photon} and n_h.

2. Plot $1/\lambda$ (ordinate) *vs* $1/n_h^2$ (abscissa) on linear graph paper.

3. Instructor's approval of graph. _____

4. From the plot determine
   a. the value of R, the Rydberg constant _____

   b. the value of $R/2^2$ _____

5. a. The accepted value of the Rydberg constant is (see instructor) _____

    b. The percent error in R is _____ %

C. Unknown Spectrum (slide no.) _____

Lines in the spectrum (nm) _____, _____, _____, _____,

_____, _____, _____, _____.

Unknown element: _____.

QUESTIONS

1. Why does a mercury light appear blue, even though yellow and green lines appear in the spectrum?

2. Explain why the color for ions in the flame tests differ.

3. Will the use of a spectroscope change the results of this experiment? Explain.

4. The Cl—Cl bond energy is 242 kJ/mol. What wavelength of electromagnetic radiation is necessary to break the bond? In what region of the spectrum does this wavelength appear?

EXPERIMENT 13

CHEMICAL PERIODICITY

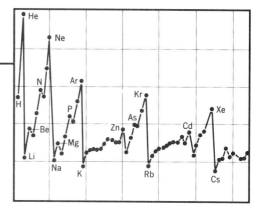

Every chemistry lecture auditorium and lab, as well as every chemist's office, has on the wall at least one seemingly necessary article of decor. This "table" appears to answer nearly all questions posed to the chemist, and is clearly a part of everyday life. What is so "magic" or what is so significant about this table? Is it really that important?

This table, called the **periodic table** of the elements, is an organized set of data stemming from years of research. The data contained therein has been used to predict the properties of elements and to predict the existence and the synthesis of other elements. Much of the data was organized in 1869 by the Russian Dimitri Mendeleev, a professor of chemistry at the University of St. Petersburg. In organizing his lectures of inorganic chemistry, he found it apparent that inorganic chemistry did not have the system and order that was evident in organic chemistry at that time. In his attempt to organize the elements, he used the accumulated data of their known chemical properties and placed them in columns and rows whereby adjacent elements were similar. Recognizing that such an organization was also related to the atomic masses of the elements led to the current organization of the periodic table.

Mendeleev was outspoken on his views over academic freedom. Mendeleev, in an outward expression of his independent and somewhat radical ideas, chose to cut his hair only once each year, in the Spring. Because he remained friendly with the Czar and his colleagues in spite of his views, the dates for many of the scientific meetings in Russia were held in the springtime.

OBJECTIVES

- To observe the physical appearance of thirteen common elements
- To observe the chemical reactivity of several elements in the third period and in Group VIIA
- To predict the physical and chemical properties of other elements

PRINCIPLES

Malleable: the property of being formed into very thin sheets.

Amorphous: the property of a solid having no orderly arrangement of atoms, molecules, or ions.

Many chemical and physical properties of an element can be predicted from its location in the periodic table. For example, elements at the left of the table are shiny solids, conduct electricity, and are malleable; elements at the right of the table are amorphous, nonconductors of electricity, and may even be gases; the metals at the left form salts with the nonmetals at the right. Elements of a **group** (vertical column) have similar chemical and physical properties with each successive element showing a gradual change in reactivity. Successive elements in a **period** (horizontal row) show more dramatic differences in properties, progressing from metallic properties at the left of the period to nonmetallic properties at the right.

In this experiment, we will look at the similarities and differences in the chemical and physical properties of several elements of Period 3 and Group VIIA (the halogens). These elements and some of their parameters are listed in Table 13.1.

Table 13.1. Measured Parameters of Some Elements

Periodic Properties	IA	IIA	IIIA	IVA	VA	VIA	VIIA
	Li	**Be**	**B**	**C**	**N**	**O**	**F**
Atomic Radius (*pm*)	145	105	85	70	65	60	50
Ionization Energy (*kJ/mol*)	519	900	799	1090	1400	1310	1680
Electronegativity	1.0	1.5	2.0	2.5	3.0	3.5	4.0
Density (*g/cm³*)	0.53	1.85	2.47	2.25[1]	1.25g/L	1.43g/L	1.70g/L
Melting Point (°C)	181	1285	2030	3370s	-210	-218	-220
Boiling Point (°C)	1347	2470	3700	---	-196	-183	-188
	Na	**Mg**	**Al**	**Si**	**P**	**S**	**Cl**
Atomic Radius (*pm*)	180	150	125	110	100	100	100
Ionization Energy (*kJ/mol*)	494	736	577	786	1060	1000	1260
Electronegativity	0.9	1.2	1.5	1.8	2.1	2.5	3.0
Density (*g/cm³*)	0.97	1.74	2.70	2.33	1.82	2.09	3.21g/L
Melting Point (°C)	98	650	660	1410	44[2]	115	-101
Boiling Point (°C)	883	1100	2350	2620	280	445	-34
	K	**Ca**	**Ga**	**Ge**	**As**	**Se**	**Br**
Atomic Radius (*pm*)	220	180	130	125	125	115	115
Ionization Energy (*kJ/mol*)	418	590	577	762	966	941	1140
Electronegativity	0.8	1.0	1.6	1.8	2.0	2.4	2.8
Density (*g/cm³*)	0,86	1.53	5.91	5.32	5.78	4.81	3.12
Melting Point (°C)	64	840	30	937	613s	220	-7
Boiling Point (°C)	774	1490	2070	2830	---	685	59
	Rb	**Sr**	**In**	**Sn**	**Sb**	**Te**	**I**
Atomic Radius (*pm*)	235	200	155	145	145	140	140
Ionization Energy (*kJ/mol*)	402	548	556	707	833	870	1010
Electronegativity	0.8	1.0	1.7	1.8	1.9	2.1	2.5
Density (*g/cm³*)	1.53	2.58	7.29	7.29[3]	6.69	6.25	4.95
Melting Point (°C)	39	770	157	232	631	450	114
Boiling Point (°C)	688	1380	2050	2720	1750	990	184
	Cs	**Ba**	**Tl**	**Pb**	**Bi**		
Atomic Radius (*pm*)	266	215	190	180	160		
Ionization Energy (*kJ/mol*)	376	502	812	920	1040		
Electronegativity	0.7	0.9	1.8	1.8	1.9		
Density (*g/cm³*)	1.87	3.59	11.87	11.34	8.90		
Melting Point (°C)	28	710	304	328	271		
Boiling Point (°C)	678	1640	1460	1760	1650		

[1]Graphite   [2]White phosphorus   [3]White tin

Definitions for the parameters in Table 13.1 are

- **Atomic radius:** the radius of an atom, expressed in picometers where $1\ pm = 1 \times 10^{-12}\ m$
- **Ionization Energy:** energy required to remove one mole of electrons from a mole of the gaseous element, expressed in kJ/mol
- **Electronegativity:** an atom's relative attraction for the electrons used for bonding to another atom, expressed on a scale relative to fluorine being assigned a number of 4.0
- **Density:** the ratio of the mass of substance to its volume, expressed as g/cm^3 for solids and liquids and g/L at STP[1] for gases
- **Melting point:** the temperature at which the solid phase becomes a liquid
- **Boiling point:** the temperature at which the liquid phase becomes a gas[2]

TECHNIQUES

- Techniques 6a, c, page 6   Heating Liquids
- Technique 11, page 10   Testing with Litmus
- Technique 18, page 18   Graphing Techniques

PROCEDURE

A. Physical Properties

Allotrope: more than one combination of atoms of the same element.

1. Samples of Na, Mg, Al, Si, P, and S are on the reagent table. Note that Na metal is stored under a nonaqueous liquid because of its reactivity with water and O_2 in air (Part B.1). Your lab instructor will cut a piece of Na; quickly notice its luster and other metallic characteristics. Use steel wool to polish the pieces of Mg and Al for better viewing. Two allotropic forms of phosphorus exist: white P_4 is so reactive with O_2 that it ignites in air; therefore white P_4 is stored under water. When stored under water for a long time, white P_4 slowly changes to the more stable red allotrope of phosphorus, which does not ignite in air. Record your observations on the Data Sheet.

2. **Preparation of Cl_2.** Place $^1/_2$ mL (10 drops) of 5% NaClO (household laundry bleach) in a 75 mm test tube and add 5 drops of toluene (**Caution:** *do not inhale*). Which layer is toluene and what is its color? Add 5 drops of 6 M HCl (**Caution:** 6 M HCl *is corrosive; wash it immediately from skin and clothes*). Agitate the solution by holding the upper part of the test tube with your thumb and index finger and tapping the lower part with your "pinky" finger (Figure 13.1). Observe the color of the toluene layer. What is the color of Cl_2?

3. **Preparation of Br_2 and I_2.**
   a. Mix equal pea-size portions of the solids KBr and MnO_2; transfer a portion of the mixture to a dry, clean 150 mm test tube until a depth of 3 mm ($\approx ^1/_8$ inch) is reached.[3] Add 3 drops of *conc* H_2SO_4 (**Caution:** *don't let it touch your skin*). *Gently* and very *carefully* warm the mixture over a low flame to initiate the reaction. What evidence of a reaction has occurred? Allow the test tube to cool, add 10 drops of water, and 5 drops of toluene, and agitate. What is the color of Br_2?

---

[1]Standard temperature and pressure are 0°C (273K) and 1 atm (101.325 kPa)

[2]More specific definitions of melting and boiling points will be given later in the course.

[3]Share the unused portion of the mixture with your neighbor—don't waste it!

b. Repeat the procedure in Part A.3, substituting KI for the KBr. What is the color of I_2?

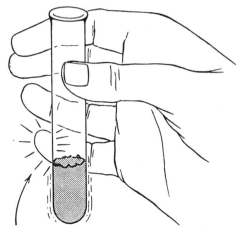

Figure 13.1.
Agitating a solution in a test tube with the "pinky" finger

Tap—tap—tap
w/pinky

4. Plot on graph paper the atomic radius (ordinate) *vs* atomic number (abscissa) and on the *same* graph, the ionization energy *vs* atomic number for the elements of Period 3. Label each axis and title the graph. Connect the data points with straight lines. Have your instructor approve your graph.

B. Chemical Properties

1. Na/H_2O. **Demonstration Only. (Caution:** *never allow Na to touch the skin; it causes a severe skin burn*)
   a. Wrap a *pea-sized* (no larger!) piece of freshly cut Na metal in aluminum foil. Fill a 200 mm test tube with water and invert it into an 800 mL beaker $^3/_4$ filled with water; test the water with litmus. Punch 5 pin-sized holes in the aluminum foil, grasp it with crucible tongs, and place it beneath the mouth of the water-filled test tube (Figure 13.2).

Gas evolution: the escape of gas from a system.

   b. After gas evolution has ceased, remove the test tube from the beaker, keeping it inverted. Place the mouth of the test tube over a Bunsen flame. What is the evolved gas? Don't be so alarmed as to drop the test tube—it may cost you 70¢ to replace it! Perform the litmus test on the water in the beaker. Is the water now acidic or basic? Based upon the properties of the gas and the acidity (or basicity) of the solution in the beaker, write a balanced equation for the reaction of Na with H_2O.

   c. **Demonstration Only.** Na/H_2O and CH_3OH (methanol). Set up a 150 mL beaker containing 20 mL of methanol, CH_3OH, and a 150 mL beaker containing 20 mL of water behind a safety shield (Yes, *behind a safety shield*—if you don't have one, proceed to Part B.3). Cut two "BB"-sized (no larger!) pieces of Na metal and, with tongs or tweezers, place one into each beaker. *Immediately* cover each beaker with a watch glass. Describe the reaction. How are water and CH_3OH similar? How do they differ?

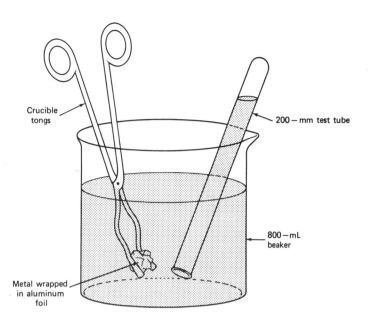

Crucible
tongs

200 – mm test tube

800 – mL
beaker

Metal wrapped
in aluminum
foil

Figure 13.2
**Collection of the H$_2$(g) evolved
from the reaction of Na with H$_2$O**

2. Mg and Al/H$_2$O and HCl
   a. Place 5 mL of distilled (or deionized) water into separate 150 mm test tubes. Place these test tubes in a half-filled 250 mL beaker of water and heat to boiling.

   b. Polish, with steel wool or sand paper, 2 cm strips of Mg and Al to remove their oxide coatings. Quickly add each metal to a separate test tube of hot water. Maintain the temperature at near boiling for 10 minutes. What is observed?

   c. Remove the test tubes containing any of the remaining Mg and Al samples. Perform a litmus test on the water in each test tube. Add 2 mL of 6 M HCl (**Caution:** *avoid skin contact*). Record your observations. What is the evolved gas?

   d. How does the reactivity of Na, Mg, and Al change in proceeding across Period 3?

Pinch: a volume no greater than a grain of rice.

3. Cl$_2$, Br⁻, and I⁻
   a. Refer again to Part A.2: prepare Cl$_2$ using the 5% NaClO and 6 M HCl solutions with the added toluene. Add a pinch of KBr to the Cl$_2$/toluene/H$_2$O mixture and agitate. Account for your observation.

   b. Repeat Part B.3a substituting a pinch of KI to the Cl$_2$/toluene/H$_2$O mixture and agitate. What chemical reaction has occurred?

4. Cl⁻, Br$_2$, and I⁻
   a. Dissolve a pinch of KCl in $1/2$ mL of water in a 75 mm test tube; add 5 drops of toluene. In the hood, add 5 drops of 2% Br$_2$/H$_2$O (**Caution:** *Br$_2$ is corrosive and causes severe skin burns*) to the test tube and agitate. What happens? Does the color of the Br$_2$ in the toluene layer disappear?

   b. Repeat Part B.4a, substituting KI for the KCl. Is the occurrence of a chemical reaction evident? Write a balanced equation to represent your observation.

Disposal Information:	Dispose of the test chemicals from Parts 3 and 4 in the "Waste Halogens" container.

Notes and Observations

CHEMICAL PERIODICITY

Date _____ Name _____ Lab Sec. _____ Desk No. _____

1. a. What is a group of elements?

    b. What is a period of elements?

2. Refer to Table 13.1 to answer the following.

    a. Which element in Period 3 is most dense?_____ least dense?_____

    b. Which element in Group VIIA has the highest electronegativity?_____ the lowest
    electronegativity?_____

    c. Which element in Group IA has the highest melting point?_____ the lowest boiling point?_____

    d. In general, the densities of the elements in a group _____ as the atomic number increases.

    e. In general, the ionization energies of the elements in a period _____ as the atomic number increases.

    f. The electronegativity of the elements in a group _____ and in a period _____ as the atomic
    number increases.

3. The relative chemical reactivity for a number of elements is studied in today's experiment.

    a. List the elements.

    b. List the element whose reactivity will be demonstrated by the instructor.

4. What commercially available compound is used to generate Cl_2 in the experiment?_____

5. a. Describe the technique for testing a solution with litmus.

    b. Describe the *proper* technique for heating a solution in a test tube (Technique 6a).

    c. What is the purpose of placing a glass stirring rod in a beaker that is being used to heat water?

CHEMICAL PERIODICITY

Date _____ Name _____ Lab Sec. _____ Desk No. _____

A. Physical Properties

Element	Symbol	Atomic Number	Atomic Mass	Physical State (g, l, s)	Color	Comments
sodium						
magnesium						
aluminum						
silicon						
phosphorus						
sulfur						

2. **Preparation of Cl_2.** Which is the toluene layer?_____ What color is

   toluene?_____ What is the color of Cl_2?_____

3. **Preparation of Br_2 and I_2**
   Evidence for the preparation of Br_2.

   Evidence for the preparation of I_2.

   Color of Br_2 _____ Color of I_2 _____

4. Instructor's approval of graph._____

   From the graph what general statement can you make about the relationship between atomic radii
   and ionization energies for a period of elements.

   Is the same statement valid for a group of elements?_____ If not, what statement holds true?

B. Chemical Properties

1. Na /H_2O
    a. Gas evolved. _____

    b. Litmus test, acidic or basic, before: _____ ; after: _____

    Write a balanced equation for the reaction of Na with H_2O.

    c. Na/H_2O and CH_3OH (methanol)
       What similarities exist in the reactions of Na with H_2O and CH_3OH?

    Is Na more reactive in H_2O or CH_3OH?_____

2. Mg and Al/H_2O and HCl

	Observation	Gas evolved	Litmus test
Mg/H_2O			
Al/H_2O			
Mg/HCl			
Al/HCl			

Compare the relative reactivity of Na, Mg, and Al with H_2O and with HCl.

3. Cl_2, Br^-, and I^-

	Observation	Balanced equation for the reaction
Cl_2 + Br^-		
Cl_2 + I^-		

4. Cl^-, Br_2, and I^-

	Observation	Balanced equation for the reaction
Cl^- + Br_2		
Br_2 + I^-		

List the halogens in order of decreasing activity. _____ > _____ > _____ Explain your listing.

QUESTIONS

1. What tool did your lab instructor use to cut the Na metal?_____ What property does Na exhibit, one that we don't normally associate with a metal, that allowed the lab instructor to use that tool?

2. a. Metallic oxides, like Na_2O and BaO, dissolved in water turn red litmus blue. What chemical property does this indicate about metallic oxides?

   b. Predict the effect on litmus when nonmetallic oxides, like SO_2 and CO_2, are dissolved in water. Explain.

3. Cl_2 is used extensively as a bleaching agent and as a disinfectant. Without regard to any adverse effects, would Br_2 be a more or less effective as a bleaching agent and disinfectant? Explain.

4. Predict the reactivity of Cs in methanol relative to that of Na. Explain.

5. Predict the reactivity of Si in water relative to that of Na, Mg, and Al. Explain.

6. Predict the chemical reactivity of fluorine gas, F_2, relative to other Group VIIA elements. Explain.

EXPERIMENT 14

SPECTRO-PHOTOMETRIC IRON ANALYSIS

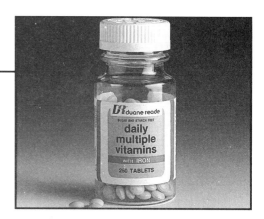

The exchange of oxygen with carbon dioxide during respiration could not occur without iron being a part of the protein hemoglobin. A healthy adult body contains about 3 g of iron with a loss of about 1 mg/day through sweat, feces, and hair. A deficiency of iron leads to a condition called anemia, resulting in symptoms of complacency and tiredness.

Scientists use a number of laboratory techniques to analyze for the amount of iron in a sample. Previously in this laboratory manual we used the balance to assist in our analyses. Many substances have color and the intensity of that color relates to the amount of substance present. For example, the more drops of blue ink that are added to water, the more intense the blue color. Quantitative and qualitative measurements based upon the property of a substance to absorb visible light are used extensively in chemical laboratories. An iron analysis is one of those applications.

Every chemical species (atoms, molecule, or ion) possesses a characteristic set of electronic, vibrational, and rotational energy states. Because they are characteristic, energy transitions between these states are often used to identify the presence and/or concentration of a substance in a mixture. This unique set of energy states for a chemical is therefore analogous to the unique set of fingerprints possessed by each person—both can be used for characteristic identifications.

OBJECTIVES

- To determine an unknown iron(II) ion concentration in solution
- To develop techniques for the use and operation of a spectrophotometer
- To determine the amount of iron in a vitamin tablet

PRINCIPLES

The principle underlying a spectrophotometric method of analysis involves the interaction of electromagnetic (EM) radiation with matter. The ultraviolet, visible, and infrared regions of the EM spectrum are the most common used in analyses; in this experiment the visible region is used. The wavelength range for the visible spectrum is from about 400 nm to 700 nm; the 400 nm radiation approximates a violet color while the 700 nm region has a red color.

The absorption of EM radiation from the visible spectrum is a result of the excitation of an electron from a lower to a higher state. The energy of the radiation that is absorbed is equal to the difference between these two energy states. The species (atom, molecule, or ion) that absorbs the radiation is in an **excited state**. The EM radiation that is not absorbed, and therefore passes through the sample, is detected by an EM detector (either our own eye or an instrument). The absorbed energy, E, is related to the wavelength, λ, of the EM radiation.

$$E = h\,\frac{c}{\lambda}$$

where h is Planck's constant and c is the speed of light.

When our eye is the detector, the color that we see is the EM radiation which the sample does not absorb. The appearance of the sample is that of the complementary color of the absorbed radiation. For example, if our sample solution absorbs energy in the yellow region of the visible spectrum, then the remaining wavelengths are transmitted to our eye and the sample appears violet. The greater the concentration of the yellow absorbing substance, the darker is the violet appearance. Table 14.1 lists the wavelengths of the visible spectrum.

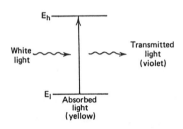

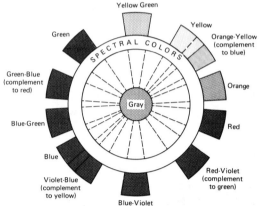

Table 14.1. Color and Wavelengths in the Visible Region of the Electromagnetic Spectrum

Color	Wavelength (nm^*)	Color Transmitted
red	750-610	blue-green
orange	610-595	blue
yellow	595-580	violet
green	580-500	purple
blue	500-435	orange
violet	435-380	yellow

*1 nanometer = 1×10^{-9} meter

In this experiment visible radiation is used to determine the concentration of the iron(II), Fe^{2+}, ion in an aqueous solution. The wavelength at which a maximum absorption of EM radiation occurs is set on the spectrophotometer, an instrument that measures light intensities with a photosensitive cell (similar to our eye) at specific (but adjustable) wavelengths (Figure 14.1).

Several factors affect the amount of visible radiation that the Fe^{2+} absorbs.

- the concentration of the Fe^{2+} in solution
- the thickness of the solution through which the visible radiation passes (this is determined by the diameter of the cuvet)
- the extent to which the Fe^{2+} absorbs the radiation at a particular wavelength (this is called its molar absorptivity). This factor is constant at a set wavelength.

The ratio of the intensities of the transmitted radiation, I_t, to the incident radiation, I_0, is the sample's **transmittance, T,** or expressed as a percent, **%T**

$$\%T = \frac{I_t}{I_0} \times 100$$

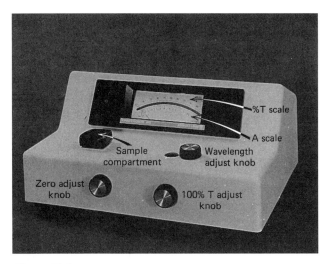
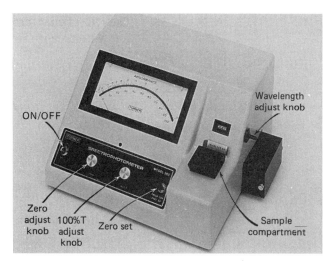

Figure 14.1
Common laboratory visible spectrophotometers

α: means "proportional to".

Measuring the Fe²⁺ Concentration

Frequently a chemist is more interested in knowing the amount of radiation that the Fe^{2+} absorbs rather than the amount that it transmits, the absorption being directly proportional to the Fe^{2+} concentration. The **absorbance, A,** of the radiation is related to the percent transmittance by the equation

$$A = -\log \frac{I_t}{I_0} \times 100 = \log \frac{100}{\%T}$$

The absorbance of radiation is directly proportional to the molar concentration of the Fe^{2+} by the equation

$$A = a \cdot b \cdot [Fe^{2+}]$$

a is called the molar absorptivity for Fe^{2+} and **b** is the thickness of the solution—both of which are constants at a given wavelength in a given cuvet. This equation is commonly referred to as Beer's Law, the important relationship being that **A α [Fe²⁺]**.

In this experiment an iron sample is dissolved in solution and all of the dissolved iron ions are reduced to Fe^{2+}. The Fe^{2+} forms a red-orange complex ion with 1,10-phenanthroline (also called ortho-phenanthroline and abbreviated *o*-phen), $[Fe(o\text{-phen})_3]^{2+}$. The absorbance of this ion is determined at a wavelength where the complex ion has a maximum molar absorptivity, called λ_{max}. The λ_{max} for $[Fe(o\text{-phen})_3]^{2+}$ is determined in Part A from a plot of **A** *vs* λ for a standard solution of $[Fe(o\text{-phen})_3]^{2+}$.

The iron(III) ion, Fe^{3+}, does not form the red-orange complex ion with *o*-phen; therefore, all iron(III) ion is reduced to iron(II). The reducing agent is hydroxylamine hydrochloride, $NH_3OH^+Cl^-$.

$$2\,Fe^{3+}(aq) + 2\,NH_3OH^+Cl^-(aq) \rightarrow$$
$$2\,Fe^{2+}(aq) + N_2(g) + 2\,H_2O + 4\,H^+(aq) + 2\,Cl^-(aq)$$

The 1,10-phenanthroline is then added to the solution to form the red-orange complex ion with Fe^{2+}.

$$Fe^{2+}(aq) + 3\ o\text{-phen}(aq) \rightarrow [Fe(o\text{-phen})_3]^{2+}(aq)$$

In addition, since the $[Fe(o\text{-phen})_3]^{2+}$ ion is most stable in the pH range from 2 to 9, sodium acetate, $NaCH_3CO_2$, is added, reacting with the H^+ from the $NH_3OH^+Cl^-$ reaction to form an $CH_3CO_2^-/CH_3COOH$ combination that maintains the pH of the solution between 4 and 6.

$$NH_3OH^+Cl^-(aq) + CH_3CO_2^-(aq) \rightarrow CH_3COOH(aq) + NH_2OH(aq) + Cl^-(aq)$$

TECHNIQUES

- Technique 2, page 1   Clean Glassware and Lab Bench
- Technique 5, page 3   Transferring Liquid Reagents
- Technique 8, page 7   Reading a Meniscus
- Technique 9, page 7   Pipetting a Liquid or Solution
- Technique 13, page 11   Using the Laboratory Balance
- Technique 14b, page 12   Separation of a Liquid from a Solid
- Technique 18, page 18   Graphing Techniques

In addition, you will develop the techniques for operating a spectrophotometer and handling cuvets. Also, you will gain additional experience in graphing data.

PROCEDURE

You and a partner are using an expensive instrument in today's experiment. Follow your instructor's suggestions. You will need three 1 mL pipets and two 10 mL graduated (Mohr) pipets from the stockroom. Read ahead and be prepared.

A. Setting the λ_{max} on the Spectro–photometer

1. Obtain (and clean thoroughly with soap and water) three 1 mL pipets and one 10 mL pipet, graduated at 1 mL increments (a Mohr pipet), and five 200 mm (8-inch) test tubes[1]. Use the graduated 10 mL pipet for the standard Fe^{3+} solution.

2. a. When using pipets, rinse the pipet twice with the reagent and then discard the reagent before preparing the test solution. This technique is especially important when using the 10 mL pipet in this experiment.

Disposal Information: Dispose of the test solutions in the "Waste Iron Salts" container

---

[1]10 mL volumetric flasks may be substituted for the 200 mm test tubes.

b. Pipet 5.00 mL of the standard Fe^{3+} solution (≈10 µg Fe^{3+}/mL) into a clean 200 mm test tube. Using 1 mL pipets add 1 mL of 10% $NaCH_3CO_2$ and 1 mL of 10% $NH_3OH^+Cl^-$. Agitate the solution periodically for at least 10 minutes to complete the reduction of Fe^{3+} to Fe^{2+}. In the meantime continue to Part A.4 and the preparation of the solutions in Part B.2.

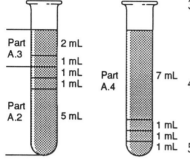

3. Use the third 1 mL pipet to add 1 mL of 0.1% *o*-phen to the solution and with the second 10 mL pipet add 2.0 mL of water. Stir, with a stirring rod, the solution for several minutes. This is test solution #4 in Table 14.2.

4. To prepare a blank solution[2], pipet 1 mL of 10% $NaCH_3CO_2$, 1 mL of 10% $NH_3OH^+Cl^-$, 1 mL of 0.1% *o*-phen and 7 mL of water into a 200 mm test tube and stir. This is test solution #1 in Table 14.2.

5. Prepare two cuvets: rinse one cuvet twice with the blank solution and then fill to the $^3/_4$ level; rinse a second cuvet (also twice) with the $[Fe(o\text{-phen})_3]^{2+}$ solution (from Parts A.2 and A.3) and similarly fill. Carefully dry the outside of each cuvet with a clean Kimwipe to remove fingerprints and water droplets. Thereafter, handle only the lip of the cuvets.[3]

6. Set the λ on the spectrophotometer at 400 nm. Insert the blank solution into the sample holder and set the meter at 100%T with the 100%T knob. Remove the blank and (without any cuvet in the sample holder) set the meter to read 0%T with the zero adjust knob. Repeat until no further adjustments are necessary.

7. Place the cuvet containing the $[Fe(o\text{-phen})_3]^{2+}$ solution in the sample holder, read the meter, and record the %T.

8. Repeat Step 7 at 20 nm intervals between 400 nm and 600 nm. Make additional %T measurements at 5 nm intervals in the region of λ_{max}. Periodically repeat Step 6.
   Save these solutions for Part B of the experiment.

9. Convert all of your %T readings to absorbance values. Plot the data as **A** (ordinate) *vs* λ (abscissa) on linear graph paper. Draw the best smooth curve through the data points. Have your instructor approve the graph.

B. Constructing the Standard Curve

1. Set the spectrophotometer at the λ_{max} determined from the graph in Part A.

2. Prepare the following solutions in the same manner as described in Parts A.2 through A.4. Be sure to add the 0.1% *o*-phen *after* the Fe^{3+} has been reduced to Fe^{2+} (again, allow at least 10 min for the reduction reaction). Use the 10 mL pipet, calibrated at 1 mL intervals, for transferring the

---

[2]A **blank** solution is used to calibrate the spectrophotometer; the solution corrects for all substances that absorb EM radiation *except* the one of interest, in this case, $[Fe(o\text{-phen})_3]^{2+}$.

[3]Foreign material on the cuvet affects the intensity of the transmitted EM radiation.

standard Fe^{3+} solution. Be sure to the rinse the pipet twice before preparing the test solution. Use the second 10 mL graduated pipet for the dilution with water.

Table 14.2. Preparation of the Test Solutions for the Construction of the Standard Curve

Test sol'n	10 µg Fe^{3+}/mL sol'n (mL)	10% $NaCH_3CO_2$ sol'n (mL)	10% $NH_3OH^+Cl^-$ sol'n (mL)	0.1% o-phen sol'n (mL)	Volume of water (mL)
1*	0.0	1.0	1.0	1.0	7.0
2	1.0	1.0	1.0	1.0	6.0
3	3.0	1.0	1.0	1.0	4.0
4*	5.0	1.0	1.0	1.0	2.0
5	7.0	1.0	1.0	1.0	0.0

*Prepared in Part A

3. Use the blank solution (from Part A.4, test solution #1) to repeat the calibration of the spectrophotometer (see Part A.6).

4. Read and record the %T of the five solutions at the λ_{max}. Follow the same techniques for handling the cuvets as indicated in Part A.5. Calculate the absorbance for each solution.

5. Prepare a standard curve by plotting on linear graph paper **A** (ordinate) vs Fe^{2+} concentration (µg/mL) (abscissa). Draw the best straight line through the five points on the graph. Have your instructor approve your graph.

C. Preparation of a Water Sample for Iron Analysis

1. Obtain at least 20 mL of a water sample known to contain dissolved iron. It may be an unknown that has already been prepared for the experiment or it may be obtained from a drinking water supply, reservoir, or river. If there is any evidence of cloudiness, filter the sample.

2. Pipet 5.00 mL of the water sample into a 100 mL volumetric flask and dilute to the "mark" with water. Pipet 1.0 mL of this solution into a 200 mm test tube, add 1.0 mL each of the $NaCH_3CO_2$, $NH_3OH^+Cl^-$, and o-phen solutions and 6.0 mL of water as in Parts A and B. Proceed to Part E.

D. Preparation of a Vitamin Tablet for Iron Analysis

1. Crush a vitamin tablet with a mortar and pestle. Transfer the powder to a clean, dry, 100 mL beaker of known mass (±0.01 g). Measure the combined mass of the beaker and crushed tablet. Add 25 mL of 6 M HCl into the beaker and stir to dissolve (≈10 to 20 min).[4]

2. Quantitatively transfer the solution to a 250 mL volumetric flask. Wash the solid remaining in the beaker with several portions of distilled (or deionized) water and add the washings to the flask. Dilute to the mark with distilled water and agitate for several minutes; if cloudiness persists, filter the solution.

---

[4]Some of the tablet's binder may not dissolve; heat the solution gently in the hood to dissolve the tablet, but do not boil.

3. Pipet 1.0 mL of the (filtered) solution into a 200 mm test tube and add 1.0 mL each of the $NaCH_3CO_2$, $NH_3OH^+Cl^-$, and o-phen solutions and 6.0 mL of water. Proceed to Part E.

E. Determination of the Fe^{2+} Concentration in the Sample

1. Visually compare the color intensity of this prepared solution with the standard solutions prepared in Part B; if the red-orange color is within the range of the standards, o.k.; if not, discard this solution and prepare a second sample solution (quantitatively) in which the second dilution has a color intensity that lies within that of the standards. You will need to make some judgment on the amount to use.

2. Record the %T of the solution on the spectrophotometer and calculate its absorbance. It may be advisable at this point to again check the calibration of the spectrophotometer with the blank solution; if it needs recalibration, the %T of the solution will need to be repeated.

3. Use the standard curve prepared in Part B to determine the Fe^{2+} concentration ($\mu g/mL$) in the prepared sample. Calculate the quantity of iron in the original sample; be sure to account for the dilution of the original sample.

Disposal Information: Dispose of the test solutions in the "Waste Iron Salts" container

Notes, Observations, and Calculations

EXPERIMENT 14

LAB PREVIEW

SPECTROPHOTOMETRIC IRON ANALYSIS

Date ____________Name ____ Lab Sec. ______Desk No. ________

1. List the purpose for each of these substances in the iron analysis.

   a. 1,10-phenanthroline (*o*-phen)

   b. $NaCH_3CO_2$

   c. $NH_3OH^+Cl^-$. Write a balanced equation for its reaction with Fe^{3+}. The oxidation half-reaction of NH_3OH^+ is

$$2\ NH_3OH^+(aq) \rightarrow N_2(g) + 2\ H_2O(l) + 4\ H^+(aq) + 2\ e^-$$

2. A concentration of 2 ppm Fe^{2+} means 2 g Fe^{2+} in 10^6 g solution. Assuming the density of the solution is 1.0 g/mL, express 2 ppm Fe^{2+} in

   a. μg Fe^{2+}/mL solution

   b. mg Fe^{2+}/L solution

3. What is the color of a solution that absorbs 600 nm EM radiation?

4. If $[Fe(o\text{-phen})_3]^{2+}$ appears red-orange, what is the approximate wavelength for its maximum absorption?

5. A sample containing 18.0 mg of iron dissolves in 250 mL of solution. One milliliter is then withdrawn and diluted to 10 mL. What is the iron concentration in the diluted sample? Express your answer in µg Fe/mL and in ppm Fe.

6. A test solution was prepared by pipetting 2.0 mL of an original sample into a 250 mL volumetric flask and then diluting to the mark. The concentration of iron in the test solution was determined (from a calibration curve) to be 1.66 µg/mL or 1.66 ppm. What was the iron concentration in the original sample? Express your answer in µg Fe/mL *and* in ppm Fe.

SPECTROPHOTOMETRIC IRON ANALYSIS

Date _____ Name _____ Lab Sec. _____ Desk No. _____

A. Setting the λ_{max} on the Spectrophotometer

$\lambda(nm)$	%T	A	$\lambda(nm)$	%T	A	$\lambda(nm)$	%T	A

Plot the data, A (ordinate) *vs* λ(abscissa) on linear graph paper.

Instructor's approval of graph. _____

B. Constructing the Standard Curve

Test sol'n	Fe^{2+} Concentration (total volume = 10 mL)		%T	A
	µg Fe^{2+}/mL	ppm Fe^{2+}		
1				
2	*			
3				
4				
5				

* Sample calculation for Fe^{2+} (µg Fe^{2+}/mL) for test solution #2 (show work here).

Plot the data, A (ordinate) *vs* Fe^{2+} concentration (µg Fe^{2+}/mL) (abscissa) on linear graph paper.

C/E. Iron Concentration in Water Sample

	Trial 1	Trial 2
1. %T		
2. Absorbance		
3. Iron concentration in diluted sample from calibration curve ($\mu g\ Fe/mL$)		
4. Iron concentration in original sample ($\mu g\ Fe/mL$) ($ppm\ Fe$)		

Show sample calculation.

D/E. Iron Concentration in Vitamin Tablet

	Trial 1	Trial 2
1. Mass of 100 mL beaker (g)		
2. Mass of 100 mL beaker + crushed tablet		
3. Mass of tablet (g)		
4. %T		
5. Absorbance		
6. Iron concentration in diluted sample from calibration curve ($\mu g\ Fe/mL$)		
7. Iron concentration in original sample ($\mu g\ Fe/mL$)		
8. mg Fe/gram tablet		
9. %Fe in tablet		

Show sample calculation.

1. This experiment measures the iron concentration in the 0.1 ppm to 10 ppm range. Explain how a solution having a higher iron concentration can be determined, and still remain within the limits of this sensitivity range.

2. How does an unfiltered sample affect the absorbance value for a solution?

3. If not enough time is allowed for the following reactions, how will it affect the reported iron concentration in the sample?

   a. the reduction of the Fe^{3+}.

   b. the formation of $[Fe(o\text{-phen})_3]^{2+}$.

4. State the purpose for the blank solution in measuring the %T of the $[Fe(o\text{-phen})_3]^{2+}$ solutions. Why is distilled water not a suitable blank?

5. a. If a 0.9 cm cuvet is mistakenly substituted for a 0.8 cm cuvet in one of the measurements, will the %T reading be higher or lower than it should be for that solution? Explain.

   b. Will the reported iron concentration be high or low for that solution?

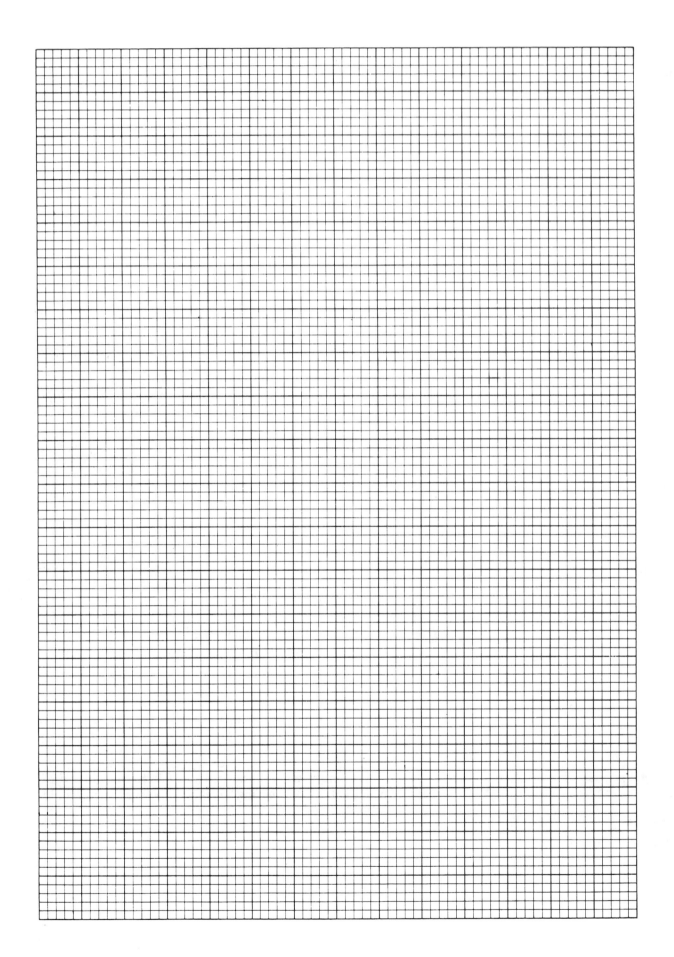

MOLECULAR GEOMETRY

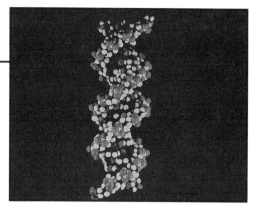

Atoms bond. Molecules and ion-pairs form. Our bodies are three-dimensional, or 3-D, and, since our bodies are made of compounds, the molecules and ion-pairs that form the compounds must also be 3-D. For that matter "everything" is 3-D so an understanding of the structure of molecules is very important to the chemist. Certainly the structure of the DNA molecule is vital to our existence—any change in its structure changes its chemical activity and alters the genetic information used in subsequent cell replication. These changes can result cancerous cells, birth defects, terminal illnesses.

Much of the early explanations and interpretations of chemical bonding is attributed to G. N. Lewis, a native of Massachusetts and later, a professor of chemistry at the Massachusetts Institute of Technology and the University of California, Berkeley. Lewis advanced the concept of the electron pair in ionic and covalent bonding. What is amazing about Lewis' theories of bonding (first proposed in 1916) is that they preceded the quantum theory of the atoms. It is from this Lewis concept of the chemical bond that the VSEPR (valence shell electron pair repulsion) theory of molecular geometry was proposed.

OBJECTIVES

- To construct models for covalently bonded molecules and polyatomic ions
- To apply the Lewis theory of bonding for predicting the three-dimensional shapes of molecules and polyatomic ions
- To predict the polarity of molecules

PRINCIPLES

Quantum theory: recognizes the wave-like properties of electrons in atoms.

Valence electrons: electrons in the highest energy level (outer shell) of the atom.

Isoelectronic: "the same electron configuration".

The structure of a molecule is basic in explaining its chemical and physical properties. For example, the facts that water is a liquid at room temperature, dissolves innumerable salts and sugars, is more dense than ice, boils at a relatively high temperature, and has a low vapor pressure can be explained through an understanding of its bonding and the bent arrangement of its atoms in the molecule.

Lewis proposed that for an atom to be stable it must gain, lose, or share valence electrons with other atoms until each electron is paired. In addition he proposed that atoms tend to be most stable in a compound when they achieve a valence number of electrons equal to that of a nearby noble gas in the Periodic Table. With the exception of helium, this number of electrons is eight and is called the **Octet Rule**. The bond formed is ionic or covalent depending upon whether the electrons are transferred or shared (respectively). Lewis' Octet Rule is quite effective in explaining bonding, especially for bonding between the representative (main group) elements.

For water, the Lewis structure shows that when hydrogen and oxygen *share* valence electrons; each atom is isoelectronic with a noble gas—hydrogen with helium and oxygen with neon.

$$\overset{\cdot\cdot}{\underset{\cdot\cdot}{O}}$$
$$H \qquad H$$

The Lewis structure only accounts for the bonding of each atom, not the 3-D structure of the molecule or polyatomic ion. Various bonding theories address the 3-D structure of a molecule; these theories account for some of the chemical and physical properties as well.

Valence Shell Electron Pair Repulsion (VSEPR) Theory

The valence shell electron pair repulsion, VSEPR, theory proposes that the geometry of a molecule or polyatomic ion is a result of a repulsive interaction of electron pairs in the valence shell of an atom; the most significant atom for determining its 3-D structure is the *central* atom. The orientation of the atoms is such that there is minimal interaction between the valence shell electron pairs; this, in turn, maximizes the distance between the electron pairs and therefore also between the atoms in the molecule.

Table 15.1. The VSEPR Structures for Molecules and Ions

Valence Electron Pairs	Bonding Electron Pairs	Nonbonding Electron Pairs	VSEPR Formula	Approx. Bond Angles	3-D Structure	Examples
2	2	0	AX_2	180°	linear	$HgCl_2$, $BeCl_2$
3	3	0	AX_3	120°	trigonal planar	BF_3, $In(CH_3)_3$
	2	1	AX_2E	<120°	V–shaped	$SnCl_2$, $PbBr_2$
4	4	0	AX_4	109.5°	tetrahedral	CH_4, $SnCl_4$
	3	1	AX_3E	<109.5°	trigonal pyramidal	NH_3, PCl_3, H_3O^+
	2	2	AX_2E_2	<109.5°	V–shaped	H_2O, OF_2, SCl_2
5	5	0	AX_5	90°/120°	trigonal bipyramidal	PCl_5, $NbCl_5$
	4	1	AX_4E	>90° <120°	irregular tetrahedral	SF_4, $TeCl_4$
	3	2	AX_3E_2	<90°	T–shaped	ClF_3
	2	3	AX_2E_3	180°	linear	ICl_2^-, XeF_2
6	6	0	AX_6	90°	octahedral	SF_6
	5	1	AX_5E	>90°	square pyramidal	BrF_5
	4	2	AX_4E_2	90°	square planar	ICl_4^-, XeF_4

In methane, CH_4, the four valence electron pairs on the carbon atom (the central atom of the molecule) repel; this positions the electron pairs (and, for CH_4, the four hydrogen atoms) at the corners of a tetrahedron in a 3-D arrangement. This positioning of electron pairs can be generalized to include all molecular systems having four valence shell electron pairs on the central atom.

$$H : \overset{\displaystyle ..}{\underset{\displaystyle ..}{C}} : H$$

Since the oxygen atom in water also has four valence shell electron pairs, they, too, are arranged tetrahedrally—two electron pairs bond the hydrogen atoms to the oxygen and two electron pairs are nonbonding. This causes water to have a V-shaped arrangement of the H–O–H atoms in the molecule.

Methane: the simplest of the hydrocarbons, better known as the major component of natural gas.

Methane has the VSEPR formula, AX_4, where A represents the central atom and X represents an attached atom; water has the VSEPR formula, AX_2E_2, where E represents a nonbonding electron pair. The H–C–H bond angles in the tetrahedral arrangement of the four electron pairs in methane is 109.5°; therefore, we can predict that all AX_4 molecules have similar bond angles and that molecules with AX_3E and AX_2E_2 formulas (also four electron pairs in the valence shell of the central atom), such as ammonia and water respectively, have bond angles slightly less than 109.5°.

A multiple bond (more than one electron pair being shared between two atoms in the molecule), as in an $X=AX_2$ or $X\equiv AX$ molecule, holds the multiple-bonded atom in the same position as a single bond. Therefore, the multiple bond is treated as a single bond for predicting the structure of a molecule. For example, carbon dioxide, which has the Lewis structure $\ddot{\text{O}}::\text{C}::\ddot{\text{O}}:$, has the VSEPR formula AX_2 and is a linear molecule.

In summary, the arrangement of valence shell electron pairs, both bonding and nonbonding pairs, around the central atom gives rise to the corresponding 3-D structures of molecules and polyatomic ions. Predicting the structure for molecules with multiple bonds are determined in the same fashion as single bonds. The 3-D structures and bond angles for molecules and polyatomic ions with a central atom having from two to six valence shell electron pairs are listed in Table 15.1.

Dipoles and Polarity

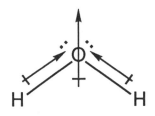

Figure 15.1
The bond polarity of the OH bonds contribute to the resultant polarity (↑) of the H_2O molecule

Once the 3-D structure of a molecule is known, its polarity can be ascertained. A **polar** molecule has a nonuniform distribution of electrons over its entire structure. This nonuniform distribution results both from atoms in the molecule having different electronegativities and from their relative positions in space. Atoms that are highly electronegative, such as fluorine, tend to strongly attract electrons, creating a center of negative charge and, therefore, a center of positive charge elsewhere in the molecule. The separation of opposite charges by a distance is called a molecular **dipole**. For example, H–Cl is a polar molecule because chlorine, being more electronegative than hydrogen, has a slightly higher concentration of negative charge than hydrogen in the molecule; H–F is even more polar because fluorine is more electronegative than chlorine.

For molecules with more than one bond, one needs to look at its 3-D structure to determine the relative orientation of each polar bond and how it contributes to the overall polarity of the molecule. For H_2O, each O–H bond is polar and their relative orientations determine the polarity of the molecule. If the bond angle were 180°, the bond polarities would cancel and the molecule would be nonpolar, but since the bond angle is 104.5°, the bond polarities do not cancel, but rather enhance each other to give the (overall) molecule a resultant dipole (Figure 15.1). A look at the 3-D structure of a molecule will enable us to predict whether a molecule is polar or nonpolar.

PROCEDURE

Check out a set of molecular models. Your instructor will assign you a number of molecules or polyatomic ions from the lists on the Data Sheet. For each molecule or polyatomic ion, you will be required to do the following:

- Write its Lewis structure
- Determine the number of bonding electron pairs on the central atom
- Determine the number of nonbonding electron pairs on the central atom
- Predict the approximate bond angle on the central atom
- Predict the 3-D structure of the molecule or polyatomic ion
- Predict if the molecule is polar or nonpolar

MOLECULAR GEOMETRY

Date _____ Name _____ Lab Sec. _____ Desk No. _____

1. a. A molecule that has two bonding pairs and two nonbonding pairs of electrons has a VSEPR formula of

   _____ . Its approximate bond angle is _____°.

   b. A molecule that has four bonding pairs and one nonbonding pair of electrons has a

   _____ structure. Its approximate bond angle is _____°.

   c. A molecule that has three bonding pairs and one nonbonding pair of electrons has a VSEPR formula of

   _____ . Its approximate bond angle is _____°.

2. a. A molecule that has two bonding pairs and one nonbonding pair of electrons has a VSEPR formula of

   _____ . Its approximate bond angle is _____°.

   b. A molecule that has two bonding pairs and three nonbonding pairs of electrons has a

   _____ structure. Its approximate bond angle is _____°.

   c. A molecule that has four bonding pairs and no nonbonding pairs of electrons has a VSEPR formula of

   _____ . Its approximate bond angle is _____°.

3. The essential amino acid, phenylalanine, one of the amino acids of NutraSweet, has the structure

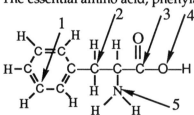

   Identify the VSEPR formula and approximate bond angles at the five designated atoms.

4. a. The number of nonbonding electron pairs in SF_4 is _____.

   b. The VSEPR formula for SF_4 is _____.

   c. The 3-D structure of SF_4 is _____.

5. a. The number of bonding electron pairs in BrF_3 is _____.

   b. The number of nonbonding electron pairs in BrF_3 is _____.

6. a. A molecule having bond angles of 120° and a trigonal planar structure has an _____ VSEPR formula.

   b. A molecule having bond angles of 90° and an octahedral structure has an _____ VSEPR formula.

7. Explain why CO_2 is a nonpolar molecule, but CO is a polar molecule.

MOLECULAR GEOMETRY

Date _____ Name _____ Lab Sec. _____ Desk No. _____

Your instructor will assign molecules and/or ions from the following lists for you to characterize according to the format shown below. The central atom of the molecule/ion is italicized and in bold face. The central atom of the molecules/ions marked with an asterisk (*) do *not* obey the octet rule.

Molecule or Ion	Lewis Structure	Bonding electron pairs	Nonbonding electron pairs	Approx. bond angle (°)	3-D Structure	Polar or nonpolar
CH_4	H $\ddot{}$ H : C : H $\ddot{}$ H	4	0	109.5°	tetrahedral	nonpolar

1. Complete the table (as outlined above) for the following molecules/ions.

   a. CF_3Cl d. H_2O g. H_3O^+ j. XeF_4*
   b. NH_3 e. AsF_5*   h. ClO_2^- k. ICl_2^+
   c. NH_4^+ f. AsF_3 i. BF_4^- l. ICl_2^-*

2. Complete the table (as outlined above) for the following molecules/ions.

   a. SF_2 d. SF_5^+*   g. SbH_3 j. OF_2
   b. SF_4*   e. BrF_2^+ h. BrF_4^-*   k. $BeCl_2$
   c. SF_6*   f. BrF_2^-*   i. BrF_4^+*   l. XeF_2*

3. Complete the table (as outlined above) for the following molecules/ions.

   a. GaI_3*   d. PF_3 g. SbF_6^-*   j. SiH_4
   b. CH_3^- e. PF_5*   h. SnF_4 k. SnF_6^{2-}*
   c. CH_3^+*   f. PF_4^+ i. SnF_2*   l. SO_4^{2-}

4. Complete the table (as outlined above) for the following molecules/ions.
   a. $OPCl_3$ d. N_2H_4 g. $COCl_2$
   b. H_2CCH_2 e. H_2O_2 h. $OCCCO$
   c. $CH_3NH_3^+$ f. $ClCN$

QUESTIONS

1. A double bond does not affect the 3-D structure of a molecule. What 3-D structure do you predict for SO_2 and CO_3^{2-}?

MOLAR MASS OF VOLATILE COMPOUND

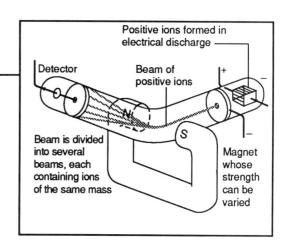

Gas molecules move at incredible velocities; for example, helium atoms travel at 1930 m/s while CO_2 molecules travel at 410 m/s at 25°C. The greater velocity of the helium atoms is attributed to its smaller mass. Even further, the molecules of a cologne travel from its source to the nose in a time span that is different from those of an orange blossom. All molecules have mass, but how do chemists measure the mass of a molecule? One modern instrumental technique employs the use of a mass spectrometer, an instrument that vaporizes the molecules at a high vacuum, ionizes them, and then subjects the molecular ions to electric and magnetic field for mass selection. This technique requires the molecule to be stable while in the gaseous state.

In this experiment we also take advantage of the volatility of a compound, but instead use the basic principles of gas phase behavior to measure the molar mass.

OBJECTIVE

• To determine the molar mass and density of a volatile compound

PRINCIPLES

The Dumas method (John Dumas, 1800–1884) for determining the molar mass of a volatile compound requires the use of the ideal gas law equation, **PV = nRT**.

Pressure: the force exerted by a gas over a surface area. Common laboratory units are atmospheres and torr.

In this experiment a compound is vaporized at a measured temperature, **T**, into a measured volume, **V**, of an Erlenmeyer flask. After the barometric pressure, **P**, is read from the laboratory barometer, the moles of gas, **n**, are calculated from the ideal gas law equation. When the pressure is recorded in atmospheres, the volume in liters, and temperature in kelvins, then the gas constant, **R**, equals 0.0821 L atm/(mol K).

Molar mass units: grams/mol (g/mol).

The mass of the vapor, **m**, is measured from the difference between the empty flask and the vapor-filled flask. The molar mass is calculated from the equation,

$$\text{molar mass} = \frac{m}{n}$$

The density of a gas is generally recorded at STP conditions.[1] A correction of the measured volume of the gas in the Erlenmeyer flask to STP, V_{corr}, must first be made before the density of the gas can be reported.

$$\text{density of gas} = \frac{m}{V_{corr}}$$

---

[1]STP (standard temperature and pressure) conditions are 273 K and 1 atm pressure.

TECHNIQUES

- Technique 2, page 1   Clean Glassware and Lab Bench
- Technique 5, page 3   Transferring Liquids
- Technique 6c, page 4   Heating Liquids
- Technique 8, page 7   Reading a Meniscus
- Technique 13, page 11   Using the Laboratory Balance

PROCEDURE

You are to complete two trials in today's experiment. Obtain no more than 15 mL of an unknown liquid compound from your laboratory instructor and check out a 110°C thermometer from the stockroom. Record the number of your unknown.

1. Set up the apparatus shown in Figure 16.1. Clean a 125 mL Erlenmeyer flask with soap and water, rinse thoroughly with distilled (or deionized) water, and *dry* either in a drying oven or by inverting the flask over a paper towel and allowing it to air-dry. Fit the flask with aluminum foil and place a rubber band around the foil and the neck of the flask. Measure the mass (±0.001 g) of the *dry* flask and Al foil/rubber band.

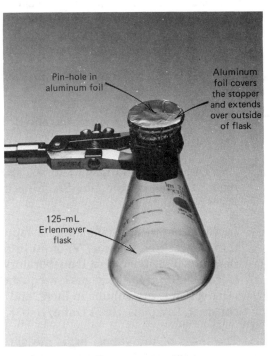

Pin-hole in aluminum foil

Aluminum foil covers the stopper and extends over outside of flask

125-mL Erlenmeyer flask

Figure 16.1
Foil-covered flask to contain the volatile liquid

2. Half-fill a 600 mL beaker with tap water, add 2 or 3 drops of 6 M HCl[2], and add a boiling stone. Support the beaker of water on a wire gauze and heat the water to boiling. While waiting for the water to boil, add approximately 6 mL of your unknown liquid to the flask and cover with the foil/rubber band.

3. Remove the heat from the hot water bath. Insert and secure the Erlenmeyer flask, containing your unknown liquid, with a utility clamp in the hot water bath; be certain the flask does not touch the beaker wall (Figure 16.2). Adjust the water level high on the neck of the flask; you may need to add or remove water. Punch 2–3 pin-sized holes in the aluminum foil. Resume the heating of the hot water bath (**Caution:** *most unknowns are flammable; use a moderate flame for heating*).

---

[2]The HCl prevents the buildup of mineral deposits on the beaker as the water evaporates.

164   **Molar Mass of a Volatile Compound**

4. As the unknown liquid is heated, some vapor escapes through the pinholes. When vapors are no longer visible[3], continue heating for another 10 minutes. The flask should now be filled with the vapor of your unknown—no liquid should remain in the flask. Record the temperature (±0.1°C) of the boiling water.

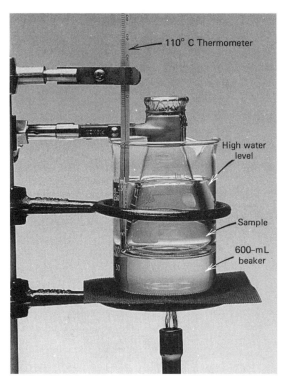

Figure 16.2

Setup for the molar mass determination of a volatile liquid

5. Remove the flask from the hot water bath. Allow to cool to room temperature. Sometimes the vapor remaining inside the flask condenses; that's o.k. Dry the outside of the flask and measure its mass along with the aluminum foil/rubber band and the condensed vapor, using the same balance that was used earlier.

6. To repeat the procedure, add 5 mL of the unknown liquid to the Erlenmeyer flask and repeat Parts 3 through 5.

Disposal Information: **Discard the condensed vapor in the "Waste Organic Liquids" container.**

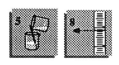

Barometer: a laboratory instrument designed to measure the pressure of a gas.

7. Fill the empty 125 mL Erlenmeyer flask to the brim with tap water. Measure the volume (±0.1 mL) of the flask by transferring portions of the water to a 50 mL graduated cylinder until it is all transferred. Sum the volumes of water and record the total volume.

8. Read the laboratory barometer and record the barometric pressure.

---

[3]To see vapors escaping through the pinholes, look across the top of the aluminum foil toward a lighted area. The vapor causes a diffraction of the lighted area.

This is a commercially available borosilicate glass bulb with a drawn tip that is used for determining the vapor density and the molar mass of volatile compounds by the Dumas method.

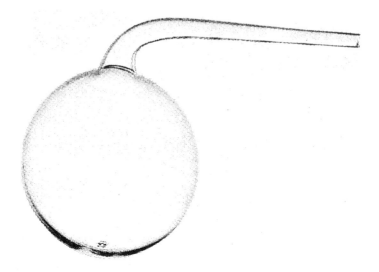

MOLAR MASS OF A VOLATILE COMPOUND

Date _____ Name _____ Lab Sec. _____ Desk No. _____

1. The vapor from an unknown compound occupies a 269 mL Erlenmeyer flask at 98.7°C and 748 torr. The mass of the vapor is 0.791 g.
    a. How many moles of vapor are present?

    b. What is the molar mass of the compound?

    c. What is the density of the vapor at STP conditions?

2. a. If the atmospheric pressure is mistakenly recorded as 760 torr in Question #1, what is the molar mass of the compound?

    b. What is the percent error in the reported molar mass caused by this assumed pressure?

3. Explain how the mass of the vapor is measured in today's experiment.

4. a. How many trials are to be completed in this experiment? _____

    b. What volume of unknown compound should be obtained from your laboratory instructor?_____

5. List the equipment that needs to be checked out from the stockroom.

6. What is the purpose of placing a boiling stone into the beaker of a hot water bath?

MOLAR MASS OF A VOLATILE COMPOUND

Date _____ Name _____ Lab Sec. _____ Desk No. _____

Unknown Number_____	Trial 1	Trial 2
1. Mass of dry flask + Al foil/rubber band (*g*)		
2. Mass of dry flask + Al foil/rubber band, and condensed vapor (*g*)		
3. Mass of vapor unknown, **m** (*g*)		
4. Temperature of boiling water, **T** (°C)		
5. Volume of 125 mL flask _____+_____+_____ , **V** (*mL*)		
6. Barometric pressure, **P** (*atm*)		
7. Moles of vapor, **n** (*mol*)		
8. Molar mass, m/n (*g/mol*)	*	
9. Average molar mass (*g/mol*)		
10. Density of gas at STP (*g/L*)		

*Show your calculations for Trial 1.

QUESTIONS

1. If the mass of the flask is measured after the liquid has been vaporized (in Part 5) but before the outside of the flask is dried, will the molar mass of the unknown be too high or too low? Explain.

2. a. Suppose the barometric pressure during today's experiment is assumed to be 760 torr instead of the value you recorded. Would the molar mass of the unknown be reported too high or too low? Explain.

    b. What would be your percent error for the molar mass if the assumption had been made?

3. If the volume of the vapor is assumed to be 125 mL instead of the measured volume, what would be the percent error for the molar mass of the unknown liquid compound? Show your work.

4. If all of the unknown liquid compound does not vaporize into the 125 mL Erlenmeyer flask in Part 4, will the reported molar mass be too high or too low? Explain.

EXPERIMENT 17

ACIDS, BASES, AND SALTS

Fresh water lakes are acidic, salt water bays and estuaries are basic. The foods and fluids that we consume are acidic (as are our stomachs) but our intestinal tract and blood are basic. Acids are corrosive, but bases remove layers of skin. The extent of such effects depend upon the strength of the acid or base. All mixtures in which water is the solvent are generally either acidic or basic, few are "neutral". Since aqueous systems are so important to our very existence then it becomes obvious as to why we should become familiar with the sources and properties of "home" and laboratory acids and bases, many of which exist because of a particular ion in a salt.

OBJECTIVE

- To become familiar with the chemical and physical properties of acids, bases, and salts.

PRINCIPLES

Acids are substances that, when dissolved in water, produce hydronium ion, H_3O^+. For example, pure HCl is a gas at room temperature and pressure, but dissolved in water, it releases H^+ to a water molecule, forming the hydronium ion, H_3O^+.

$$HCl(g) + H_2O(l) \rightarrow H_3O^+(aq) + Cl^-(aq)$$

Bases dissolved in water produce hydroxide ion, OH^-. Barium hydroxide, $Ba(OH)_2$, is a solid at room temperature and pressure, but in water, it dissociates into Ba^{2+} and OH^- ions.

$$Ba(OH)_2(s) \xrightarrow{H_2O} Ba^{2+}(aq) + 2 OH^-(aq)$$

When aqueous solutions of an acid and a base are mixed, a reaction occurs producing water and a salt as products. This is a **neutralization reaction**. For example, a mixture of HCl and $Ba(OH)_2$ produce water and the salt barium chloride, $BaCl_2$, as products.

$$2 HCl(g) + Ba(OH)_2(s) \xrightarrow{H_2O} 2 H_2O(l) + BaCl_2(aq)$$

A **salt** therefore is any ionic compound that is a neutralization product of an acid-base reaction.

Ionic compound: a compound that consists of ratio of cations and anions resulting in a neutral compound.

But in water, hydrogen chloride actually exists as H_3O^+ and Cl^- and barium hydroxide as Ba^{2+} and OH^-. Barium chloride is a soluble, ionic compound—in an aqueous solution it dissociates into Ba^{2+} and Cl^-. Therefore, a better representation of the reaction between HCl and $Ba(OH)_2$ is

$$2 H_3O^+(aq) + 2 Cl^-(aq) + Ba^{2+}(aq) + 2 OH^-(aq) \rightarrow$$
$$4 H_2O(l) + Ba^{2+}(aq) + 2 Cl^-(aq)$$

or written as a net ionic equation,

$$H_3O^+(aq) + OH^-(aq) \rightarrow 2\,H_2O(l)$$

Electrolytes are substances that dissociate in water to produce ions and conduct an electrical current. Some compounds are *strong* electrolytes in water—these substances completely dissociate ($\approx100\%$) into ions. Hydrochloric acid, HCl, barium hydroxide, $Ba(OH)_2$, and barium chloride, $BaCl_2$, are strong electrolytes. **Nonelectrolytes** are substances that dissolve in water but do not form ions ($\approx0\%$ ionization). Sugar and alcohol are nonelectrolytes. A large number of compounds only partially dissociate in an aqueous solution; these are called *weak* electrolytes. Acetic acid is a weak electrolyte.

Acidic solutions result from the action of water on (1) nonmetallic hydrides, such as HCl and HBr, (2) nonmetallic oxides, such as CO_2 and SO_3, or (3) compounds of hydrogen, oxygen, and one other element (usually a nonmetal), such as H_2SO_4 and HNO_3.

$$HCl(g) + H_2O(l) \rightarrow H_3O^+(aq) + Cl^-(aq)$$
$$CO_2(g) + H_2O(l) \rightarrow H_3O^+aq) + HCO_3^-(aq)$$
$$H_2SO_4(l) + H_2O(l) \rightarrow H_3O^+(aq) + HSO_4^-(aq)$$

Sulfuric acid, also called oil of vitriol, is perhaps the most versatile of all inorganic industrial chemicals. There is hardly a chemical industry that does not use sulfuric acid in its manufacturing process. In 1991, when it ranked number one in chemical usage, nearly 87 billion pounds of H_2SO_4 were produced in the United States, nearly twice that of the second ranked chemical. Other acids that ranked among the "Top 50" chemicals in production were phosphoric acid, H_3PO_4, nitric acid, HNO_3, and hydrochloric acid, HCl (called muriatic acid). See Table 17.1.

In addition, many organic acids that rank in the "Top 50", such as acetic, adipic, and oleic acids, are useful and important to the chemical industry.

Table 17.1. Acids and Bases Ranked Among the "Top 50" in Production for 1991 in the United States from *Chemical and Engineering News* , April 13, 1992, p. 16.

Rank	Chemical	Formula	Billions of Pounds (1991)
1	sulfuric acid	H_2SO_4	86.62
5	ammonia	NH_3	34.04
6	lime	CaO	33.60
7	phosphoric acid	H_3PO_4	24.68
8	sodium hydroxide	NaOH	24.39
11	sodium carbonate	Na_2CO_3	20.50
13	nitric acid	HNO_3	15.05
27	hydrochloric acid	HCl	5.60
34	acetic acid	CH_3COOH	3.61
46	adipic acid	$C_4H_8(COOH)_2$	1.56

Metallic oxides: also called basic anhydrides.

Three sources of basic solutions are: (1) metallic hydroxides, such as $Ba(OH)_2$ and $NaOH$, (2) metallic oxides, such Na_2O, or (3) a select number of polyatomic anions, such as PO_4^{3-} and CO_3^{2-}.

$$Ba(OH)_2(s) \xrightarrow{-H_2O} Ba^{2+}(aq) + 2\,OH^-(aq)$$
$$Na_2O(s) + H_2O(l) \rightarrow 2\,Na^+(aq) + 2\,OH^-(aq)$$
$$PO_4^{3-}(aq) + H_2O(l) \rightarrow HPO_4^{2-}(aq) + OH^-(aq)$$

The strong bases such as $NaOH$ (called caustic soda or lye) and KOH (called caustic potash) are also known as **alkalis**.

The most notable bases ranked in the "Top 50" are ammonia, NH_3, sodium hydroxide, calcium oxide, CaO (also called quicklime or, more simply, lime), and sodium carbonate, Na_2CO_3 (also known as soda ash). Again, refer to Table 17.1.

TECHNIQUES

- Technique 3, page 2   Microscale Analyses
- Technique 6a, page 4   Heating Liquids
- Technique 11, page 10   Testing with Litmus
- Technique 13, page 11   Using the Laboratory Balance
- Technique 17a,b, page 16   Handling Gases

PROCEDURE

During this experiment, **STOP** at each numbered superscript (i.e., [1]) and record your observation(s) or conclusion(s) on the Data Sheet.

Caution: *Dilute and concentrated (conc) acids and bases cause severe skin burns and irritation to mucous membranes. Be very careful in handling these chemicals. Clean up all spills immediately with excess water, followed by a covering of baking soda, $NaHCO_3$. Notify your instructor if a spill occurs. Read the "Laboratory Safety" section in Experiment 1.*

A. Conductivity

The apparatus shown in Figure 17.1 is used to determine the strength or extent of dissociation of various electrolytes. When the two electrodes are submerged in a solution that is a good conductor of current (a strong electrolyte), the circuit is completed and the bulb shines brightly; if the solution is a poor conductor, the bulb burns dimly; or if a nonconductor, not at all. Therefore the brightness of the bulb is a qualitative measure of the degree of dissociation of a substance into ions.

1. Connect the apparatus to an electrical outlet. Half-fill a 150 mL beaker with distilled (or deionized) water and place it in contact with the electrodes. Does the bulb glow?[1] Add about 0.5 g of NaCl to the water and stir to dissolve. What happens to the conductivity of the solution?[2] Is NaCl a strong, weak, or nonelectrolyte?[3] Remove the solution and rinse the electrodes with distilled (or deionized) water from your wash bottle.

2. Place about 25 mL of conc acetic acid (read the **Caution** for handling acids) into a clean 150 mL beaker. Place the acid in contact with the electrodes. Classify its strength as an electrolyte.[4] While keeping the acetic acid in contact with the electrodes, add distilled (or deionized) from your wash bottle and swirl the solution. Observe. Continue to add water (up to 100 mL), swirl, and observe the glow of the bulb. What happens to the conductivity as the acetic solution is diluted?[5] Rinse the electrodes with distilled (or deionized) water and discard the rinse.

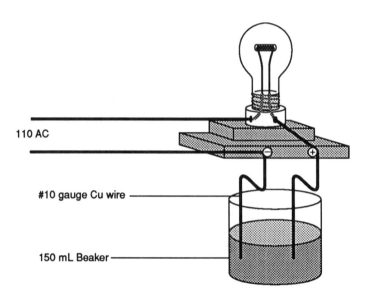

110 AC

#10 gauge Cu wire

Figure 17.1
Apparatus for testing the conductivity of a solution

150 mL Beaker

3. Test the conductivity of the following solutions.* Be sure to rinse the electrodes after each test to avoid contamination of the solutions.

   0.1 M CH_3COOH, 0.1 M HCl, 0.1 M NH_3, 0.1 M NaOH, 0.1 M $NaNO_3$, 0.1 M NH_4Cl, 0.1M $C_{12}H_{22}O_{11}$, and 0.1 M $CO(NH_2)_2$. Record your observations on the Data Sheet. Classify the solutions as strong, weak, or nonelectrolytes.[6]

4. Place 25 mL of 0.1 M H_2SO_4 in a 150 mL beaker; submerge the electrodes.[7] Obtain about 15 mL of 0.2 M $Ba(OH)_2$. Slowly add the 0.2 M $Ba(OH)_2$; keep the mixture stirred. What happens to the conductivity of the mixture? Why?[8] Continue adding 0.2 M $Ba(OH)_2$ until the glow disappears. Now add additional $Ba(OH)_2(aq)$. Explain your observations.[9]

B. Concentrated Acids

1. Several bottles of concentrated acids (purchased from the chemical supplies distributor) are on the reagent shelf. Select one of the acids and closely read the label. Answer the questions on the Data Sheet.

C. Acids—Sulfuric, Nitric, Hydrochloric, and Acetic Acids

1. Arrange 1 x 4 array of microcells containing 10 drops of water. Place a finger flush on the bottom of the first microcell and add 5 drops of conc HCl; is noticeable heat produced?[10] Test the solution with litmus paper.[11] Repeat the tests with 5 drops of conc H_2SO_4, conc HNO_3, and conc CH_3COOH (**Caution**: *do not allow the concentrated acids to touch the skin*).[12]

2. Some acids can be prepared by the action of another acid on a salt. H_2SO_4 and H_3PO_4 are often used as the acids for the preparation. Place about 0.1 g of NaCl in a microcell. Add 3 or 4 drops of conc H_2SO_4 to the solid salt and hold both red and blue moistened litmus papers over the reaction. Explain the effect on litmus.[13] Write a balanced equation for the reaction.[14]

---

*
To avoid waste, the test solutions can be shared with other students, especially if care is taken to avoid contamination.

3. Place a small piece of polished (with steel wool) Mg, Zn, Fe, and Cu in a 1 x 4 array of microcells. Add about 10 drops of 6 M HCl to each metal. Note any reaction.[15] Repeat the test with 6 M HNO₃.[16]

4. **Instructor Demonstration.** Test the dehydrating effects of conc H_2SO_4 by placing a few drops on a wood splint and, in an evaporating dish, on a small amount (≈1 g) of sugar. Repeat the test with conc HCl and conc HNO₃. Record your observations for each acid on the Data Sheet.[17-19]

5. Set up a 1 x 3 array of microcells. Pipet 1 mL of 0.1 M NaOH into each microcell and add 1 drop of phenolphthalein indicator. Add drops of 0.5 M HCl to microcell (1) until a color change (from pink to colorless) occurs (swirl after each drop). Record the number of drops.[20]

   Repeat the determination, substituting 0.5 M H_2SO_4 and 0.5 M HNO₃ for the 0.5 M HCl.[21,22] What can you conclude about the available acidity of the three acids?[23] Write balanced equations for the reactions.[24]

Disposal Information: **Dispose of all acid test solutions in the "Waste Acids" container.**

D. Bases—Sodium Hydroxide, Calcium Oxide, and Sodium Carbonate, and Ammonia

1. In four successive microcells, place a small BB-sized piece of NaOH, a sample of oven cleaner or solid drain cleaner, a fresh sample of CaO, and a sample of Na_2CO_3. Place the tip of your finger flush on the bottom of the microcell and add 1 mL of water to the first microcell. Note the heat generated in the dissolving of the NaOH.[25] Repeat the test with the other samples. Test each solution with litmus.[26]

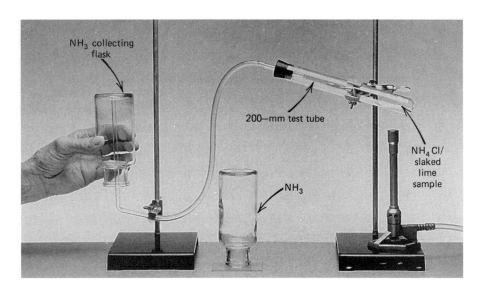

Figure 17.2
Collection of NH₃(*g*) by air displacement

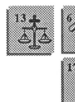

2. Set up the apparatus shown in Figure 17.2. Place 3 g (± 0.01 g) of a 2:1 mixture (by volume of solid) of NH_4Cl and $Ca(OH)_2$ into the 200 mm test tube. Heat the mixture gently; begin at the mouth of the test tube and gradually extend the heat over the entire test tube. As soon as NH_3 gas evolves freely and the apparatus is free of air, collect *two* bottles of the NH_3 by air displacement.§ Write balanced equations for the reaction of NH_4Cl with $Ca(OH)_2$ and for the reaction of HCl with NH_3 (see footnote).[27] Note the color and odor of the NH_3.[28]

Phenolphthalein: an acid-base indicator that is colorless in acidic solutions but pink/red in basic solutions.

3. To test for the solubility of NH_3 in water, place about 300 mL of water in an 800 mL beaker. Add 1 mL of phenolphthalein to the water. Place the mouth of a bottle of NH_3 under the surface of the water (Figure 17.3). Allow it to remain there for several minutes; be sure that the mouth of the bottle stays submerged. What happens to the water level in the bottle?[29] What happens to the color of the solution? Explain.[30]

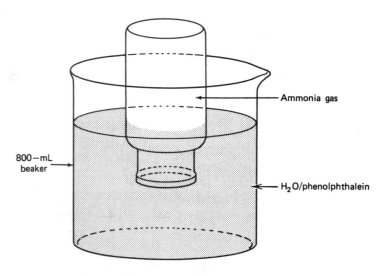

Ammonia gas

800−mL beaker

H_2O/phenolphthalein

Figure 17.3
A setup for testing the solubility of $NH_3(g)$ in water

Disposal Information:	Dispose of the base test solutions in the "Waste Base" container.

---

§To test if a bottle is filled with NH_3, suspend a drop of conc HCl from the tip of a glass rod at the mouth of the bottle; the conc HCl fumes strongly when ammonia is present. Consult the instructor on this technique.

ACIDS, BASES, AND SALTS-LAB PREVIEW

Date _____ Name _____ Lab Sec. _____ Desk No. _____

1. a. What is an acid?

   b. What is a base?

   c. What is a salt?

   d. What is an alkali?

2. Distinguish between a strong electrolyte and a weak electrolyte.

3. a. What is a neutralization reaction?

   b. In what part of the experimental procedure is a neutralization reaction conducted?

4. a. What is the color of litmus paper in an acidic solution?_____

   b. What is the color of litmus paper in a basic solution?_____

   c. What is the color of phenolphthalein in an acidic solution?_____

   d. What is the color of phenolphthalein in a basic solution?_____

5. Briefly describe the technique for testing a solution with litmus paper.

6. How are gases that are less dense than air collected?

7. Describe the technique for heating a test tube with a direct flame (Part D.2).

8. Describe the technique for testing the odor of a chemical.

9. List the common (or commercial) names and the correct chemical names associated with these compounds.

	Common (commercial) Names	Chemical Names
a. Na_2CO_3		
b. HCl		
c. H_2SO_4		
d. NaOH		
e. CaO		
f. $Ca(OH)_2$		
g. $CaCO_3$		
h. NH_3		
i. $NaHCO_3$		

10. How should acids spills be cleaned up in the laboratory?

ACIDS, BASES, AND SALTS

Date _____Name_____ Lab Sec. _____Desk No._____

A. Conductivity

	Compound	Observation	Electrolyte		
			Strong	Weak	Non–
#1	H_2O				
#2,3	NaCl				
#4	conc CH_3COOH				
#5	dil CH_3COOH				
#6	0.1 M CH_3COOH				
	0.1 M HCl				
	0.1 M NH_3				
	0.1 M NaOH				
	0.1 M $NaNO_3$				
	0.1 M NH_4Cl				
	0.1 M $C_{12}H_{22}O_{11}$				
	0.1 M $CO(NH_2)_2$				
#7	0.1 M H_2SO_4				
#8,9	H_2SO_4 + $Ba(OH)_2$				

B. Concentrated Acids

1. What is the percent (range) by mass of the acid in the bottle?_____

   What is its major impurity?_____

2. List two **Danger** warnings printed on the label.

   a. _____

   b. _____

3. List two **First Aid** remedies for the acid.

   a. _____

   b. _____

C. Acids

	Acid	Heat Change	Litmus Test
#10,11	HCl		
#12	H_2SO_4		
	HNO_3		
	CH_3COOH		

#13 Litmus test on vapors_____

#14 Balanced equation: $NaCl + H_2SO_4 \rightarrow$ _____

	Acid/Metal	Mg	Zn	Fe	Cu
#15	HCl				
#16	HNO_3				

	Acid	H_2SO_4	HCl	HNO_3
#17-19	wood splint			
	sugar			

	Acid	HCl	H_2SO_4	HNO_3
#20-22	Drops of NaOH			

#23 Conclusion on available acidity

#24 Balanced equations: $NaOH + HCl \rightarrow$ _____

$NaOH + H_2SO_4 \rightarrow$ _____

$NaOH + HNO_3 \rightarrow$ _____

D. Bases

		Heat Change	Litmus Test
#25	NaOH		
#26	Oven/Drain Cleaner		
	CaO		
	Na_2CO_3		

#27 Balanced equations: $NH_4Cl + Ca(OH)_2 \rightarrow$ _____

$HCl + NH_3 \rightarrow$ _____

#28 Color of NH_3 _____ . Does NH_3 have an odor?_____

#29 Is NH_3 soluble in water?_____

#30 Why does the color of the solution change?

EXPERIMENT 18

HARD WATER ANALYSIS

What does it mean when we say that water is "hard"? Hard water is obviously not hard in the same sense that wood or metal is hard, nor difficult to understand as is the creation of the universe. Hard water contains the dissolved salts of calcium, magnesium, and iron ions which are called hardening ions. In low concentrations these ions are not considered harmful for domestic use, but at higher concentrations these ions interfere with the cleansing action of soaps and accelerate the corrosion of steel pipes, especially those carrying hot water.

The source of hardening ions is the result of slightly acidic rainwater flowing over mineral deposits of varying compositions; the acidic rainwater reacts with the *very* slightly soluble carbonate salts of calcium and magnesium and with various iron-containing rocks. A partial dissolution of these salts releases the ions into the drinking water supply, which may be surface water from a lake or ground water.

OBJECTIVES

- To learn the cause and effects of hard water
- To determine the hardness of a water sample

PRINCIPLES

Hardening ions, such as Ca^{2+}, Mg^{2+}, and Fe^{3+}, form insoluble compounds with soaps. Soaps, sodium salts of fatty acids such as sodium stearate, $C_{17}H_{35}CO_2^- Na^+$, are very effective cleansing agents so long as they remain soluble; the presence of the hardening ions however causes the formation of a grey, insoluble soap scum such as $(C_{17}H_{35}CO_2)_2Ca$.

$$2 C_{17}H_{35}CO_2^-Na^+(aq) + Ca^{2+}(aq) \rightarrow (C_{17}H_{35}CO_2)_2Ca(s) + 2 Na^+(aq)$$

This grey precipitate appears as a "bathtub ring" and it also clings to clothes, causing white clothes to appear grey.

Hard water is also responsible for the appearance and undesirable formation of "boiler scale" on tea kettles and pots used for heating water. The boiler scale is a poor conductor of heat and thus reduces the efficiency of transferring heat. Boiler scale also builds on the inside of hot water pipes to decrease the flow of water; in extreme cases, this buildup causes the pipe to break. Boiler scale consists primarily of the carbonate salts of the hardening ions and is formed according to

$$Ca^{2+}(aq) + 2 HCO_3^-(aq) \xrightarrow{\Delta} CaCO_3(s) + CO_2(g) + H_2O(l)$$

Ground water becomes hard as it flows through underground limestone ($CaCO_3$) deposits; generally, the water from deep wells has a higher hardness than that from shallow wells because of a longer time of contact with the limestone. Surface water similarly accumulates hardening ions as a result of it flowing over limestone deposits. In either case the CO_2 dissolved in rainwater[1] solubilizes limestone deposits.

$$CO_2(aq) + H_2O(l) + CaCO_3(s) \rightarrow Ca^{2+}(aq) + 2\,HCO_3^{-}(aq)$$

Surface water: water that is collected from a watershed, e.g., lakes, rivers, and streams.

Notice that this reaction is just the reverse of the reaction for the formation of boiler scale.

Because of the relative large natural abundance of limestone deposits and other calcium minerals, such as gypsum, $CaSO_4 \cdot 2H_2O$, it is not surprising that Ca^{2+} ion, in conjunction with Mg^{2+}, is a major component of the dissolved solids in water. A general classification of hard waters is listed in Table 18.1.

Table 18.1 Hardness Classification of Water

Hardness (*ppm* $CaCO_3$)	Classification
< 15 ppm	very soft water
15 ppm - 50 ppm	soft water
50 ppm - 100 ppm	medium hard water
100 ppm - 200 ppm	hard water
>200 ppm	very hard water

ppm: parts per million by mass, e.g., 15 ppm $CaCO_3$ is 15 g $CaCO_3$ in 10^6 g solution.

The concentration of the hardening ions in a water sample is commonly expressed as though the hardness is due exclusively to $CaCO_3$. The units for hardness is mg $CaCO_3$/L, which is also ppm $CaCO_3$.[2]

In this experiment a titration technique is used to measure the combined Ca^{2+} and Mg^{2+} concentrations in a water sample. The titrant is the disodium salt of ethylenediaminetetraacetic acid (abbreviated Na_2H_2Y)[3].

In aqueous solution Na_2H_2Y dissociates into Na^+ and H_2Y^{2-} ions. The H_2Y^{2-} ion reacts with the hardening ions, Ca^{2+} and Mg^{2+}, to form very stable complex ions, especially in a solution buffered at a pH of about 10. An ammonia–ammonium ion buffer is often used for this pH adjustment in the analysis.

A special indicator is used to detect the endpoint in the titration. Called Eriochrome Black T (EBT), it forms complex ions with the Ca^{2+} and Mg^{2+}

---

[1]CO_2 dissolved in rainwater makes rainwater slightly acidic.

$$CO_2(g) + 2\,H_2O(l) \rightarrow H_3O^+(aq) + HCO_3^{-}(aq)$$

[2]ppm means "parts per million"—1 mg of $CaCO_3$ in 1 000 000 mg solution is 1 ppm $CaCO_3$. Assuming the density of the solution is 1 g/mL, then 1 000 000 mg solution = 1 L solution.

[3]Ethylenediaminetetraacetate is often simply referred to as EDTA.

ions, but binds more strongly to Mg^{2+} ions. Because only a small amount of EBT is added, only a small quantity of Mg^{2+} is complexed; no Ca^{2+} ion is complexed to EBT—therefore, most of the hardening ions remain "free" in solution. The EBT indicator is sky-blue in solution but the $[Mg-EBT]^{2+}$ complex ion is wine-red.

$$Mg^{2+}(aq) + EBT(aq) \rightarrow [Mg-EBT]^{2+}(aq)$$
$$\text{sky-blue} \quad\quad \text{wine-red}$$

Therefore even before any H_2Y^{2-} titrant is added for the analysis, the solution is wine-red. As H_2Y^{2-} titrant is added, it complexes with the "free" Ca^{2+} and Mg^{2+}.

$$Ca^{2+}(aq) + H_2Y^{2-}(aq) \rightarrow CaY^{2-}(aq) + 2H^+(aq)$$
$$Mg^{2+}(aq) + H_2Y^{2-}(aq) \rightarrow MgY^{2-}(aq) + 2H^+(aq)$$

Once the H_2Y^{2-} complexes all of the "free" Ca^{2+} and Mg^{2+} from the water sample, it then removes the Mg^{2+} from the $[Mg-EBT]^{2+}$ complex; the solution turns from wine-red back to the sky-blue color of the indicator, and the endpoint is reached.

$$[Mg^{2+}-EBT]^{2+}(aq) + H_2Y^{2-}(aq) \rightarrow MgY^{2-}(aq) + 2H^+(aq) + EBT(aq)$$
$$\text{wine-red} \quad\quad\quad\quad\quad\quad\quad\quad\quad\quad\quad\quad\quad\quad\quad \text{sky-blue}$$

Equimolar: an equal number of moles.

For the endpoint to appear, Mg^{2+} must be present; therefore a small amount of MgY^{2-} is usually added to the buffer solution. The added Mg^{2+} does not affect the amount of H_2Y^{2-} used in the analysis because an equimolar amount of Na_2H_2Y is also added.

From the balanced equations, it is apparent that once the molar concentration of the Na_2H_2Y solution is known, the moles of hardening ions in a water sample can be calculated.

volume H_2Y^{2-} x molar concentration of H_2Y^{2-} = moles H_2Y^{2-}
= moles hardening ions

The hardening ions, for reporting purposes, are assumed to be exclusively Ca^{2+} from the dissolving of $CaCO_3$. Since one mole of Ca^{2+} is from one mole of $CaCO_3$, the equivalent hardness expressed as mg $CaCO_3$ per liter of water sample is

moles hardening ions = moles Ca^{2+} = moles $CaCO_3$

$$\text{ppm } CaCO_3 \left(\frac{\text{mg } CaCO_3}{\text{L sample}}\right) = \frac{\text{mol } CaCO_3}{\text{L sample}} \times \frac{100.1 \text{ g } CaCO_3}{\text{mol}} \times \frac{\text{mg}}{10^{-3}\text{g}}$$

TECHNIQUES

- Technique 2, page 1   Clean Glassware and Lab Bench
- Technique 8, page 12   Reading a Meniscus
- Technique 9, page 13   Pipetting a Liquid or Solution
- Technique 10, page 14   Titrating a Solution
- Technique 13, page 16   Using the Laboratory Balance
- Technique 14b, page 5   Separation of a Liquid from a Solid

PROCEDURE

A. A Standard 0.01 M Disodium Ethylenediaminetetraacetate, Na$_2$H$_2$Y, Solution

Three trials should be completed for the standardization of the Na$_2$H$_2$Y solution. To save time, prepare the three Erlenmeyer flasks in Part A.3 at the same time.

1. Measure about 0.5 g (±0.01 g) of Na$_2$H$_2$Y•2H$_2$O (molar mass = 372.24 g/mol) on weighing paper; transfer it to a 250 mL volumetric flask containing about 100 mL of distilled (or deionized) water and stir to dissolve. Dilute to the "mark" on the volumetric flask with distilled (or deionized) water.

2. Prepare a buret for titration. Rinse the buret with the Na$_2$H$_2$Y solution and then fill. Record the volume (±0.01 mL) of the solution.

3. Obtain about 80 mL of the standard Ca^{2+} solution from the reagent shelf and record its molar concentration. Pipet 25.0 mL into a 125 mL Erlenmeyer flask, add 1 mL of the buffer (pH = 10) solution, and 2 drops of EBT indicator.

4. Titrate the standard Ca^{2+} solution with the Na$_2$H$_2$Y solution; swirl continuously. Near the endpoint, slow the rate of addition to drops; the last few drops should be added at 3–5 s intervals. The solution changes from wine-red to purple to blue—no tinge of the wine-red color should remain; the solution is *blue* at the endpoint.

5. Repeat the titrations on the other two samples and then calculate the molar concentration of the Na$_2$H$_2$Y solution.

B. Analysis of Water Sample

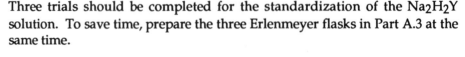

Complete three trials in your analysis. The first trial is an indication of the hardness of your water sample—the hardness values may vary tremendously, depending upon your water sample. You may want to adjust the volume of water in the analysis for the second and third trials.

1. a. Obtain about 100 mL of a water sample from your instructor. You may want to bring your own water sample or simply use the tap water in the laboratory.

    b. If the water sample is from a lake, stream, or ocean, you will need to gravity filter the sample before the analysis.

    c. If your sample is acidic, add 1 M NH$_3$ until it is basic to litmus.

2. Pipet 25.0 mL of your (filtered, if necessary) water sample[4] into a 125 mL Erlenmeyer flask, add 1 mL of the buffer (pH = 10) solution, and 2 drops of EBT indicator.

3. Repeat Parts A.4 and A.5 to determine the hardness of your water sample.

Disposal Information: Dispose of the analyzed solutions in the "Waste EDTA" container.

---

[4]If your water is known to have a high hardness, decrease the volume of the water proportionally until it takes about 15 mL of Na$_2$H$_2$Y titrant for your second and third trials. Similarly if your water sample is known to have a low hardness, increase the volume of the water proportionally.

HARD WATER ANALYSIS

Date _____ Name _____ Lab Sec. _____ Desk No. _____

1. What ions are responsible for water hardness?

2. Which one of the hardening ions binds more tightly to (forms a stronger complex ion with) the anion of the disodium salt of ethylenediaminetetraacetate, H_2Y^{2-}?

3. a. How do hardening ions cause soap to be less effective?

    b. What is and what causes boiler scale?

4. a. While gravity filtering a solution, the funnel should be no more than _____ full at any time.

    b. Two folds are made on a piece of filter paper that is to be fit into a gravity filtering funnel; describe the two folds.

5. The analysis procedure in this experiment requires that the titration be conducted at a pH = 10. How is the solution adjusted to this pH?

6. A 50.0 mL water sample required 5.14 mL of 0.0100 M Na_2H_2Y to reach the endpoint for the titration procedure described in this experiment.

   a. What is the name of the indicator used in the titration?

   b. What is the color change at the endpoint?

   c. Calculate the moles of hardening ions in the water sample.

   d. Assuming the hardness is due exclusively to $CaCO_3$, express the hardness concentration in mg $CaCO_3$/L sample.

   e. What is this hardness concentration expressed in ppm $CaCO_3$?

   f. Classify the hardness of this water according to Table 18.1.

7. The hardness of a water sample is known to be 500 ppm $CaCO_3$. In an analysis of a 50 mL water sample, what volume of 0.0100 M Na_2H_2Y is needed to reach the endpoint?

HARD WATER ANALYSIS

Date _____ Name _____ Lab Sec. _____ Desk No. _____

A. A Standard 0.01 M Disodium Ethylenediaminetetraacetate, Na_2H_2Y, Solution

	Trial 1	Trial 2	Trial 3
1. Mass of weighing paper + Na_2H_2Y (g)			
2. Mass of weighing paper (g)			
3. Mass of Na_2H_2Y (g)			
4. Volume of standard Ca^{2+} solution (mL)	25.0	25.0	25.0
5. Concentration of standard Ca^{2+} solution			
6. Mol Ca^{2+} = mol Na_2H_2Y (mol)			
7. Buret reading, final (mL)			
8. Buret reading, initial (mL)			
9. Volume of Na_2H_2Y titrant (mL)			
10. Molar concentration of Na_2H_2Y solution (mol/L)			
11. Average molar concentration of Na_2H_2Y solution (mol/L)			

B. Analysis of Water Sample

	Trial 1	Trial 2	Trial 3
1. Volume of water sample (mL)			
2. Buret reading, final (mL)			
3. Buret reading, initial (mL)			
4. Volume of Na_2H_2Y titrant (mL)			
5. Mol Na_2H_2Y = mol Ca^{2+} (mol)			
6. Mass of equivalent $CaCO_3$ (g)			
7. ppm $CaCO_3$ (mg $CaCO_3$/L sample)			
8. Average ppm $CaCO_3$			

1. Because the chemist couldn't accurately detect the endpoint, the addition of Na_2H_2Y titrant was discontinued before the endpoint was reached. How will this affect the reported hardness of the water sample? Explain.

2. Explain what might happen in the analysis if the indicator had been omitted from the procedure.

3. Washing soda, $Na_2CO_3 \cdot 10H_2O$, is often used to "soften" hard water, i.e., to remove hardening ions. Assuming hardness is due to Ca^{2+}, the CO_3^{2-} ion precipitates the Ca^{2+}.

$$Ca^{2+}(aq) + CO_3^{2-}(aq) \rightarrow CaCO_3(s)$$

How many grams and pounds of washing soda are needed to remove the hardness from 500 gallons of water having a hardness of 150 ppm $CaCO_3$ (1 gal = 4 qt; see Appendix B for additional conversion factors)?

REDOX; ACTIVITY SERIES

Man! Look at that old car! What a piece of junk! One sign of an "old" car is its external appearance. The door panels are beginning to rust (and perhaps disappear!) and the paint is losing its luster. Oftentimes only professional care and repair will return the exterior to its original beauty. All one needs to do is to visit a junk yard of old cars to see what happens when the finish is exposed to the harsh conditions of rain, salt, various air contaminants, and sunlight; the surface and underlying steel alloys become "oxidized" and the coating looks dull and ugly.

All metals have a tendency to oxidize; some are easily oxidized by water, others require harsher chemicals such as nitric acid. In this experiment we will use oxidizing agents of various strengths to differentiate the ease of oxidation of a series of metals. For example, should we use magnesium instead of iron to guard against the corrosion of our automobile? Or would zinc be better?

OBJECTIVES

- To learn rules for determining the oxidation numbers of the elements in compounds
- To develop a general understanding of redox reactions
- To determine the relative chemical reactivity of several metals

PRINCIPLES

Oxidize: to combine with oxygen or to increase in oxidation number.

Valence electrons: electrons in the highest principal energy level of the atom.

Reduce: to remove oxygen from a compound or to decrease the oxidation number of an element.

Most elements exist in chemical combination with other elements, forming any number of compounds. An element that chemically combines with another element to form a compound no longer has the same physical or chemical properties because of changes in its electronic structure—it has either lost or gained valence electrons to form an ion or has shared its valence electrons to form a molecule.

An atom in its use of its valence electrons to form a molecule (or ion) assumes an "apparent" charge. This apparent charge, the charge the atom would have if all the electrons in each bond were assigned to the more electronegative element, may be either positive (+) or negative (–). This apparent charge is called the **oxidation number** (or **oxidation state**) of the atom in the molecule. When the atom appears to have *lost* valence electrons in the bond formation, its oxidation number has increased and we say that it has been *oxidized*; but when the atom appears to have *gained* valence electrons in the bond formation, its oxidation number has decreased, it has been *reduced*.

Oxidation numbers are very handy "bookkeeping" devices for keeping track of valence electrons when elements combine to form compounds. From a few rules about oxidation numbers, we can not only write the correct chemical formulas for a large number of compounds, but also determine what is oxidized and what is reduced in a chemical reaction. These rules are:

1. Any element in the "free" state (not combined with any other element) has an oxidation number of zero, regardless of the complexity of the molecule in which it occurs. Each atom in Ne, O_2, P_4, and S_8 has an oxidation number of zero.

2. The oxidation number of any ion (monoatomic or polyatomic) equals the charge on the ion. The Ca^{2+}, NH_4^+, S^{2-}, and PO_4^{3-} ions have oxidation numbers of 2^+, 1^+, 2^-, and 3^- respectively.

3. Oxygen has an oxidation number of 2^- (except 1^- in peroxides, e.g., H_2O_2, and 2^+ in OF_2). The oxidation number of oxygen is 2^- in $KMnO_4$, Fe_2O_3, CaO, and N_2O.

4. Hydrogen has an oxidation number of 1^+ (except 1^- in metal hydrides, e.g., NaH or CaH_2). Its oxidation number is 1^+ in HCl, $NaHCO_3$, and NH_3.

5. Some elements exhibit only one oxidation number in certain compounds.
   a. Group IA elements (Li, Na, K, Rb, and Cs) always have a 1^+ oxidation number in compounds.
   b. Group IIA elements (Be, Mg, Ca, Sr, and Ba) always have a 2^+ oxidation number in compounds.
   c. Boron and aluminum always possess a 3^+ oxidation number in compounds.
   d. Group VIA elements (O, S, Se, and Te) exhibit a 2^- oxidation number in all *binary* compounds with *metals*.
   e. Group VIIA elements (F, Cl, Br, and I) exhibit a 1^- oxidation number in all *binary* compounds with *metals*.

6. The oxidation number of any other element does not follow set rules, it may vary depending upon the elements with which it combined to form the compound. Generally one or more of the elements in a compound follow Rules 1–5; since the compound must be neutral, the oxidation number of the element in question can be calculated.

   Example 1. What is the oxidation number of Fe in Fe_2O_3?
   Since each oxygen is known to have a 2^- oxidation number (for a total of 6^-) and the sum of the oxidation numbers of Fe_2O_3 is zero, the combined oxidation numbers of the two iron atoms must be 6^+, or *each* Fe atom must be 3^+.

7. The oxidation number of polyatomic ions is the charge of the ion (see Rule 2). The oxidation numbers of the constituent atoms in the ion are determined as in Rule 6.

   Example 2. What is the oxidation number of Cr in the dichromate ion, $Cr_2O_7^{2-}$?
   Each oxygen has a 2^- oxidation number (for a total of 14^-) and the oxidation number of $Cr_2O_7^{2-}$ is 2^-. Since $2\ Cr + 14^- = 2^-$, the combined oxidation number of the two chromium atoms must be 12^+, or *each* Cr atom is 6^+.

Additional practice questions are available in the Lab Preview.

The relative tendency for a metal to acquire a positive oxidation number by losing valence electrons (referred to as its chemical activity) is measured in this experiment. The metals listed in order of decreasing activity comprise an abbreviated **activity series**.

| **Metal and Water—Rate of Hydrogen Gas Evolution** | Very reactive metals are oxidized by water resulting in the evolution of H_2 gas and heat. |

oxidation: $M(s) \rightarrow M^{n+}(aq) + n\ e^-$

reduction: $2\ H_2O(l) + 2\ e^- \rightarrow 2\ OH^-(aq) + H_2(g)$

The rate of evolution of H_2 gas is not an absolute criterion of the chemical activity of the metal because the reaction involves several variables, e.g., particle size of a metal. It is, however, a qualitative method for observing reactivity.

| **Metal and Hot Water—Rate of Disappearance of Metal** | Less active metals in hot water cause a similar reaction to occur. The rate of evolution of H_2 gas is often so slow that the rate of disappearance of the metal is used as the measure of activity, even though the particle size of the metal does influence the reaction rate. |

| **Metal and Nonoxidizing Acid—Rate of Hydrogen Gas Evolution** | By comparison, some metals evolve H_2 gas only with the addition of a nonoxidizing acid, such as $HCl(aq)$ or $H_2SO_4(aq)$. The metal is oxidized and the H^+ is reduced. |

oxidation: $M(s) \rightarrow M^{n+}(aq) + n\ e^-$

reduction: $2\ H^+(aq) + 2\ e^- \rightarrow H_2(g)$

| **Metal and Oxidizing Acid—Rate of Gas Evolution** | Metals that can only be oxidized by an oxidizing acid have a very low activity. Common oxidizing acids include nitric acid, $HNO_3(aq)$, and perchloric acid, $HClO_4(aq)$. For conc nitric acid, the reactions are: |

oxidation: $M(s) \rightarrow M^{n+}(aq) + n\ e^-$

reduction: $2\ H^+(aq) + NO_3^-(aq) + e^- \rightarrow NO_2(g) + H_2O(l)$

| **Metal and More Reactive Metal Cation—A Displacement Reaction** | The activity one metal relative to another can be determined by placing the metal into a solution containing the cation of the other. If metal M is more active (a greater tendency to be oxidized) than metal R, then M will displace R^{n+} from the aqueous solution. M goes into solution as M^{n+} and R^{n+} forms the metal, R. |

$$M(s) + R^{n+}(aq) \rightarrow M^{n+}(aq) + R(s)$$

For example, when iron wire is placed in a lead(II) solution, a reaction occurs because the iron is more active than lead, causing the formation of Fe^{2+} and the reduction of Pb^{2+} to Pb metal.

oxidation: $Fe(s) \rightarrow Fe^{2+}(aq) + 2\ e^-$

reduction: $Pb^{2+}(aq) + 2\ e^- \rightarrow Pb(s)$

On the other hand, when a lead strip is placed in an iron(II) solution, no reaction occurs because lead metal is less active than iron and is incapable of displacing the iron(II) ion from solution:

$$Pb(s) + Fe^{2+}(aq) \rightarrow no\ reaction$$

Since iron metal displaces lead(II) ion from solution, it is said to have a greater activity than lead.

• Technique 3, page 2   Microscale Analyses
• Technique 6a, c, page 4   Heating liquids

TECHNIQUES

PROCEDURE

This experiment determines the relative activity of Ca, Na, Pb, Cu, Zn, Fe, Al, and Mg.

You and your partner will need a number of solids and solutions for this experiment. Have them readily available so the testing can proceed quickly.

A. Metal and Water—Rate of Hydrogen Gas Evolution

1. **Demonstration Only. Ca, Na. (Caution:** *never touch* Na or Ca; *each causes a severe skin burn.*)[1]

    a. Wrap a pea-sized (*no larger!*) piece of the freshly cut calcium in aluminum foil. Fill a 200 mm Pyrex test tube with water and invert it in an 800 mL beaker $^3/_4$-filled with water. Add 2 drops of phenolphthalein.[2] Set the beaker and test tube behind a safety shield.

    b. Punch 5 pin-sized holes in the aluminum foil. With a pair of tongs or tweezers, place the wrapped calcium in the mouth of the test tube, keeping it under water (Figure 19.1).

2. Repeat the test for sodium and compare the rates of $H_2(g)$ evolution. Record these observations.

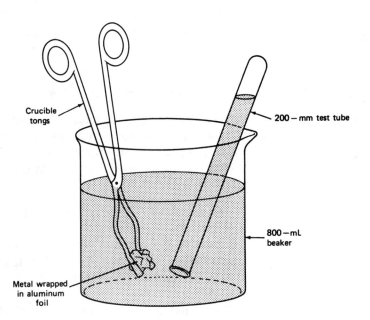

Figure 19.1

Collection of $H_2(g)$ from the reaction of an active metal with water

---

[1]It is strongly recommended that this be completed as a laboratory demonstration by the laboratory instructor.

[2]Phenolphthalein is an acid-base indicator that has a pink color in a basic solution but colorless in acidic and neutral solutions. A color change from colorless to pink indicates the generation of the hydroxide ion, OH^-.

B. Metal and Hot Water—Rate of Disappearance of Metal

Oxide coating: the dull finish on a metal caused by its reaction with the oxygen in air.

1. Pb, Cu, Zn, Fe, Al, and Mg. Thoroughly clean a 2 cm strip of each metal with steel wool to remove any oxide coating. This is especially critical for Al and Mg since they form tough, protective oxide coatings. Obtain a 24-cell plate, rinse six microcells with boiling, distilled (or deionized) water, and then half-fill with boiling water; add 1 drop of phenolphthalein to each cell.

2. *Quickly* place each metal into the microcells, Al and Mg first (Figure 19.2). Look for $H_2(g)$ evolution, discoloration of the metal surface, a color change of the solution (due to the phenolphthalein), or a disappearance of the metal. Some reactions may not be immediately apparent. Record your observations.

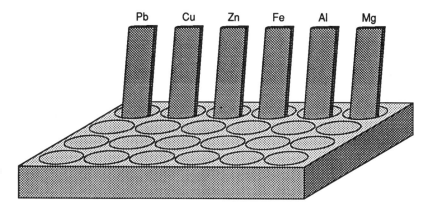

Figure 19.2
Arrangement of metals for testing reactivity

C. Metal and Nonoxidizing Acid—Rate of Hydrogen Gas Evolution

1. Add 5 drops of conc HCl (**Caution:** *avoid skin contact*) to the metals from Part B in which no reaction was observed. Swirl the solutions; allow 10–15 minutes for evidence of a reaction. Record.

D. Metal and Oxidizing Acid—Rate of Gas Evolution

1. If a metal shows no reaction in Parts B or C, draw off the water/acid solution with a Beral pipet and discard it. Add 1 mL of 3 M HNO_3 (**Caution:** *avoid skin contact*). If a reaction is now observed, record your observation.

E. Metal and More Reactive Metal Cation—A Displacement Reaction

1. Pb, Cu, Zn, and Fe. Place a small amount of a freshly cleaned metal in four consecutive microcells of the 24-cell plate—one metal in each microcell. Half-fill each microcell with 0.1 M $Pb(NO_3)_2$. Any tarnishing or dulling of the metal or changing of the color of the solution indicates a reaction. Allow 5–10 minutes for a reaction to be observed.

2. Repeat the procedure, using the following 0.1 M test solutions on each metal: $Cu(NO_3)_2$, $Zn(NO_3)_2$, and $Fe(NH_4)_2(SO_4)_2$. In each case the same metal strip may be reused if it remains unreacted from the previous test, rinsed with distilled water, and cleaned with steel wool. Record.

F. Establishment of Activity Series

Using the observations from Parts A through E, list the eight metals in order of decreasing activity.

Disposal Information: Dispose of the metals in the "Waste Solids" container and the solutions in the "Waste Salt Solution" container

Activity Series for Hydrogen and Some Metals

Element	Reduced State	Oxidized State
Cesium	Cs	Cs^+
Rubidium	Rb	Rb^+
Potassium	K	K^+
Barium	Ba	Ba^{2+}
Strontium	Sr	Sr^{2+}
Calcium	Ca	Ca^{2+}
Sodium	Na	Na^+
Magnesium	Mg	Mg^{2+}
Aluminum	Al	Al^{3+}
Manganese	Mn	Mn^{2+}
Zinc	Zn	Zn^{2+}
Chromium	Cr	Cr^{3+}
Iron	Fe	Fe^{2+}
Cadmium	Cd	Cd^{2+}
Cobalt	Co	Co^{2+}
Tin	Sn	Sn^{2+}
Lead	Pb	Pb^{2+}
Hydrogen	H_2	H^+
Copper	Cu	Cu^{2+}
Silver	Ag	Ag^+
Mercury	Hg	Hg^{2+}
Gold	Au	Au^{3+}

REDOX; ACTIVITY SERIES

Date _____ Name _____ Lab Sec. _____ Desk No. _____

1. Indicate the oxidation number of the bold faced element.

   a. **C**O _____   d. **N**$_2$O _____   g. **P**Cl$_3$ _____   j. **P**F$_5$ _____   m. **Cu**$_2$SO$_4$ _____

   b. **C**O$_2$ _____   e. O**F**$_2$ _____   h. **S**O$_3$$^{2-}$ ____   k. H**P**O$_3$$^-$ ___   n. K$_2$**Mn**O$_4$ _____

   c. **I**F$_7$_____   f. **N**I$_3$ _____   i. **I**O$_3$$^-$ _____   l. **Au**Cl$_3$ ____   o. **Cr**O$_3$ _____

2. From the following displacement reactions, arrange, at right, these hypothetical metals in order of increasing activity.

   $A + B^+ \rightarrow A^+ + B$ _____ least reactive
   $C + B^+ \rightarrow$ no reaction   _____
   $D + C^+ \rightarrow D^+ + C$ _____
   $F + G^+ \rightarrow$ no reaction   _____
   $A + F^+ \rightarrow$ no reaction   _____
   $F + D^+ \rightarrow F^+ + D$ _____ most reactive
   $D + B^+ \rightarrow$ no reaction
   $G + A^+ \rightarrow G^+ + A$

3. a. List four methods by which the chemical activity of a metal can be determined.

   _____   _____

   _____   _____

   b. Which method is used to determine the most reactive metals?

   _____

   c. Which method is used to determine the least reactive metals?

   _____

4. Cesium is a very reactive metal, reacting with cold water to produce H$_2$(g).
   a. Write a balanced equation for the reaction.

   b. The oxidizing agent is _____; the substance oxidized is _____.

5. Considering the trend in ionization energies of the Group IA elements, predict their relative reactivity in cold water?

6. Would you predict a reaction to occur when
   a. Zn metal is placed in a solution containing Ag^+? (Consider the relative activity of Zn *versus* Ag.)

   b. Au metal is placed in a solution containing Zn^{2+}?

7. Distinguish by definition an oxidizing acid from a nonoxidizing acid. Give an example of each.

8. What is Technique 6c?

REDOX; ACTIVITY SERIES-DATA SHEET

Date _____ Name _____ Lab Sec. _____ Desk No. _____

A. Metal and Water—Rate of Hydrogen Gas Evolution

Ca and Na

1. From the rate of $H_2(g)$ evolution, arrange Ca and Na in order of decreasing activity.

2. Write a balanced equation for each reaction of the metal with water.

B. Metal and Hot Water—Rate of Disappearance of Metal

Pb, Cu, Zn, Fe, Al, and Mg

1. Which metals show a definite reaction with hot water?

2. If possible, arrange the metals that do react in order of decreasing activity.

3. Write a balanced equation for each reaction.

C. Metal and Nonoxidizing Acid—Rate of Hydrogen Gas Evolution

Pb, Cu, Zn, Fe, Al, and Mg

1. Which remaining metals show a definite reaction with HCl?

2. If possible, arrange these metals that do react in order of increasing activity.

3. Write a balanced equation for each reaction.

D. Metal and Oxidizing Acid—Rate of Gas Evolution

1. Which of the remaining metals react with only HNO_3?

E. Metal and More Reactive Metal Cation—A Displacement Reaction

Pb, Cu, Zn, and Fe

Complete the table with NR (no reaction) or R (reaction) where appropriate.

For all reactions observed, write a balanced net–ionic equation. Use additional paper if necessary.

Test Reagents	Pb	Cu	Zn	Fe
HCl				
$Pb(NO_3)_2$	none			
$Cu(NO_3)_2$		none		
$Zn(NO_3)_2$			none	
$Fe(NH_4)_2(SO_4)_2$				none

Balanced Equations

F. Establishment of Activity Series

List the eight metals studied in this experiment in order of *decreasing* activity.

QUESTIONS

1. Tons of aluminum and magnesium are used annually where light-weight construction is required (e.g., for airplanes) even though each reacts rapidly with oxygen. Explain.

2. Zinc is used as the protective covering of steel on "galvanized iron". Explain its function in terms of chemical activity. Is zinc or iron more reactive?

3. a. Does hydrochloric acid added to copper metal generate $H_2(g)$? Explain.

   b. Which metals generate $H_2(g)$ in a reaction with HCl?

4. Magnesium metal is used as a sacrificial metal to reduce the corrosion of underground storage tanks (made of iron, actually steel which is an iron/carbon alloy). As a result the sacrificial metal corrodes instead of the iron. Describe the chemical reaction that occurs.

5. Circle the metals that react with
   a. Na^+: Pb, Cu, Mg
   b. Fe^{2+}: Na, Mg, Cu
   c. Cu^{2+}: Ca, Zn, Pb

6. Write a balanced equation for only one (if any) of the reactions that occurs in each part of Question 5.

   a. $Na^+ +$ _____ $\rightarrow$

   b. $Fe^{2+} +$ _____ $\rightarrow$

   c. $Cu^{2+} +$ _____ $\rightarrow$

EXPERIMENT 20

VITAMIN C ANALYSIS

The human body does not synthesize vitamins; therefore the vitamins that we need for catalyzing specific biochemical reactions are gained only from the food that we eat. We are generally aware that vitamin C can be obtained from citrus fruits, but it can also be obtained from a variety of fresh fruit and vegetables. However, storage and processing causes vegetables to lose a part of their vitamin C content. Cooking (boiling or steaming) leaches the water soluble vitamin C from the vegetables; in addition, the high temperatures accelerates its degradation by air oxidation. Therefore to maximize the intake of vitamin C, only freshly harvested fruits or vegetables should be consumed; their natural protective coverings (e.g., orange peel) should only be removed just before its consumption.

OBJECTIVES

- To prepare and standardize a sodium thiosulfate solution
- To determine the vitamin C concentration in a vitamin tablet, a fresh fruit, or a fresh vegetable sample

PRINCIPLES

Vitamin C, also called **ascorbic acid**, is one of the more abundant and easily obtained vitamins in nature. It is a colorless, water-soluble acid that, in addition to its acidic properties, is a powerful biochemical reducing agent, meaning it readily undergoes oxidation, even from the oxygen of the air.

Table 20.1 lists the concentration ranges of ascorbic acid for various vegetables.

Table 20.1 Ascorbic Acid in Foods

<10 mg/100 g	beets, carrots, eggs, milk
10-25 mg/100 g	asparagus, cranberries, cucumbers, green peas, lettuce, pineapple
25-100 mg/100 g	Brussels sprouts, citrus fruits, tomatoes, spinach
100-350 mg/100 g	chili peppers, sweet peppers, turnip, greens

Even though ascorbic acid is an acid, its *reducing* properties are used in this experiment to analyze its concentration in various samples. There are many other acids present in foods (e.g, citric acid) that would interfere with an acid analysis and not permit us to selectively determine the ascorbic acid content. The equation for the oxidation of ascorbic acid is

$$\text{or } C_6H_8O_6(aq) + H_2O(l) \rightarrow C_6H_8O_7(aq) + 2H^+(aq) + 2\,e-$$

Vitamin C Analysis

In analyzing for ascorbic acid, the sample is dissolved in water and treated with a measured excess of periodate ion, IO_3^-, a strong oxidizing agent; in an acidic solution containing an excess of I^-, IO_3^- converts to (red-brown) I_3^-, a milder oxidizing agent (Step 1).

$$IO_3^-(aq) + 8\,I^-(aq) + 6\,H^+(aq) \rightarrow 3\,I_3^-(aq) + 3\,H_2O(l) \quad \text{(Step 1)}$$

Some of the I_3^- then oxidizes the ascorbic acid (Step 2) in the sample.

$$I_3^-(aq) + C_6H_8O_6(aq) + H_2O(l)$$
$$\rightarrow C_6H_8O_7(aq) + 3\,I^-(aq) + 2\,H^+(aq) \quad \text{(Step 2)}$$

The rest of the I_3^- (or the excess, "xs") is titrated with a standard thiosulfate, $S_2O_3^{2-}$, solution, producing the colorless I^- and $S_4O_6^{2-}$ ions (Step 3).

$$\text{(xs)}\ I_3^-(aq) + 2\,S_2O_3^{2-}(aq) \rightarrow 3\,I^-(aq) + S_4O_6^{2-}(aq) \quad \text{(Step 3)}$$

Therefore the difference between the I_3^- generated initially (from the IO_3^-) and that which is titrated in excess is a measure of the ascorbic acid content of the sample.

Stoichiometric point: when the moles of reactants in the reaction mixture are in the same ratio as that expressed by the balanced equation.

The stoichiometric point is detected using starch as an indicator. Just prior to the disappearance of the red-brown I_3^- in the titration (Step 3), starch is added; this forms the deep-blue ion, $[I_3 \bullet \text{starch}]^-$.

$$I_3^-(aq) + \text{starch}(aq) \rightarrow [I_3 \bullet \text{starch}]^-(aq)$$

The addition of the $S_2O_3^{2-}$ titrant is continued until the $[I_3 \bullet \text{starch}]^-$ has been reduced to I^-; the solution appears colorless at the endpoint.

$$[I_3 \bullet \text{starch}]^-(aq) + 2\,S_2O_3^{2-}(aq) \rightarrow 3\,I^-(aq) + \text{starch}(aq) + S_4O_6^{2-}(aq)$$

Standardization of Na₂S₂O₃ Solution

A standard $Na_2S_2O_3$ solution is prepared using solid KIO_3 as a primary standard. A quantitative amount of KIO_3 is dissolved in an acidic solution containing KI. The generated I_3^- (see equation above) is titrated with your prepared solution of $Na_2S_2O_3$ using starch as the indicator. The endpoint is a color change from deep-blue to colorless. The equations for the standardization of the $Na_2S_2O_3$ solution (Steps 1 and 3) and for the analysis of vitamin C are exactly the same, except that there is no vitamin C (Step 2) to react with any of the I_3^- in this standardization procedure.

TECHNIQUES

- Technique 2, page 1   Clean Glassware and Lab Bench
- Technique 6d, page 4   Heating Liquids
- Technique 8, page 7   Reading a Meniscus
- Technique 9, page 7   Pipetting a Liquid or Solution
- Technique 10, page 8   Titrating a Solution
- Technique 13, page 11   Using the Laboratory Balance
- Technique 14c, d, page 12   Separation of a Liquid from a Solid

PROCEDURE

In this experiment you are required to prepare and standardize a $Na_2S_2O_3$ solution using solid KIO_3 as a primary standard. Three trials are necessary. This standard solution is then used to analyze for ascorbic acid in a sample assigned by your instructor. Quantitative, reproducible data are objectives of this experiment; practice good laboratory techniques.

A. A Primary Standard 0.01 M KIO₃ Solution

1. Measure about 0.5 g (±0.001 g) of KIO_3 (dried at 110°C for one hour) on weighing paper, transfer the solid to a 250 mL volumetric flask, dissolve and dilute to the mark. Calculate and record the molar concentration of the KIO_3 solution.

B. A Standard 0.1 M Na₂S₂O₃ Solution

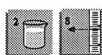

1. The $Na_2S_2O_3$ solution[1] should be prepared about one week in advance because of the unavoidable decomposition of the sodium thiosulfate. Dissolve about 6 g (±0.01 g) of $Na_2S_2O_3 \cdot 5H_2O$ with freshly boiled, distilled (or deionized) water and dilute to 250 mL. Agitate until the salt dissolves.

2. Properly prepare a clean, 50 mL buret for titration. Fill it with your $Na_2S_2O_3$ solution, drain the tip of air bubbles, and, after 30 s, read and record the volume (±0.02 mL).

3. Pipet 25 mL of the standard KIO_3 solution into a 125 mL Erlenmeyer flask and add about 1 g (±0.01 g) of solid KI. Add about 5 mL of 0.5 M H_2SO_4 and 0.1 g of $NaHCO_3$.[2]

4. Immediately begin titrating with the $Na_2S_2O_3$ solution. When the red-brown solution (due to I_3^-) changes to a pale yellow color, add 2 mL of starch solution. Stirring constantly, continue titrating slowly until the blue color disappears.

5. Repeat the procedure twice by rapidly adding the $Na_2S_2O_3$ titrant until 1 mL before the endpoint point. Add the starch solution and continue titrating until the solution is colorless.

6. Refill the buret with the $Na_2S_2O_3$ solution in preparation for the sample analysis in Part D. Read and record the volume (±0.02 mL) of $Na_2S_2O_3$ titrant in the buret.

---

[1]Ask your instructor about this solution; it may have already been prepared.

[2]The $NaHCO_3$ reacts in the acidic solution to produce $CO_2(g)$, providing an inert atmosphere above the solution and minimizing the possibility of the air oxidation of I^- ions (and in Part D, the ascorbic acid).

C. Sample Preparation

1. **Vitamin C tablet.** Read the label to determine the approximate mass of vitamin C in each tablet. Measure (±0.001 g) the fraction of the total mass of a tablet that corresponds to 100 mg of ascorbic acid. Dissolve it in a 250 mL Erlenmeyer flask with 40 mL of 0.5 M H_2SO_4[3] and then add about 0.5 g $NaHCO_3$. Kool-Aid™, Tang™, or Gatorade™ may be substituted as dry samples, even though their ascorbic acid concentrations are much lower. Fresh Fruit™ has a very high concentration of ascorbic acid. Proceed immediately to Part D.

2. **Fresh fruit sample.** Filter 125–130 mL of freshly squeezed juice through several layers of cheesecloth (or vacuum-filter). Measure the mass (±0.01 g) of a clean, dry 250 mL Erlenmeyer flask. Add about 100 mL of filtered juice and again determine the mass. Add 40 mL of 0.5 M H_2SO_4 and 0.5 g $NaHCO_3$. Concentrated fruit juices may also be used as samples. Proceed immediately to Part D.

3. **Fresh vegetable sample.** Measure about 100 g (±0.01 g) of a fresh vegetable. Transfer the sample to a mortar[4] and grind. Add 5 mL of 0.5 M H_2SO_4 and continue to pulverize the sample. Add another 15 mL of 0.5 M H_2SO_4, stir, and filter through several layers of cheesecloth (or vacuum-filter). Add 20 mL of 0.5 M H_2SO_4 to the mortar, stir, and pour through the same filter. Repeat the washing of the mortar with 20 mL of freshly boiled, distilled (or deionized) water. Combine all of the washings in a 250 mL Erlenmeyer flask and add 0.5 g $NaHCO_3$. Proceed immediately to Part D.

D. Vitamin C Analysis

1. Pipet 25.0 mL of the standard KIO_3 solution (from Part A) into the sample solution from Part C and add 1 g of KI. Add about 5 mL of 0.5 M H_2SO_4 and 0.1 g of $NaHCO_3$. Titrate the excess I_3^- in the sample with the standard $Na_2S_2O_3$ solution as described in Part B.4. Read and record the final buret reading (±0.02 mL).

2. Repeat the analysis twice in order to complete the three trials.

---

[3]Remember that vitamin tablets contain binders and other material that may be insoluble in water—do not heat in an attempt to dissolve the tablet!

[4]A blender may be substituted for the mortar and pestle.

VITAMIN C ANALYSIS

Date _____ Name _____ Lab Sec. _____ Desk No. _____

1. Explain why cooked fruits and vegetables have a lower vitamin C content than fresh fruits and vegetables.

2. a. What is the oxidizing agent in Part B of the procedure?

   b. What will be the color change of the starch indicator that signals the end of the titration in today's experiment?

   c. Vitamin C is an acid (ascorbic acid) and a reducing agent. Which property is utilized for its analysis in the experiment?

3. Six ounces (1 fl. oz. = 29.57 mL) of a well-known vegetable juice contains 35% of the recommended daily allowance of vitamin C (equal to 60 mg). How many milliliters of the vegetable juice will provide 100% of the recommended daily allowance?

4. a. Which finger should be used to control the delivery of a solution from a pipet?

   b. How much time should elapse between the time that the stopcock on a buret is closed and the time that a volume reading should be made and recorded?

   c. What criterion is used to determine if a buret is clean?

   d. What does the phrase "Properly prepare a clean, 50 mL buret for titration" mean? (see Part B.2)

5. A 25.0 mL volume of 0.010 M KIO_3, containing an excess of KI, is added to a 0.246 g sample of a Real Lemon™ solution containing vitamin C. The red-brown solution, caused by the presence of excess I_3^-, is titrated to a colorless starch endpoint with 10.7 mL of 0.100 M $Na_2S_2O_3$.

a. How many moles of KIO_3 were added to the Real Lemon™ solution?

b. Calculate the moles of I_3^- that are generated from the KIO_3?

c. How many moles of I_3^- reacted with the 0.100 M $Na_2S_2O_3$ in the titration?

d. How many moles (of the total moles) of I_3^- had reacted with the vitamin C in the Real Lemon™ sample?

e. Calculate the moles and grams of vitamin C, $C_6H_8O_6$, in the sample.

f. Calculate the percent (by mass) of vitamin C in the Real Lemon™ sample.

VITAMIN C ANALYSIS

Date _____ Name _____ Lab Sec. _____ Desk No. _____

A. A Primary Standard 0.01 M KIO_3 Solution

    1. Mass of KIO_3 + weighing paper (*g*)   ~~~~~~~~~~~~~~~~~~~~~~~~~~~~~~~~

    2. Mass of weighing paper (*g*)   ~~~~~~~~~~~~~~~~~~~~~~~~~~~~~~~~

    3. Mass of KIO_3 (*g*)   ~~~~~~~~~~~~~~~~~~~~~~~~~~~~~~~~

    4. Moles of KIO_3 (*mol*)   ~~~~~~~~~~~~~~~~~~~~~~~~~~~~~~~~

    5. Molar concentration of standard KIO_3 solution (*mol/L*)   ~~~~~~~~~~~~~~~~~~~~~~~~~~~~~~~~

B. A Standard 0.1 M $Na_2S_2O_3$ Solution

	Trial 1	Trial 2	Trial 3
1. Volume of KIO_3 solution (*mL*)	25.0	25.0	25.0
2. Moles of KIO_3 titrated (*mol*)			
3. Moles of I_3^- generated (*mol*)			
4. Buret reading, final (*mL*)			
5. Buret reading, initial (*mL*)			
6. Volume of $Na_2S_2O_3$ added (*mL*)			
7. Moles of $Na_2S_2O_3$ added (*mol*)			
8. Molar concentration of $Na_2S_2O_3$ solution (*mol/L*)			

    9. Average molar concentration of $Na_2S_2O_3$ solution (*mol/L*)   ~~~~~~~~~~~~~~~~~~~~~~~~

C. Sample Preparation

    Sample Name:

1. Mass of sample (*g*)			

D. Vitamin C Analysis

	Trial 1	Trial 2	Trial 3
1. Volume of KIO_3 added (mL)	25.0	25.0	25.0
2. Moles of IO_3^- added (mol)			
3. Moles of I_3^- generated, total (mol)			
4. Buret reading, final (mL)			
5. Buret reading, initial (mL)			
6. Volume of $Na_2S_2O_3$ added (mL)			
7. Moles of $S_2O_3^{2-}$ added (mol)			
8. Moles of I_3^- titrated with $S_2O_3^{2-}$ (mol)			
9. Moles of I_3^- reduced by $C_6H_8O_6$ (mol)			
10. Moles of $C_6H_8O_6$ in sample (mol)			
11. Mass of $C_6H_8O_6$ in sample (g)			
12. Percent of $C_6H_8O_6$ in sample (%)			

13. Average percent of $C_6H_8O_6$ in sample (%) _____

QUESTIONS

1. How will the addition of 2 g of KI to the sample in Part B.3, instead of the 1 g, affect the molar concentration of the $Na_2S_2O_3$ solution? Explain.

2. If the blue color does not appear when the starch solution is added during the titration, should you continue titrating or discard the sample (or leave for a cup of coffee)? Explain (but with a serious response).

3. After adding the KIO_3 solution and KI to the sample (in Part D) and stirring, the sample solution remains colorless; what modification of the procedure can be made to correct for this unexpected observation in order to complete the analysis.

EXPERIMENT 21

BLEACH ANALYSIS

Whiter that white! How can that be? "To remove that stubborn strawberry stain from baby's new white shirt, just apply some of our new and improved liquid detergent, now with bleach." Chemical companies do extensive advertising in an attempt to convince the consumer that their detergent is stronger, gets clothes whiter, and is safe to the environment. Of all the formulations that have been placed on the market that claim to remove stains from most clothing, perhaps the most effective is still simply bleach, such as Chlorox, Purex, or some similar generic brand of bleach. In addition many laundry detergents also add "whiteners" which actually fluoresce in the presence of ultraviolet radiation, giving the appearance of a whiter wash.

OBJECTIVES

- To prepare and standardize a sodium thiosulfate solution
- To determine the percent "available Cl_2" in a bleach

PRINCIPLES

Commercial bleaching agents contain the hypochlorite ion, ClO^-, as the "active ingredient" for making clothes whiter or removing stains. This ion is generally in the form of the sodium salt, $NaClO$, or the calcium salt, $Ca(ClO)_2$.

Oxidant: an oxidizing agent.

Normally when the strengths of various bleaches are compared, the standard oxidant for bleaching is assumed to be Cl_2. The oxidizing strength of the bleaching agent is rated according to an equivalent mass of Cl_2 per unit volume of the solution (or mass of the powder). This rating is called the **available Cl_2** in the bleach. Occasionally, its strength is expressed as percent chlorine by mass.

Liquid laundry bleach, generally a 5.25% (by mass) $NaClO$ solution, is prepared by the electrolysis of a cold, stirred $NaCl$ solution, producing Cl_2 at the anode and OH^- ion at the cathode. Cl_2 and OH^- react to form the ClO^- ion.

oxidation, anode:	$2\,Cl^-(aq) \rightarrow Cl_2(g) + 2\,e^-$
reduction, cathode:	$2\,e^- + 2\,H_2O(l) \rightarrow 2\,OH^-(aq) + H_2(g)$
stirred solution:	$Cl_2(g) + 2\,OH^-(aq) \rightarrow ClO^-(aq) + Cl^-(aq) + H_2O(l)$

net overall reaction: $Cl^-(aq) + H_2O(l) \rightarrow ClO^-(aq) + H_2(g)$

Bleaching powder—a mixture of $CaCl_2$, $Ca(ClO)Cl$, and $Ca(ClO)_2$—is prepared by the reaction of Cl_2 and slaked lime $Ca(OH)_2$.

Substances (fabrics and stains) have a color because energy from visible light excites electrons to a higher energy level in the molecule, causing the unabsorbed visible radiation to be transmitted and detected with the eye. In both dry and liquid bleaches, the hypochlorite ion removes these electrons from the substance.

$$\text{ClO}^-(aq) + \text{H}_2\text{O}(l) + 2\,e^- \rightarrow \text{Cl}^-(aq) + 2\,\text{OH}^-(aq)$$

Since the electrons are no longer present, visible radiation is not absorbed and, because the eye detects only visible light, the clothes appear whiter.

Bleach Analysis

In this experiment the oxidation-reduction analysis of a bleach involves the reaction of the hypochlorite ion with an excess of iodide ion.

$$\text{ClO}^-(aq) + 3\,\text{I}^-(aq) + \text{H}_2\text{O}(l) \rightarrow \text{I}_3^-(aq) + \text{Cl}^-(aq) + 2\,\text{OH}^-(aq)$$

The solution is acidified and the triiodide ion, I_3^-, is titrated with a standardized sodium thiosulfate $\text{Na}_2\text{S}_2\text{O}_3$, solution until the yellow color of I_3^- (or $\text{I}_2 \bullet \text{I}^-$)[1] nearly disappears just prior to the stoichiometric point.

$$\text{I}_3^-(aq) + 2\,\text{S}_2\text{O}_3^{2-}(aq) \rightarrow 3\,\text{I}^-(aq) + \text{S}_4\text{O}_6^{2-}(aq)$$

Starch solution is added to form a soluble, deep-blue complex ion, $[\text{I}_3 \bullet \text{starch}]^-$, with the remaining I_3^- ion.

$$\text{I}_3^-(aq) + \text{starch}(aq) \rightarrow [\text{I}_3 \bullet \text{starch}]^-(aq, \text{deep-blue})$$

At the stoichiometric point titration the I_3^- in the complex ion is reduced and the blue color disappears.

$$2\,e^- + [\text{I}_3 \bullet \text{starch}]^-(aq) \rightarrow 3\,\text{I}^-(aq) + \text{starch}(aq, \text{colorless})$$

To summarize the analysis: a measured amount of $\text{Na}_2\text{S}_2\text{O}_3$ reacts with a generated quantity of I_3^-—this I_3^- is generated from the ClO^- in the bleach; therefore, since the moles of $\text{Na}_2\text{S}_2\text{O}_3$ that react are known, the moles of ClO^- in the bleach can be determined.

The available Cl_2 is calculated as if free Cl_2 per unit volume (or mass, if the bleach is a powder) reacts with the I^- ion. The chlorine is assumed to be analyzed with $\text{S}_2\text{O}_3^{2-}$, just as the hypochlorite ion:

$$\text{Cl}_2(aq) + 3\,\text{I}^-(aq) \rightarrow 2\,\text{Cl}^-(aq) + \text{I}_3^-(aq)$$
$$\text{I}_3^-(aq) + 2\,\text{S}_2\text{O}_3^{2-}(aq) \rightarrow 3\,\text{I}^-(aq) + \text{S}_4\text{O}_6^{2-}(aq)$$

Net equation: $\text{Cl}_2(aq) + 2\,\text{S}_2\text{O}_3^{2-}(aq) \rightarrow 2\,\text{Cl}^-(aq) + \text{S}_4\text{O}_6^{2-}(aq)$

This equation serves as the basis for determining the strengths of bleaching agents in this experiment. Notice that both 1 mol ClO^- and 1 mol Cl_2 react with 2 mol $\text{S}_2\text{O}_3^{2-}$.

Standardization of a Na$_2$S$_2$O$_3$ Solution

A standard $\text{Na}_2\text{S}_2\text{O}_3$ solution is prepared using solid KIO_3 as a primary standard. A measured amount of KIO_3 is dissolved in an acidic solution containing an excess of KI.

---

[1]The yellow color is due to the presence of I_2; the excess I^- combines with I_2 to form a soluble I_3^- ion. It is common to refer to $\text{I}_2(aq)$ and $\text{I}_3^-(aq)$ as the same substance.

$$IO_3^-(aq) + 8 I^-(aq) + 6 H^+(aq) \rightarrow 3 I_3^-(aq) + 3 H_2O(l)$$

The generated I_3^- is titrated with your prepared solution of $Na_2S_2O_3$ using starch as the indicator.

$$3 I_3^-(aq) + 3 \text{ starch}(aq) \rightarrow 3 [I_3 \bullet \text{starch}]^-(aq, \text{ deep-blue})$$
$$3 [I_3 \bullet \text{starch}]^-(aq) + 6 S_2O_3^{2-}(aq) \rightarrow 9 I^-(aq) + 3 S_4O_6^{2-}(aq) + 3 \text{ starch}(aq)$$

Again, the endpoint is a color change from deep-blue to colorless. The equations and the analysis procedure are exactly the same as that described above for the bleach analysis.

TECHNIQUES

- Technique 2, page 1   Clean Glassware and Lab Bench
- Technique 8, page 7   Reading a Meniscus
- Technique 9, page 7   Pipetting a Solution
- Technique 10, page 8   Titrating a Solution
- Technique 13, page 11   Using the Laboratory Balance

PROCEDURE

The strengths of two bleaching agents are determined in this experiment; two trials are required for each determination. Therefore, obtain about 12 mL of each liquid bleach (or 5 g of each powdered bleach) from your instructor.

A. A Primary Standard 0.01 M KIO₃ Solution

1. Measure on weighing paper about 0.2 g ($\pm$0.002 g) of KIO_3 (dried at 110°C), transfer the solid to a 100 mL volumetric flask, dissolve and dilute to the mark. Calculate and record the molar concentration of the KIO_3 solution.

B. A Standard 0.1 M Na₂S₂O₃ Solution

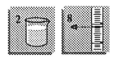

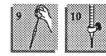

1. Dissolve about 6 g ($\pm$0.01 g) of $Na_2S_2O_3 \bullet 5H_2O$ in freshly boiled, distilled (or deionized) water and dilute to about 250 mL. Swirl the solution until the salt dissolves. This solution[2] should be prepared about one week in advance because of the inevitable decomposition of the thiosulfate solution.

2. Properly prepare a clean, 50 mL buret for titration. Fill it with your $Na_2S_2O_3$ solution, drain the tip of air bubbles, and, after 30 s, read and record the volume ($\pm$0.02 mL).

3. Pipet 25 mL of the standard KIO_3 solution into a 125 mL Erlenmeyer flask and add about 2 g ($\pm$0.01 g) of solid KI. Add about 10 mL of 0.5 M H_2SO_4 and 0.5 g of $NaHCO_3$.[3] Immediately begin titrating with the $Na_2S_2O_3$ solution. When the red-brown solution (due to I_3^-) changes to a pale yellow color, add 2 mL of starch solution. Stirring constantly, continue titrating slowly until the blue color disappears.

4. Repeat the procedure twice by *rapidly* adding the $Na_2S_2O_3$ titrant until 1 mL before the stoichiometric point. Add the starch solution and continue titrating until the solution is colorless.

---

[2]Ask your instructor about this solution; it may have already been prepared.

[3]The $NaHCO_3$ reacts in the acidic solution to produce $CO_2(g)$, providing an inert atmosphere above the solution and minimizing the possibility of the air oxidation of I^- ions.

5. Refill the buret with the $Na_2S_2O_3$ solution in preparation for the sample analysis in Part E. Read and record the volume (±0.02 mL) of $Na_2S_2O_3$ titrant in the buret.

C. Preparation of Liquid Bleach

1. Pipet 10.0 mL of bleach[4] into a 100 mL volumetric flask and dilute to the mark with boiled, distilled (or deionized) water. Mix thoroughly. Pipet 25.0 mL of this diluted bleach solution into a 250 mL Erlenmeyer flask and add 20 mL of distilled water, 2 g of KI, and 10 mL of 3 M H_2SO_4. A yellow color indicates the presence of I_2 as I_3^-. Proceed immediately to Part E.

D. Preparation of a Powdered Bleach

Glacial acetic acid: also called concentrated acetic acid.

1. Place about 5 g of powdered bleach into a mortar and grind. Measure the exact mass (±0.01 g) of the pulverized sample on weighing paper and transfer it to a 100 mL volumetric flask fitted with a funnel. Dilute to 100 mL with distilled (or deionized) water. Mix the solution thoroughly.[5] Pipet 25.0 mL of this solution into a 250 mL Erlenmeyer flask and add 20 mL of distilled (or deionized) water, 2 g of KI, and 15 mL glacial acetic acid. A yellow color indicates the presence of I_2 as I_3^-. See footnote #1. Proceed immediately to Part E.

E. Titration of Bleach

1. *Immediately* titrate the liberated I_2 with your standardized $Na_2S_2O_3$ solution until the yellow color is *almost* gone. Consult your instructor at this point if you are uncertain. Add 2 mL of starch solution; this will form the deep-blue $[I_3 \bullet starch]^-$ ion.[6] While swirling the flask, continue titrating *slowly* until the deep-blue color disappears. Record the final volume of the titrant in the buret.

2. Repeat the experiment with the same bleaching agent. Calculate the amount of available chlorine in the sample.

F. A Comparative Determination

1. Analyze a similar bleaching agent to determine the amount of available Cl_2. Compare the percent Cl_2 per unit price to determine the best buy.

---

[4]The density of liquid bleach is 1.084 g/mL.

[5]All of the powder may not dissolve because of the insoluble abrasive in the bleach.

[6]If you are analyzing a powdered bleach, the 2 mL of starch may be added after the glacial acetic acid solution in Part D.

BLEACH ANALYSIS

Date _____Name _____ Lab Sec. _____ Desk No. _____

1. a. Write the formula for the hypochlorite ion.

    b. In today's experiment, what substance is oxidized by the hypochlorite ion?

    c. Write the equation for the reaction in **b**.

2. Define available Cl_2.

3. a. The number of moles of ClO^- that react with 1 mol of I^- is _____.

    b. The number of moles of Cl_2 that react with 1 mol of I^- is _____.

    c. The number of moles of ClO^- equivalent to 1 mol of available Cl_2 is _____.

4. a. What is the color change at the stoichiometric point in today's analysis?

    b. What is the cause of the color change?

5. Sodium thiosulfate is the standardized titrant for the bleach analysis.

    a. Write the formula for sodium thiosulfate. _____

    b. Does sodium thiosulfate serve as an oxidizing agent or a reducing agent? _____

    c. What does sodium thiosulfate oxidize (or reduce)? _____

    d. Write a balance equation for its reaction in today's experiment.

    e. Each mole of the sodium thiosulfate titrant is equivalent to _____ moles of ClO^- and _____ moles of Cl_2 in the bleach.

6. A 0.684 g sample of a powdered bleach is analyzed according to the procedure in this experiment. A volume of 31.7 mL of 0.100 M $Na_2S_2O_3$ is required to reach the endpoint.

   a. How many moles of I_2 (or I_3^-) did the powdered bleach produce?

   b. Calculate the moles and grams of available Cl_2 in the bleach.

   c. What is the percent (by mass) of available Cl_2 in the bleach?

7. A 10.0 mL bleach sample is diluted to 100 mL in a volumetric flask. A 25.0 mL portion of this solution is analyzed according to the procedure in this experiment. If 11.3 mL of 0.30 M $Na_2S_2O_3$ is needed to reach the endpoint, what is the percent (by mass) of available Cl_2 in the *original* sample? Assume that the density of the bleach solution is 1.084 g/mL.

8. Explain the technique for transferring an insoluble solid in one beaker into a second beaker or Erlenmeyer flask.

9. a. Where on a meniscus should the volume of solution be read in a buret?

   b. When reading a meniscus, how does the use of a black mark on a white card better define the volume of a solution?

10. When a solution is delivered from a pipet, some of the solution invariably remains in the tip. What should be done with that solution: should you force it out or leave it in the pipet? Explain.

BLEACH ANALYSIS

Date _____ Name _____ Lab Sec. _____ Desk No. _____

A. A Primary Standard 0.01 M KIO_3 Solution

1. Mass of KIO_3 + weighing paper (*g*) _____

2. Mass of weighing paper (*g*) _____

3. Mass of KIO_3 (*g*) _____

4. Moles of KIO_3 (*mol*) _____

5. Molar concentration of standard KIO_3 solution (*mol/L*) _____

B. A Standard 0.1 M $Na_2S_2O_3$ Solution

	Trial 1	Trial 2	Trial 3
1. Volume of KIO_3 solution (*mL*)	25.0	25.0	25.0
2. Moles of KIO_3 titrated (*mol*)			
3. Moles of I_3^- generated (*mol*)			
4. Buret reading, final (*mL*)			
5. Buret reading, initial (*mL*)			
6. Volume of $Na_2S_2O_3$ added (*mL*)			
7. Moles of $Na_2S_2O_3$ added (*mol*)			
8. Molar concentration of $Na_2S_2O_3$ solution (*mol/L*)			

9. Average molar concentration of $Na_2S_2O_3$ solution (*mol/L*) _____

C. Preparation of Liquid Bleach

Sample and unit price (¢/*g*)				
	Trial 1	Trial 2	Trial 1	Trial 2
1. Volume of original liquid bleach titrated (*mL*)	2.5	2.5	2.5	2.5

D. Preparation of a Powdered Bleach

Sample and unit price (¢/g)

1. Mass of weighing paper and bleach(g)

2. Mass of weighing paper (g)

3. Mass of bleach (g)

E. Titration of Bleach

1. Buret reading, final (mL)

2. Buret reading, initial (mL)

3. Volume of $Na_2S_2O_3$ added (mL)

4. Moles of $Na_2S_2O_3$ (mol)

5. Moles of available Cl_2(mol)

6. Mass of available Cl_2(g)

7. Percent (by mass) available Cl_2 (assume the density of liquid bleach to be 1.084 g/mL)

8. Average % available Cl_2

9. Best buy (%Cl_2/unit price)

QUESTIONS

1. The starch solution can be added at the same time as the KI solution in Part D. However, the [I_3•starch]⁻ ion does not readily dissociate as the $Na_2S_2O_3$ titrant is added. What difficulties might you have encountered if the starch solution had been added at the same time as the KI solution in Part D?

2. If an air bubble initially trapped in the tip of the buret is released during the titration, will the reported percent available Cl_2 be too high or too low? Explain.

EXPERIMENT 22

ALUM FROM SCRAP ALUMINUM

Ahhh! What a cool, refreshing taste! But now what are you going to do with that aluminum can? Look at the can for a moment: the aluminum in that can is 98-99% aluminum, the remainder being magnesium; the plastic lining on the inner wall and the advertisement on the outside wall contribute the rest of the mass of the can.

Aluminum is the most abundant metal in the earth's crust (8.1% by mass), ranking just ahead of iron (5.0% by mass). Because of its low density, high tensile strength, and resistance to corrosion, it is widely used for the manufacture of airplanes and automobiles as well as for aluminum cans. The recycling of aluminum cans and other aluminum products (such as lawn furniture and window screens) is a very positive contribution to saving energy and saving our natural resources. Most of the recycled aluminum is melted and recast into other aluminum metal products or used in the production of various aluminum compounds, the most common of which are the alums, used in papermaking, tanning of leather products, and in foodstuffs.

The average American uses more than 20 aluminum cans per month.

OBJECTIVE

• To prepare an alum from scrap aluminum

PRINCIPLES

Double salt: a salt that contains two cations and generally the same anion.

In this experiment, a scrap piece of aluminum is used to prepare potassium **alum**, a hydrated double salt with the formula, $KAl(SO_4)_2 \cdot 12H_2O$. This alum is used for water purification, for sewage treatment and in fire extinguishers.

Aluminum metal rapidly reacts in a hot aqueous KOH solution producing the soluble potassium aluminate, $KAl(OH)_4$, salt.

$$2\,Al(s) + 2\,KOH(aq) + 6\,H_2O(l) \rightarrow 2\,KAl(OH)_4(aq) + 3\,H_2(g)$$

When the potassium aluminate solution is treated with sulfuric acid, insoluble $Al(OH)_3$ initially forms, but dissolves with heat and a small excess of H_2SO_4.

$$2\,KAl(OH)_4(aq) + H_2SO_4(aq) \rightarrow 2\,Al(OH)_3(s) + K_2SO_4(aq) + 2\,H_2O(l)$$
$$2\,Al(OH)_3(s) + 3\,H_2SO_4(aq) \xrightarrow{\Delta} Al_2(SO_4)_3(aq) + 6\,H_2O(l)$$

Dodeca: a prefix meaning twelve.

Saturated solution: The dissolution of the maximum amount of solute in a fixed mass of solvent at a given temperature.

The alum, called potassium aluminum sulfate dodecahydrate, crystallizes when the nearly saturated solution is cooled.

$$K_2SO_4(aq) + Al_2(SO_4)_3(aq) + 12\,H_2O(l) \rightarrow 2\,KAl(SO_4)_2 \cdot 12H_2O(s)$$

TECHNIQUES

• Technique 5, page 3	Transferring a Liquid
• Technique 6b, page 4	Heating Liquids
• Technique 13, page 11	Using the Laboratory Balance
• Technique 14b, c, page 12	Separation of a Liquid from a Solid

PROCEDURE

For the synthesis of the alum you will need to obtain a 110°C thermometer and a vacuum-filter apparatus from the stockroom. Be sure to follow the laboratory safety suggestions in today's experiment.

A. Preparation of the Alum

1. Polish a small piece of scrap aluminum with steel wool. Cut about 0.5 g (±0.01 g) of the scrap aluminum into very small pieces. Place these into a 125 mL Erlenmeyer flask.[1]

2. Add 25 mL of 4 M KOH to the Al pieces to dissolve the aluminum. (**Caution:** *wear safety glasses. Do not splatter the solution. KOH is caustic— do not permit skin contact.*) Heating the flask gently with a small flame may be necessary for the dissolution of the aluminum metal. Since $H_2(g)$ is evolved, do this step in a well-ventilated area. Continue heating until all of the Al reacts.[2]

3. Heat the solution to reduce the liquid level by one-half, to about 10–15 mL. When evidence of reaction is no longer visible, gravity filter the *warm* solution into a 200 mm test tube through very porous, low retention filter paper[3] or through a thin layer of glass wool placed in a funnel.

4. Allow the clear filtrate (solution) to cool. While swirling, slowly add 6 M H_2SO_4 (**Caution:** *avoid skin contact*) until insoluble $Al(OH)_3$ just forms in the solution.[4]

5. Gently reheat the mixture until the $Al(OH)_3$ dissolves. After the solution is clear, remove the heat, and again gravity filter, but *only* if solids are present. Cool in an ice bath for about 20 minutes; alum crystals should form. For better results, allow the crystallization to continue overnight. If crystals do not form, reheat the mixture to reduce the volume of the solution to about 20 mL (about one-fourth the volume of the test tube) or less and then cool again.

6. Vacuum filter the crystals from the solution. Wash them on the filter paper with four 5 mL portions of a 50/50 ethanol-water mixture (by volume).[5] Continue applying the vacuum until the crystals appear dry. Determine the mass (±0.01 g) of the alum crystals and show them to the laboratory instructor.

B. Melting Point of the Alum

1. Place finely ground alum to a depth of about 0.5 cm in the bottom of a melting point capillary tube (see Figure 22.1 for this technique). Attach the capillary tube to a 110°C thermometer with a rubber band or tubing and position it in a hot water bath (Figure 22.2).

---

[1]The smaller the Al pieces, the more rapid is the reaction. Aluminum foil may be used instead of scrap aluminum.

[2]Impurities, such as the plastic coating or paint covering, may remain as floating insoluble particles.

[3]Whatman No.4 or Fisher*brand* P8 are low retention filter paper brands.

[4]You will need to use some judgment here; if a precipitate has not yet formed, add a small amount of $H_2SO_4(aq)$ until it does; if the precipitate has redissolved, add a small amount of KOH(*aq*) until the $Al(OH)_3$ reforms.

[5]Alum crystals have a very low solubility in a 50/50 ethanol-water mixture.

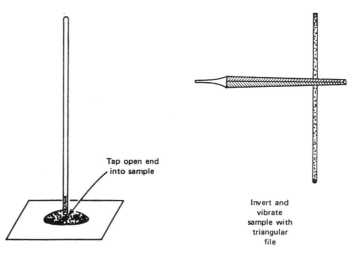

Figure 22.1
Filling a capillary tube with a solid

Tap open end into sample

Invert and vibrate sample with triangular file

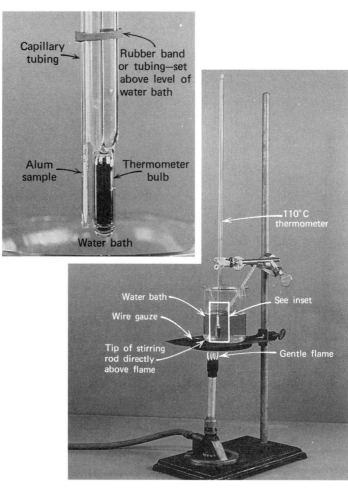

Capillary tubing

Rubber band or tubing—set above level of water bath

Alum sample

Thermometer bulb

Water bath

110°C thermometer

Water bath

See inset

Wire gauze

Tip of stirring rod directly above flame

Gentle flame

Figure 22.2
Determination of the melting point of the alum crystals

2. Slowly heat the water so that its temperature increases about 3°C per minute; carefully watch the solid during the heating. At the moment the solids melts, note the temperature. This is the melting point of the alum.

3. To again determine the melting point of the alum, allow the water bath to cool below that of the melting point. Again, reheat but at a slower rate (e.g., 1°C per minute) to check the temperature at which the alum changes from a solid to a liquid.

Shredded aluminum cans can be melted and recycled into useful aluminum products such as aluminum cans, lawn furniture, or compounds containing aluminum.

ALUM FROM SCRAP ALUMINUM

Date _____ Name _____ Lab Sec. _____ Desk No. _____

1. What does it mean when a salt is said to be a dodecahydrate?

2. An alum is a **double salt** with the general formula $M^+M^{3+}(SO_4)_2 \cdot 12H_2O$. M^+ is commonly Na^+, K^+, Tl^+, NH_4^+, or Ag^+; M^{3+} is Al^{3+}, Fe^{3+}, Cr^{3+}, Ti^{3+}, or Co^{3+}.

   a. The alum commonly used for pickling cucumbers is ammonium aluminum sulfate dodecahydrate. Write its formula.

   b. Sodium aluminum sulfate dodecahydrate is the acid commonly used in baking powders. What is its formula?

   c. Potassium chromium(III) sulfate dodecahydrate is used in tanning leather and in waterproofing fabrics. Write its formula.

3. The precipitate $Al(OH)_3$ is amphoteric. What does this mean?

4. Describe the procedure for filling a capillary tube with a solid.

5. A mass of 13.02 g $(NH_4)_2SO_4$ is dissolved in water. After the solution is heated, 27.22 g of $Al_2(SO_4)_3 \cdot 18H_2O$ is added. Calculate the theoretical yield of the resulting alum (see Question 2 for the formula of the alum). Hint: this is a limiting reactant-type problem.

6. How many milliliters of H_2 gas at STP are evolved in the reaction of 1.02 g Al with excess KOH?

7. a. While gravity filtering a solution, what is the maximum level to which a funnel should be filled with the solution?

   b. Explain how filter paper is sealed into a Büchner funnel when used for vacuum filtration.

   c. Briefly describe the technique for heating a test tube containing a solution over an open flame.

ALUM FROM SCRAP ALUMINUM

Date _____ Name _____ Lab Sec. _____ Desk No. _____

A. Preparation of the Alum

1. Mass of aluminum metal (g)   _____

2. Mass of alum, $KAl(SO_4)_2 \bullet 12H_2O$ (g)   _____

3. Theoretical yield of alum (g)*   _____

4. Percent yield (%)   _____

*Show calculation based upon pure aluminum metal.

B. Melting Point of the Alum

1. Melting point (°C)   _____

2. Melting point, second determination (°C)   _____

_____

QUESTIONS

1. Why is a 50/50 ethanol-water mixture used for washing the alum crystals after preparation rather than distilled (or deionized) water?

2. How might the reaction rate between Al and KOH be increased in Part A.2 of the procedure?

3. In the Lab Preview, it was noted that Al(OH)$_3$ is amphoteric. Where was this property observed in the experiment?

4. A greater yield with nearly perfect crystals is obtained when the alum solution is allowed to cool in a refrigerator overnight or for a few days. Explain.

5. How can you determine the mass of water hydrated to the alum crystals? (See Experiment 8).

EXPERIMENT 23

MOLAR MASS DETERMINATION

To make ice cream at home, the ice cream mix is bathed and turned in a bucket containing a mixture of salt, ice, and water. Salt is used in the bath because it lowers the temperature of the ice-water mixture below 0°C. At the lower temperature the ice cream freezes much more quickly. Salts are also applied to sidewalks and highways in the wintertime to lower the freezing temperature of the accumulated ice and snow. Could sugar be substituted for the salt in each case?

Antifreeze, or ethylene glycol, is added to cooling systems in automobiles to protect against freezing temperatures. While water freezes at 0°C, and could be used exclusively in cooling systems, the addition of the ethylene glycol appreciably lowers the freezing temperature.

In both of these instances the addition of a substance to water lowers its melting temperature; the substance also increases the boiling temperature of water as well. Try the experiment of adding salt to boiling water; the water stops boiling momentarily and *not* because the salt is cooler than the water!

OBJECTIVE

- To determine the molar mass of a nonvolatile solute by observing the difference between the freezing points of a solvent and a solution

PRINCIPLES

Colligative properties: properties of a solvent that result from the presence of the number of solute particles in the solution, and not their chemical composition.

The addition of a nonvolatile solute to a solvent produces a decrease in the freezing point and the increase in the boiling point of the solvent. These changes are called **colligative properties** of a solution, where the colligative property is due to the number of moles of solute particles dissolved in the solvent and not necessarily the kind of solute. This means that one mole of sugar, $C_{12}H_{22}O_{11}$, and one mole of ethylene glycol have the same effect on the freezing point and boiling point of water. It is important to note that two moles of solute, whether they be molecules or ions, have nearly twice the effect on the melting and boiling points as does one mole; for example, two moles of sugar and one mole of salt, NaCl, (2 moles of solute particles, $NaCl(aq) \rightarrow Na^+(aq) + Cl^-(aq)$) have approximately the same effect. Other colligative properties are the lowering of the vapor pressure of a solvent and the phenomenon of osmosis.

Osmosis: the passage of a solvent through a semipermeable membrane from a solution of lower concentration to one of higher concentration.

The freezing point change, ΔT_f, and boiling point change, ΔT_b, are proportional to the molality, **m**, of the solute in solution. The proportionality is made an equality by inserting a constant; k_f and k_b are called the molal freezing and boiling point constants for the solvent. Constants for several solvents are listed in Table 23.1. The temperature changes are related to the moles and molar mass, M, of the solute by the equations

$$\Delta T_f = k_f m = k_f \left[\frac{\text{mol solute}}{\text{kg solvent}} \right] = k_f \left[\frac{(\text{g/M}) \text{ solute}}{\text{kg solvent}} \right]$$

$$\Delta T_b = k_b m = k_b \left[\frac{\text{mol solute}}{\text{kg solvent}} \right] = k_b \left[\frac{(\text{g/M}) \text{ solute}}{\text{kg solvent}} \right]$$

Substance	Freezing Point (°C)	k_f (°C•kg/mol)	Boiling Point (°C)	k_b (°C•kg/mol)
water	0.0	1.86	100.0	0.512
benzene	5.45	4.90	80.2	2.53
cyclohexane	*	20.0	80.7	2.79
naphthalene	80.2	6.9	....	....
acetic acid	16.6	3.90	118.3	3.07
camphor	178.4	37.7	....	....

* The freezing point of cyclohexane is measured in this experiment

In this experiment you will use the freezing point colligative property of cyclohexane to determine the molar mass of a nonvolatile compound dissolved in cyclohexane. A mass of the unknown solute, added to a known mass of cyclohexane, causes a freezing point change, ΔT_f, which is measured. Since ΔT_f is proportional to the moles of solute added, the molar mass, M, of the unknown is calculated.

$$\text{molar mass, M (g/mol)} = \frac{\text{mass of solute (g)}}{\text{moles of solute (mol)}}$$

TECHNIQUES

- Technique 2, page 1   Clean Glassware and Lab Bench
- Technique 13, page 11   Using the Laboratory Balance
- Technique 18, page 18   Graphing Techniques

PROCEDURE

You and a partner should complete at least two trials in today's experiment. A 110°C thermometer and a wire stirrer are needed from the stockroom. One of you should fill a 400 mL beaker with ice.

A. Freezing Point of Cyclohexane

1. Place a clean, dry 200 mm test tube in a 250 mL beaker and determine the combined mass (±0.01 g). Add approximately 15 g (15–20 mL) of cyclohexane to the test tube and again measure the mass, using the same balance (Figure 23.1).

2. Prepare about 300 mL of an ice-water slurry in a 400 mL beaker. Place the test tube containing the cyclohexane in the ice-water bath. Maintain the ice-water slurry through Parts A.3 and B. Clamp a thermometer with a small three-pronged clamp and insert its bulb into the solvent to measure the temperature (Figure 23.2).

3. While stirring with the wire stirrer record on the Data Sheet the temperature (±0.1°C) at timed intervals (10 s or 30 s). The temperature remains virtually constant at the freezing point until the solidification is almost complete. Continue recording until the temperature begins to drop again.

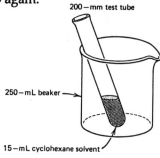

200—mm test tube

250—mL beaker

15—mL cyclohexane solvent

Figure 23.1

Determine the mass of cyclohexane in a beaker

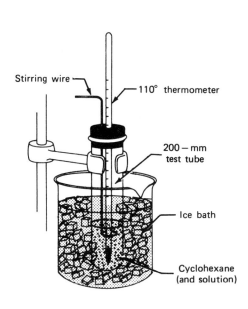

Stirring wire

110° thermometer

200 – mm test tube

Ice bath

Cyclohexane (and solution)

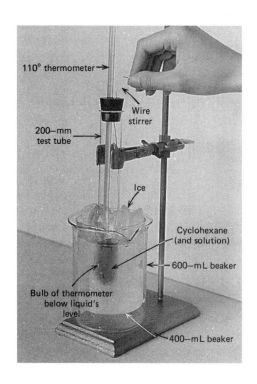

110° thermometer

Wire stirrer

200–mm test tube

Ice

Cyclohexane (and solution)

600–mL beaker

Bulb of thermometer below liquid's level

400–mL beaker

Figure 23.2

Apparatus for measuring the freezing point of cyclohexane and a cyclohexane solution

4. On linear graph paper plot the temperature (°C, ordinate) *vs* time (seconds, abscissa) to obtain the "cooling curve" for cyclohexane. (See the solid line in Figure in 23.3.) Determine the freezing point of cyclohexane from the plot. Obtain your instructor's approval for your graph.

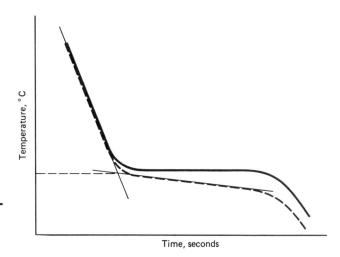

Figure 23.3

A cooling curve for the cyclohexane (——) and for the cyclohexane solution (- - - -)

B. Freezing Point of the Solution and Molar Mass of an Unknown Solute

1. Dry the outside of the test tube containing the cyclohexane and again measure its mass in the same 250 mL beaker with the same balance as in Part A.1. Obtain an unknown from your instructor. Measure approximately 0.2–0.5 g of the unknown (ask your instructor for the approximate mass to use).[1] Quantitatively transfer the unknown[2] to the cyclohexane in the 200 mm test tube.

2. Determine the freezing point of the (now) solution in the same way as that for cyclohexane in Part A.3. When the solution nears its freezing point, record the temperature at more frequent intervals (15–20 s).

3. Plot the temperature *vs* time on the *same* graph as in Part A.4 to obtain the cooling curve for the solution. (See the broken line in Figure 23.3.) The curve will show a "break" at the temperature where freezing began; this is only an approximate freezing point and is not as well defined as that for pure cyclohexane. To determine the freezing point of the solution:

   • draw a line tangent to the curve *prior* to the freezing point.
   • draw a line tangent to the curve *after* the freezing point is reached.
   • at the intersection point of the two drawn tangents, draw a line parallel to the abscissa (time axis) until it intersects the ordinate (temperature axis)–this temperature value is the freezing point of the solution.

   Again obtain your instructor's approval for your graph.

4. Warm the solution to about room temperature; repeat the temperature-time measurements and repeat the graphing until a reproducibility of the freezing point is achieved.

5. If time permits, repeat the procedure, starting with a new sample of cyclohexane (Part A) and a new mass of unknown (Part B).

Disposal Information:	Discard the test solution into a jar marked "Waste Cyclohexane Solution".

---

[1]If the unknown is a **solid**, measure its mass on weighing paper. If the unknown is a **liquid**, transfer about 3 mL from a 10 mL graduated cylinder into the cyclohexane; again measure the combined masses, now containing the liquid unknown. Calculate the mass difference for the mass of the unknown.

[2]In the transfer, be certain that *none* of the solute adheres to the wall of the test tube. If some does, roll the test tube so that the cyclohexane contacts and dissolves the solute.

MOLAR MASS DETERMINATION

Date _____ Name _____ Lab Sec. _____ Desk No. _____

1. What is the freezing point of a 1 m sugar (aqueous) solution?

2. Define a colligative property.

3. a. A 0.224 g sample of a nonvolatile solute is dissolved in 15.0 g of benzene. The solution freezes at 4.85°C. What is the molar mass of the solute?

   b. What would be the boiling point of this solution?

   c. What would be the freezing point change, ΔT_f, of a solution containing 0.224 g of the solute in 15.0 g of naphthalene?

4. a. The molar mass of *p*-nitrotoluene is 137 g/mol. How many grams must be dissolved in 15.0 g of benzene to lower its freezing point by 2.0°C?

    b. How many grams of *p*-nitrotoluene must be dissolved in 15.0 g of camphor to lower its freezing point by 2.0°C?

5. a. In plotting data, which axis is the abscissa?_____

    b. In a plot of pressure *vs* volume for a gas, which values are customarily plotted along the y-axis?

6. A solution does not have a constant freezing point. Why is this so?

MOLAR MASS DETERMINATION

Date _____ Name _____ Lab Sec. _____ Desk No. _____

A. Freezing Point of Cyclohexane

1. Mass of beaker, test tube and cyclohexane (*g*)   _____

2. Mass of beaker and test tube (*g*)   _____

3. Mass of cyclohexane (*g*)   _____

4. Freezing point, from cooling curve (°C)   _____

5. Instructor's approval of graph   _____

B. Freezing Point of the Solution and Molar Mass of an Unknown Solute

	Trial 1	Trial 2
1. Mass of beaker, test tube, and cyclohexane (*g*)		
2. Mass of cyclohexane in solution (*g*)		
3. Mass of unknown solute in solution (*g*)		
4. Freezing point of solution, from cooling curve (°C)		
5. Instructor's approval of graph		
6. k_f for cyclohexane	20.0°C kg/mol	
7. Freezing point change, ΔT_f (°C)		
8. Molar mass of unknown solute (*g/mol*)		

9. Average molar mass (*g/mol*)   _____

10. Percent (elemental) composition of unknown   _____
    (Obtain this from your instructor)

11. Empirical formula of unknown (from B.10 above)   _____

12. Empirical molar mass   _____

13. Molecular formula of unknown (compare B.9 to B.12)   _____

14. Actual molar mass of unknown (use atomic masses from   _____
    periodic table and the molecular formula from B.13)

15. Percent error, $\dfrac{\text{actual M} - \text{expt'l M}}{\text{actual M}} \times 100 =$ _____

1. a. If the freezing point of the solution is recorded 0.2°C lower than its actual freezing point, will the molar mass of the unknown be reported too high or too low? Explain.

    b. If the freezing point of cyclohexane is recorded 0.2°C lower that its actual freezing point, will the molar mass of the unknown be reported too high or too low? Explain.

2. How will the change in the freezing point, ΔT_f, of cyclohexane be affected by:
    a. the presence of a nonvolatile solute that dissociates? Explain.

    b. the presence of two solutes that react according to the equation, $A + B \rightarrow C$? Explain.

3. If a thermometer is miscalibrated to read 0.5°C higher than the actual temperature over its entire scale, how will it affect the *reported* molar mass of the solute? Explain.

4. Suppose the "pure" cyclohexane in today's experiment was initially contaminated with a nonvolatile solute.
    a. How would the ΔT_f have been affected? Explain.

    b. Would the molar mass of the unknown have been reported as being too high, too low, or unchanged? Explain.

5. Cyclohexane has a relatively high vapor pressure. If some of the cyclohexane evaporates during the experiment, how will this affect
    a. the ΔT_f in the experiment?

    b. the reported molar mass of the unknown solute? Explain.

EXPERIMENT 24

LeCHATELIER'S PRINCIPLE

The flow of traffic across a bridge seems to never cease. In the morning hours the traffic moves preferentially in one direction, but in the evening the flow is reversed. Construction and accidents often inhibit and alter its flow as well. However, traffic never seems to stop—even at midday and at midnight, the traffic continues to flow. It is evident that traffic flow is continuous and dynamic, its direction and magnitude is dependent upon the time of day and upon the conditions of the highway. The "stresses" to continuous and dynamic traffic flow are varied and many.

The movement of molecules in physical systems or in chemical reaction systems is also continuous and dynamic, and is influenced by many stresses. Suppose that two tanks (#1 and #2) of water are connected by glass capillary tubing as shown in the diagram. A stopcock is connected to each tank. If we focus on the water molecules that are moving from one tank through the glass tubing to the other, we can visualize an equal flow of water molecules. This condition is called a **dynamic equilibrium** (as contrasted to a static equilibrium) in a physical system. Now if we place a "stress" on this equilibrium condition, what can we predict?

Stress 1. Water is added to tank #1; what happens initially to the water level in tank #1 and finally in tank #2? Explain in terms of the flow of water molecules through the glass tubing.

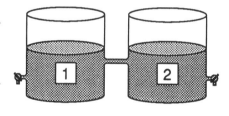

Stress 2. The stopcock of tank #2 is opened; what happens initially to the water level in tank #2 and subsequently what happens to the water level in tank #1? Again explain in terms of the flow of water molecules in the system?

OBJECTIVES

• To study the effects of concentration and temperature changes on the equilibrium position in various chemical systems
• To study the pH effect of strong acid and strong base addition on buffered and unbuffered systems

PRINCIPLES

A chemical system reaches a **dynamic equilibrium** when the reactants combine to form the products at the same the rate at which the products combine to re-form the reactants, a condition that can exist in reversible reactions. For the reaction,

$$2 \, SO_2(g) \; + \; O_2(g) \; \rightleftharpoons \; 2 \, SO_3(g) \; + \; 197 \, kJ$$

chemical equilibrium is reached when the reaction rate for SO_2 and O_2 forming SO_3 is the same as that for SO_3 molecules re-forming SO_2 and O_2.

If the amount of one of the species in the *equilibrium* system changes or the temperature changes, then the equilibrium position of the reaction system tends to shift compensating for the change. For example, when more SO_2 is added it reacts (stoichiometrically) with the O_2, increasing the amount of SO_3 until a new dynamic equilibrium is established. The new equilibrium has more SO_2 (because it was added) and SO_3 (because it formed from the reaction), but less moles of O_2 (because it reacted with the SO_2). Therefore, the addition of the SO_2 resulted in a *shift* in the position of the original equilibrium to the *right*, a shift to compensate for the added SO_2.

A general statement governing all chemical systems at dynamic equilibrium is *if an external stress (change in concentration, temperature, ...) is applied to a chemical system at equilibrium, the equilibrium shifts in the direction which minimizes the effect of that stress*—this is LeChatelier's Principle, proposed by Henri Louis LeChatelier in 1888. For instance, if more bleach is added to the laundry, more of the stains can be removed (more oxidation can occur).

Often the concentrations of the species in a chemical system at equilibrium can be quantitatively determined. From this, an equilibrium constant is calculated; its magnitude then indicates the position of the equilibrium. Equilibrium constants are determined in Experiments 25 and 26.

In this experiment, we will observe the effects of creating a stress on a number of reactions that exist at equilibrium. These stresses will involve concentration and temperature changes.

Changes in Concentration

Many salts are only slightly soluble in water. The Ag^+ ion forms slightly soluble salts with many anions. This experiment studies several equilibria involving the relative solubility of the CO_3^{2-}, Cl^-, I^-, and S^{2-} silver salts; an explanation of these equilibria follows.

Silver carbonate equilibrium. Ag_2CO_3 precipitates in the presence of excess Ag^+ and CO_3^{2-} ions in aqueous solution to establish an equilibrium.

$$Ag_2CO_3(s) \rightleftharpoons 2\,Ag^+(aq) + CO_3^{2-}(aq)$$

Addition of HNO_3 causes Ag_2CO_3 to dissolve. The H^+ ions from the HNO_3 react with the CO_3^{2-} ions, removing the CO_3^{2-} from the equilibrium. The system shifts *right* to compensate for the removal of the CO_3^{2-} ions, causing the Ag_2CO_3 to dissolve. The H^+ and CO_3^{2-} combine to form H_2CO_3 which, because of its instability at room temperature and pressure, decomposes to CO_2 and H_2O.

$$CO_3^{2-}(aq) + 2\,H^+(aq) \text{ from } HNO_3 \rightarrow H_2CO_3(aq) \rightarrow H_2O(l) + CO_2(g)$$

The Ag^+ and NO_3^- remain in solution.

Silver chloride equilibrium. The addition of Cl^- ions to the Ag^+ ions remaining in solution causes insoluble, white $AgCl$ to form.

$$Ag^+(aq) + Cl^-(aq) \rightleftharpoons AgCl(s)$$

When $NH_3(aq)$ is added to the $AgCl$ equilibrium, a complex ion of Ag^+ and NH_3, the diamminesilver ion, $Ag(NH_3)_2^+$, forms

$$Ag^+(aq) + 2\,NH_3(aq) \rightarrow Ag(NH_3)_2^+(aq)$$

The Ag^+ ion is therefore removed from the $AgCl$ equilibrium, causing dissolution of the $AgCl$. The Cl^- remains in solution.

Adding acid to this system results in an acid-base reaction with the NH_3 [$NH_3(aq) + H^+(aq) \rightarrow NH_4^+(aq)$]. This destroys the $Ag(NH_3)_2^+$ complex and releases the Ag^+, which then recombines with the Cl^- to form $AgCl(s)$.

$$Ag(NH_3)_2^+(aq) + 2H^+(aq) \rightarrow Ag^+(aq) + 2NH_4^+(aq)$$
$$Ag^+(aq) + Cl^-(aq) \rightleftharpoons AgCl(s)$$

Silver iodide equilibrium. I^- ion from KI added to the $Ag(NH_3)_2^+$ precipitates yellow AgI.

$$Ag(NH_3)_2^+(aq) + I^-(aq) \rightleftharpoons AgI(s) + 2NH_3(aq)$$

This occurs because of the greater insolubility of AgI than AgCl.

Silver sulfide equilibrium. Silver sulfide, Ag_2S, is even less soluble than AgI. Addition of S^{2-} ion (from Na_2S) precipitates Ag^+ from the AgI equilibrium. The AgI equilibrium shifts right (therefore the AgI dissolves) and the Ag_2S precipitates.

$$AgI(s) \rightleftharpoons Ag^+(aq) + I^-(aq)$$
$$2Ag^+(aq) + S^{2-}(aq) \rightarrow Ag_2S(s)$$

Common-ion effect: the effect of adding an ion or ions common to those in an existing equilibrium.

Common-ion effect. The solubility of a salt can be affected by the addition of a common-ion. In the case of ammonium chloride, a relatively soluble salt, the addition of concentrated amounts of ammonium ion or chloride ion can decrease its solubility.

$$NH_4Cl(s) \rightleftharpoons NH_4^+(aq) + Cl^-(aq)$$

The common-ion effect is also observed for the following equilibria in this experiment.

$$4Cl^-(aq) + Co(H_2O)_6^{2+}(aq) \rightleftharpoons CoCl_4^{2-}(aq) + 6H_2O(l)$$
$$CH_3COOH(aq) + H_2O(l) \rightleftharpoons H_3O^+(aq) + CH_3CO_2^-(aq)$$
$$Mg^{2+}(aq) + 2H_2O(l) + 2NH_3(aq) \rightleftharpoons Mg(OH)_2(s) + 2NH_4^+(aq)$$

Changes in Temperature

The reaction of SO_2 with O_2 producing SO_3 is exothermic by 197 kJ.

$$2SO_2(g) + O_2(g) \rightleftharpoons 2SO_3(g) + 197\,kJ$$

To favor the formation of SO_3, the reaction vessel is kept cool. Removal of heat shifts the equilibrium to the *right* favoring the formation of SO_3. Added heat results in a shift in the direction that absorbs heat—in this case, to the *left*.

This experiment examines the effect of temperature on:

$$4Cl^-(aq) + Co(H_2O)_6^{2+}(aq) \rightleftharpoons CoCl_4^{2-}(aq) + 6H_2O(l)$$

This system involves an equilibrium between the "coordination sphere" about the cobalt(II) ion which is concentration *and* temperature dependent. The tetrachlorocobaltate(II) ion, $CoCl_4^{2-}$ is more stable at the higher temperatures.

TECHNIQUES

- Technique 3, page 2
- Technique 6d, page 4

Microscale Analyses
Heating Liquids

PROCEDURE

At each superscript (i.e., [#1]) in the Procedure, **STOP**, and record your observations.

A. Silver Ion Equilibria

1. Add 10 drops of 0.1 M Na_2CO_3 to 10 drops of 0.1 M $AgNO_3$ in one microcell of a 24-cell plate.[#1] Now add drops of 6 M HNO_3 until change no longer occurs. What did you see?[#2]

2. Now add drops of 0.1 M HCl, again until no further change occurs.[#3] Allow the precipitate to settle and withdraw about one-half of the supernatant with a Beral pipet and properly discard. Add *drops* of conc NH_3 (**Caution:** *avoid inhalation and skin contact with conc* NH_3) until the precipitate dissolves.[#4] Reacidify the solution with 6 M HNO_3 and record.[#5] What happens if an excess of conc NH_3 is added? Try it.[#6]

3. After "trying it", add 10 drops of 0.1 M KI and stir.[#7]

4. To the solution containing the yellow AgI precipitate, add 10 drops of 0.1 M Na_2S. Record and explain your observations on the Data Sheet.[#8]

B. Saturated NH_4Cl Solution

1. Add an excess of solid NH_4Cl to a second microcell half-filled of H_2O and stir to dissolve. Touch the bottom of the cell with your finger.[1] Is the dissolution an endothermic or exothermic process?[#9]

2. Transfer 10 drops of the concentrated supernatant (none of the solid!) to a third microcell. Add 2–3 drops conc HCl (**Caution:** *conc HCl is a severe skin irritant*) to the saturated solution until a "first" change in appearance occurs.[#10] Write a chemical equation to explain this. Warm the test tube with your finger. What happens to the appearance of the system?[#11]

C. $Co(H_2O)_6^{2+}$ and $CoCl_4^{2-}$ Equilibrium

1. Place about 10 drops of 1 M $CoCl_2$ in a microcell. Record the color of the solution.[#12] Slowly and carefully add drops of conc HCl (**Caution:** *avoid inhalation or skin contact*) until a color change occurs.[#13] Write an equation that explains your observations. Slowly add water to the system.[#14]

2. Transfer about 1 mL of 1 M $CoCl_2$ to a 75 mm test tube; place it into a boiling water bath. Compare its color to the original 1 M $CoCl_2$ solution.[#15] Write an equation that explains your observation.

D. CH_3COOH and $CH_3CO_2^-$ Equilibrium (A Buffer System)

Pinch: about the size of a grain of rice.

1. Number a 1 x 4 array of microcells in a the 24-cell plate. Add two drops of universal indicator to each. Half-fill microcells 1 and 2 with 0.1 M CH_3COOH; note the color.[#16] Now add 0.05 g (a "pinch") of solid $NH_4CH_3CO_2$ to microcells 1 and 2 and agitate to dissolve the salt. Compare the solution's color with the pH color chart for the universal indicator.[#17]

2. Place 1 mL of distilled water in microcells 3 and 4.

---

[1] A thermometer may be inserted into the cell to determine the direction of heat flow.

3. To microcells 1 and 3, add 5–8 drops of 0.1 M NaOH; compare the colors of both solutions.[18] To cells 2 and 4, add 5–8 drops of 0.1 M HCl; compare the colors of these two solutions.[19]

Controlling the pH change in a chemical system when an acid or base is added is termed **buffer action**. Explain the effect that a buffered system has on pH when a strong acid or base is added to it.[20]

E. $Mg^{2+}(aq)$ and $Mg(OH)_2$ Equilibrium

1. Half-fill a microcell with 1 M $MgCl_2$ and add several drops of 3 M NH_3. Note what occurs.[21] Add a pinch of solid NH_4Cl and stir the mixture. What happens to the precipitate?[22]

Disposal Information:	Rinse the 24-cell plate with a minimum amount of water over a large beaker. Dispose of the test solution rinse in the "Waste Salts" container.

Notes and Observations

LeCHATELIER'S PRINCIPLE

Date _____ Name _____ Lab Sec. _____ Desk No. _____

1. Indicate the direction in which the equilibrium shifts in each of the chemical systems when the following stress is placed on each.

    a. Ag^+ is added to $Ag^+(aq) + Cl^-(aq) \rightleftharpoons AgCl(s)$ _____

    b. H_3O^+ is added to $Ag_2CO_3(s) \rightleftharpoons 2\,Ag^+(aq) + CO_3^{2-}(aq)$ _____

    c. H_3O^+ is added to $Ag^+(aq) + 2\,NH_3(aq) \rightleftharpoons Ag(NH_3)_2^+(aq)$ _____

    d. Ag^+ is removed from $Ag^+(aq) + I^-(aq) \rightleftharpoons AgI(s)$ _____

    e. heat is added to $2\,NO(g) + Cl_2(g) \rightleftharpoons 2\,NOCl(g) + 77.1\,kJ$ _____

    f. heat is removed from $2\,SO_2(g) + O_2(g) \rightleftharpoons 2\,SO_3(g) + 197\,kJ$ _____

    g. Cl_2 is added to $2\,NO(g) + Cl_2(g) \rightleftharpoons 2\,NOCl(g) + 77.1\,kJ$ _____

    h. Br^- is added to $4\,Br^-(aq) + Cu(H_2O)_4^{2+}(aq) \rightleftharpoons CuBr_4^{2-}(aq) + 4\,H_2O(l)$ _____

    i. $NaHCO_3$ is added to $CO_3^{2-}(aq) + H_2O(l) \rightleftharpoons HCO_3^-(aq) + OH^-(aq)$ _____

2. Complete the following statements with "increase", "decrease", or "no change".

    a. When Ag^+ is added to $Ag^+(aq) + Cl^-(aq) \rightleftharpoons AgCl(s)$, the amount of Cl^- _____.

    b. When Cl_2 is added to $2\,NO(g) + Cl_2(g) \rightleftharpoons 2\,NOCl(g) + 77.1\,kJ$, the temperature _____ and the

    amount of NOCl _____ .

    c. The reaction, $CH_4(g) + 2\,H_2S(g) \rightleftharpoons CS_2(g) + 4\,H_2(g)$, is endothermic. If $CH_4(g)$ is added to the

    system, the amount of H_2S _____, the amount of CS_2 _____, and temperature _____.

    d. When CO_3^{2-} is added to the $Mg^{2+}(aq) + CO_3^{2-}(aq) \rightleftharpoons MgCO_3(s)$, the amount of Mg^{2+} _____

    and the amount of insoluble $MgCO_3$ _____.

    e. Adding NH_3 to $4\,NH_3(aq) + Cu(H_2O)_4^{2+}(aq) \rightleftharpoons Cu(NH_3)_4^{2+}(aq) + 4\,H_2O$ causes the amount of

    $Cu(H_2O)_4^{2+}$ to _____ and the amount of $Cu(NH_3)_4^{2+}$ to _____ .

    f. Adding H_3O^+ to $HCN(aq) + H_2O(l) \rightleftharpoons H_3O^+(aq) + CN^-(aq)$ causes the amount of CN^- to

    _____. Adding OH^- to the same equilibrium causes the amount of CN^- to _____ and the

    amount of HCN to _____.

3. Answer the water tank questions presented as Stress 1 and Stress 2 at the beginning of the experiment.

LeCHATELIER'S PRINCIPLE

Date _____ Name _____ Lab Sec. _____ Desk No. _____

A. Silver Ion Equilibria

#1 Equation for Ag_2CO_3 equilibrium system _____

#2 Effect of 6 M HNO_3. Explain. _____

   Equation _____

#3 Effect of 0.1 M HCl. Explain. _____

   Equation _____

#4 Effect of conc NH_3. Explain. _____

   Equation _____

#5 Effect of reacidification with 6 M HNO_3. Explain. _____

   Equation _____

#6 Effect of additional conc NH_3. Explain. _____

#7 Effect of 0.1 M KI. Explain. _____

   Equation _____

#8 Effect of 0.1 M Na_2S. Explain. _____

B. Saturated NH_4Cl Solution

#9 Is the dissolution exothermic or endothermic? _____

   Equation for equilibrium system _____

#10 Effect of conc HCl. Explain. _____

#11 Effect of heat. Explain. _____

C. $Co(H_2O)_6^{2+}$ and $CoCl_4^{2-}$ Equilibrium

#12 Color of $CoCl_2(aq)$ _____

#13 Effect of conc HCl. Explain. _____

   Equation for equilibrium _____

#14 Effect of added H_2O. Explain. _____

#15　Effect of temperature increase. Explain._____

D. CH₃COOH and CH₃CO₂⁻ Equilibrium (A Buffer System)

Brønsted equation for CH_3COOH in water_____

#16　Color of universal indicator in CH_3COOH_____; pH = _____

#17　Color of universal indicator after $NH_4CH_3CO_2$ addition_____; pH = _____

Effect of $NH_4CH_3CO_2$ on the equilibrium_____

	$CH_3COOH/CH_3CO_2^-$		H_2O	
Cell number	1	2	3	4
#18 Color effect from 0.1 M NaOH		xxx		xxx
Approximate pH		xxx		xxx
Approximate change in pH		xxx		xxx
#19 Color effect from 0.1 M HCl	xxx		xxx	
Approximate pH	xxx		xxx	
Approximate change in pH	xxx		xxx	

#20　How does the magnitude of the pH change when HCl (a strong acid) or NaOH (a strong base) is added to a solution of CH_3COOH and $NH_4CH_3CO_2$ as compared to the magnitude of the pH change that occurs when HCl or NaOH is added to pure water?

F. Mg²⁺(aq) and Mg(OH)₂ Equilibrium

#21　Addition of $NH_3(aq)$ to the $MgCl_2$ solution_____

Equation_____

#22　Effect of NH_4Cl_____

QUESTIONS

1. Identify the color of each:

   a. $CoCl_4^{2-}$ ____________________

   b. $Co(H_2O)_6^{2+}$ ____________________

   c. AgI ____________________

   d. $Ag(NH_3)_2^+$ ____________________

   e. S^{2-} ____________________

   f. I^- ____________________

2. What is the color of the $Co(H_2O)_6^{2+}/CoCl_4^{2-}$ equilibrium at 80°C?

3. a. Explain why the pH change was small when HCl was added to the $CH_3COOH/CH_3CO_2^-$ system, but much larger when added to water.

   b. Explain why the pH change was small when NaOH was added to the $CH_3COOH/CH_3CO_2^-$ system, but much larger when added to water.

AN EQUILIBRIUM CONSTANT

In a manner of speaking, all natural processes tend to reach a "steady state" or to arrive at a position of equilibrium. For example, the traffic flow in both directions across a bridge appears to be constant; a ball rolls and comes to rest; the corrosion of metal eventually takes the metal back to its original state, to some equilibrium state. Chemical reactions also begin with substances that eventually react and reach some final steady state position. Industrial chemists and engineers oftentimes wage a war on chemical reactions, trying to prevent them from coming to an equilibrium position so that product can be continuously produced. Knowing this position of equilibrium is invaluable information in studying chemical reactions. Once a chemical reaction begins, they have, for the most part, a tendency to proceed in the reverse direction. For example, in the production of an industrial chemical, such as sulfur trioxide gas from the reaction of sulfur dioxide and oxygen gases,

$$2\,SO_2(g)\ +\ O_2(g)\ \rightarrow\ 2\,SO_3(g)$$

it is the chemists and chemical engineers who must contend with the undesirability of the reverse reaction from occurring. For chemists to know in advance the position that a chemical system attempts to achieve at equilibrium, they are then able to work with the system to obtain desirable results. This knowledge is available as the equilibrium constant for the reaction.

OBJECTIVES

- To develop the techniques for the use and operation of a spectrophotometer
- To determine the equilibrium constant for a soluble ionic system

PRINCIPLES

The magnitude of an equilibrium constant, K_c, expresses the equilibrium position for a chemical system. For example, a small K_c indicates that the equilibrium favors the reactants whereas a large K_c favors the product. The value of K_c is constant for a chemical system at a constant temperature.

This experiment determines K_c for a system in which all species are ionic and soluble. The equilibrium involves the hydrated iron(III) ion, the thiocyanate ion, and the iron(III)-thiocyanate complex ion:

$$Fe(H_2O)_6{}^{3+}(aq)\ +\ SCN^-(aq)\ \rightleftharpoons\ Fe(H_2O)_5NCS^{2+}(aq)\ +\ H_2O(l)$$

Because the concentration of H_2O is essentially constant in dilute aqueous solutions, we can ignore the waters of hydration and represent the equilibrium with a simplified equation:

$$Fe^{3+}(aq)\ +\ SCN^-(aq)\ \rightleftharpoons\ FeNCS^{2+}(aq)$$

Its equilibrium (or mass action) expression for the system at equilibrium is

$$K_c = \frac{[FeNCS^{2+}]}{[Fe^{3+}][SCN^-]}$$

To determine K_c for the system, known amounts of Fe^{3+} and SCN^- are mixed and react to form an equilibrium with $FeNCS^{2+}$, a deep, blood-red ion with an absorption maximum at about 447 nm. Because of its intense color the concentration of $FeNCS^{2+}$ is determined spectrophotometrically.[1]   By knowing the *initial* molar concentrations of Fe^{3+} and SCN^- and by measuring the $FeNCS^{2+}$ *equilibrium* molar concentration spectrophotometrically, the equilibrium molar concentrations of Fe^{3+} and SCN^- can be calculated.

$$[Fe^{3+}]_{\text{at equilibrium}} = [Fe^{3+}]_{\text{initial}} - [FeNCS^{2+}]_{\text{at equilibrium}}$$
$$[SCN^-]_{\text{at equilibrium}} = [SCN^-]_{\text{initial}} - [FeNCS^{2+}]_{\text{at equilibrium}}$$

Using these equilibrium concentrations, the K_c for the chemical reaction is calculated.

In Part A of this procedure, you will prepare a set of standard solutions for the $FeNCS^{2+}$ ion and then measure the percent transmittance, %T, of each solution. The absorbance, A, for each solution is then calculated[2] and plotted against its known molar concentration of $FeNCS^{2+}$ to establish a **calibration curve**. This graph is then used to determine the the molar concentration of $FeNCS^{2+}$ at equilibrium for the systems in Part B.

Absorbance: the amount of light that is absorbed when it passes through a sample; its value is directly proportional to the concentration of the absorbing substance in the sample.

In preparing the standard solutions for Part A, the Fe^{3+} concentration is added in a *large excess* relative to the SCN^- concentration; this drives the equilibrium ($Fe^{3+}(aq) + SCN^-(aq) \rightleftharpoons FeNCS^{2+}(aq)$) far to the *right*. Because of this procedure, we assume that the moles of $FeNCS^{2+}$ that form approximates the number of moles of SCN^- placed in the system, i.e., we assume that all of the SCN^- is in the form of the $FeNCS^{2+}$ ion at equilibrium.

The calculations for K_c are involved, but the questions in the Lab Preview should clarify most of these steps. The Data Sheet is also outlined in such detail as to help with the calculations.

TECHNIQUES

- Technique 2, page 1   Clean Glassware and Lab Bench
- Technique 8, page 7   Reading a Meniscus
- Technique 9, page 7   Pipetting a Liquid or Solution
- Technique 18, page 18   Graphing Techniques

In addition, the proper techniques for using a spectrophotometer for analysis and the handling of cuvets are necessary.

PROCEDURE

Five 25 mL volumetric flasks, one 10 mL graduated (1.0 mL) pipet, and one 10 mL pipet are needed for Part A of this experiment. One 5 mL pipet and two 10 mL graduated pipets are required for Part B. Ask your instructor about working with a partner(s) for Parts A and/or B. Listen carefully to your instructor's advice on the use and operation of the spectrophotometer.

---

[1]The principles of a spectrophotometric analysis are presented in Experiment 14.

[2]$A = \log \dfrac{I_o}{I_t} = a \bullet b \bullet c = \log \dfrac{100}{\%T}$

A. A Set of Standard FeNCS^{2+} Solutions

1. Prepare the solutions in Table 25.1: use the 10 mL graduated pipet to pipet 0, 1, 2, 3, and 4 mL of 0.001 M NaSCN[3] into separate, (clean) labeled 25 mL volumetric flasks. Pipet 10.0 mL of 0.2 M Fe(NO$_3$)$_3$[4] into each flask and dilute to the "mark" with 0.1 M HNO$_3$. These solutions are used to establish a calibration curve for determining the equilibrium [FeNCS^{2+}] spectrophotometrically. An additional test solution of 5 mL of 0.001 M NaSCN is suggested.

2. Set the wavelength on the spectrophotometer at 447 nm. Fill a cuvet approximately three-fourths full with the blank solution and dry the outside with a clean Kimwipe to remove water droplets and fingerprints. Handle only the lip of the cuvet henceforth. Any foreign material on the cuvet affects the amount of the transmitted light.

Table 25.1. A Set of Standard FeNCS^{2+} Solutions

Solution	0.2 M Fe(NO$_3$)$_3$ (in 0.1 M HNO$_3$)	0.001 M NaSCN (in 0.1 M HNO$_3$)	0.1 M HNO$_3$
Blank	10.0 mL	0 mL	dilute to 25 mL
1	10.0 mL	1 mL	dilute to 25 mL
2	10.0 mL	2 mL	dilute to 25 mL
3	10.0 mL	3 mL	dilute to 25 mL
4	10.0 mL	4 mL	dilute to 25 mL

3. Calibrate the spectrophotometer with the blank solution as follows: place the cuvet containing blank solution in the cuvet holder and set the spectrophotometer to read 100% T. Remove the cuvet and set the spectrophotometer to read 0%T. Repeat until no further adjustments are necessary. Be sure to position the cuvet in the cuvet holder the same way each time.

4. Empty the cuvet and rinse the cuvet with several portions Solution #1 and then fill it approximately three-fourths full and dry the outside with a clean Kimwipe. Place the cuvet in the cuvet holder and read its %T.

5. Repeat the %T measurements for Solutions #2, #3, and #4. Convert your %T readings to absorbance values, A, and plot them *vs* [FeNCS^{2+}]. Draw the best straight line through the four (or five if a fifth solution was prepared) points *and* the origin. Ask the instructor to approve your graph.

B. Set of Equilibrium Solutions

1. Prepare the test solutions in Table 25.2 in clean, *dry* 200 mm test tubes.[5] Use three pipets, one for each solution, for the volumetric measurements. Be careful not to mix the pipets. Thoroughly stir each solution with a clean stirring rod.

---

[3] Record the *exact* concentration of the NaSCN on the Data Sheet.

[4] Record the *exact* concentration of the Fe(NO$_3$)$_3$ on the Data Sheet.

[5] 10 mL volumetric flasks can be substituted for the 200 mm test tubes and the solutions can then be diluted to the mark with the 0.1 M HNO$_3$.

2. Again use the blank solution to check the calibration of the spectrophotometer (see Part A.3).

3. Rinse the cuvet with 1 or 2 portions of Solution #5 and fill three-fourths full; wipe dry the cuvet and insert it (properly aligned) into the cuvet holder. Read and record its %T. Be careful in handling the cuvet—handle only the lip of the cuvet and don't drop it!

4. From the calibration curve, determine the equilibrium molar concentration of $FeNCS^{2+}$ for each solution.

Table 25.2. A Set of Equilibrium Solutions

Solution	0.002 M Fe(NO$_3$)$_3$* (in 0.1 M HNO$_3$)	0.002 M NaSCN (in 0.1 M HNO$_3$)	0.1 M HNO$_3$
5	5 mL	1 mL	4 mL
6	5 mL	2 mL	3 mL
7	5 mL	3 mL	2 mL
8	5 mL	4 mL	1 mL
9	5 mL	5 mL	••••

* If 0.002 M Fe(NO$_3$)$_3$ is not available, dilute 1.0 mL (measure with a 1.0 mL pipet) of the 0.2 M Fe(NO$_3$)$_3$ used in Part A with 0.1 M HNO$_3$ in a 100 mL volumetric flask and then share the leftover solution with other students.

Disposal Information:	Dispose of the waste iron(III)-thiocyanate solutions in the "Waste Iron Salts" container.

AN EQUILIBRIUM CONSTANT

Date _____Name _____ Lab Sec. _____Desk No. _____

1. a. What is white light?

   b. What is monochromatic light?

2. Describe the procedure for setting 0%T and 100%T on the spectrophotometer for its calibration.

3. A maximum absorption of visible light occurs at 447 nm for the $FeNCS^{2+}$ ion. What is the dominant color of light absorbed?_____ and the dominant color of light transmitted?_____

4. What effect do fingerprints on a cuvet containing the test sample have on the transmittance (%T) of visible light through the solution?

5. A 5.0 mL volume of 0.0200 M $Fe(NO_3)_3$ is mixed with 5.0 mL of 0.00200 M NaSCN; the blood-red $FeNCS^{2+}$ ion forms and the equilibrium is established.

$$Fe^{3+}(aq) + SCN^-(aq) \rightleftharpoons FeNCS^{2+}(aq)$$

The molar concentration of $FeNCS^{2+}$ was measured spectrophotometrically and determined to be 7.0×10^{-4} mol/L at equilibrium. To calculate the K_c for the equilibrium system, proceed through the following steps.

a. moles of Fe^{3+}, initial   _____

b. moles of SCN^-, initial   _____

c. moles of $FeNCS^{2+}$ at equilibrium   _____

d. moles of Fe^{3+} reacted   _____

e. moles of SCN^- reacted   _____

f. moles of Fe^{3+} remaining unreacted, at equilibrium (a-d)   _____

g. moles of SCN^- remaining unreacted, at equilibrium (b-c)   _____

h. $[Fe^{3+}]$ at equilibrium   _____

i. $[SCN^-]$ at equilibrium   _____

j. $[FeNCS^{2+}]$ at equilibrium 7.0×10^{-4} mol/L

k. K_c of $FeNCS^{2+}$ _____

6. Using the value of K_c in Question 5, determine the equilibrium $[SCN^-]$ after 90.0 mL of 0.100 M Fe^{3+} are added to 10.0 mL of a SCN^- solution; the equilibrium $[FeNCS^{2+}]$ is 1.0×10^{-6} mol/L.

AN EQUILIBRIUM CONSTANT

Date _____ Name _____ Lab Sec. _____ Desk No. _____

A. A Set of Standard FeNCS^{2+} Solutions

Exact molar concentration of NaSCN   _____

Molar concentration of Fe(NO$_3$)$_3$ _____

Standard solutions	Blank	1	2	3	4
Volume of NaSCN (*mL*)					
Moles of SCN$^-$ (*mol*)					
[SCN$^-$] (25 mL solution)					
[FeNCS^{2+}]					
%T					
Absorbance, A					

Calibration curve, (A *vs* [FeNCS^{2+}]); instructor's approval   _____

B. Set of Equilibrium Solutions

Exact molar concentration of Fe(NO$_3$)$_3$ _____

Exact molar concentration of NaSCN   _____

Solutions	5	6	7	8	9
Volume of Fe(NO$_3$)$_3$ (*mL*)					
Moles of Fe^{3+}, initial (*mol*)					
Volume of NaSCN (*mL*)					
Moles of SCN$^-$, initial (*mol*)					
%T					
Absorbance, A					

C. Determination of K_c

Solutions	5	6	7	8	9
$[FeNCS^{2+}]$, from calibration curve					
moles of $FeNCS^{2+}$ in solution at equilibrium					

Calculation for $[Fe^{3+}]$ at equilibrium

	5	6	7	8	9
moles of Fe^{3+}, reacted					
moles of Fe^{3+}, unreacted					
$[Fe^{3+}]$, equilibrium (unreacted)*					

Calculation of $[SCN^-]$, at equilibrium

	5	6	7	8	9
moles of SCN^-, reacted					
moles of SCN^-, unreacted					
$[SCN^-]$, equilibrium (unreacted)**					
$K_c = \dfrac{[FeNCS^{2+}]}{[Fe^{3+}][SCN^-]}$					

Average K_c

*Show calculation for $[Fe^{3+}]$ at equilibrium

**Show calculation for $[SCN^-]$ at equilibrium

Standard deviation of K_c, $\sigma = \sqrt{\dfrac{d_1^2 + d_2^2 + d_3^2 + ...+ d_n^2}{(n-1)}} =$ _____

d = difference between a single K_c value and the average K_c

n = number of K_c values determined

QUESTIONS

1. What effect will a dirty cuvet (caused by fingerprints, water spots, or lint) have on the value of K_c in this experiment?

2. In our calculations the solution's thickness and the probability of light absorption by the $FeNCS^{2+}$ were not considered. See the discussion of Beer's Law in Experiment 14. Explain.

3. Why is 0.2 M $Fe(NO_3)_3$ in 0.1 M HNO_3 used as the blank solution for calibrating the spectrophotometer instead of distilled water?

4. Considering Solution #8, suppose the 1.00 mL of 0.1 M HNO_3 is not added.

a. How will this affect the %T of the solution?

b. How will this affect the reported the $FeNCS^{2+}$ concentration for the solution?

c. Will the reported K_c be greater or less than it should have been? Explain.

5. Why should the spectrophotometer be periodically recalibrated during the course of the experiment?

EXPERIMENT 26

SOLUBILITY CONSTANT AND COMMON ION EFFECT

Rocks seem not to change in appearance..."once a rock, always a rock." Many national parks, such as Bryce Canyon, Utah or Garden of the Gods, Colorado, have a large number of unusual rock formations that have maintained their appearance for centuries. Rocks are composed of various minerals, or complex salts, that have a very low solubility in water. Even though the solubility of each of these minerals is low, it is not zero. The mineral slowly dissolves with each bit of rainfall because of its very slight solubility; slowly the edges of the rock formations become rounded and the height of the rock formation is less. Geologists refer to the Appalachian Mountains as "old", but the Rocky Mountains as "young" formations because of the relative amounts of weathering and because of the slight solubility of the minerals in water.

In this experiment the solubility of a salt having a very low solubility is determined, the extent of its solubility is expressed by an equilibrium constant, called a solubility constant.

OBJECTIVES

- To determine the molar solubility and solubility constant of $Ca(OH)_2$
- To determine the molar solubility of $Ca(OH)_2$ in the presence of Ca^{2+}

PRINCIPLES

Salts which have a low solubility in water are called **slightly soluble salts**. A saturated solution of a slightly soluble salt is a dynamic equilibrium between the solid salt and a very low concentration of its ions in solution. Limestone rock appears to be insoluble in its natural environment; however, with time, the limestone slowly dissolves due to its slight solubility. The major component of limestone is calcium carbonate, $CaCO_3$. Its equilibrium with the Ca^{2+} and CO_3^{2-} favors the solid $CaCO_3$, or the equilibrium lies far to the *left*.

$$CaCO_3(s) \rightleftharpoons Ca^{2+}(aq) + CO_3^{2-}(aq)$$

The mass action expression for this system at equilibrium is a constant, called the solubility constant, K_s, for the salt. Since the concentration of a solid remains a constant, the mass action expression is

$$K_s = [Ca^{2+}][CO_3^{2-}]$$

This equation states that a product of the molar concentrations of the Ca^{2+} ion and the CO_3^{2-} ion is a constant at equilibrium; it does **not** state that the molar concentrations of the Ca^{2+} ion and the CO_3^{2-} ion must be equal in solution. If a solution has a high $[Ca^{2+}]$, then its $[CO_3^{2-}]$ must be necessarily low—it is their product that is experimentally found to be a constant.

Handbook of Chemistry and Physics: a reference book that contains chemical data.

According to the CRC's *Handbook of Chemistry and Physics*, the K_s of $CaCO_3$ is 4.95×10^{-9} at 25°C. In a saturated solution, when the $[Ca^{2+}] = [CO_3^{2-}]$, the molar concentration of each ion is

$$K_s = [Ca^{2+}][CO_3^{2-}] = 4.95 \times 10^{-9}$$

$$[Ca^{2+}] = [CO_3^{2-}] = \sqrt{4.95 \times 10^{-9}} = 7.04 \times 10^{-5} \text{ mol/L}$$

The **molar solubility** of $CaCO_3$ is also 7.04×10^{-5} mol/L because for each mole of $CaCO_3$ that dissolves 1 mol of Ca^{2+} and 1 mol of CO_3^{2-} are in solution.

Common-Ion Effect

What happens to the solubility of a slightly soluble salt when an ion, common to the salt, is added to a saturated solution? According to LeChatelier's Principle, its equilibrium shifts to compensate for the added (common) ions and favor the formation of more solid salt, reducing the salt's molar solubility.

> **Example.** Suppose 0.0010 mol CO_3^{2-} is added to a saturated $CaCO_3$ solution. What happens?
> The molar solubility decreases because the added CO_3^{2-} shifts the equilibrium, $CaCO_3(s) \rightleftharpoons Ca^{2+}(aq) + CO_3^{2-}(aq)$, to the *left*, meaning that less moles of $CaCO_3$ can now dissolve. The new molar solubility of $CaCO_3$ equals the $[Ca^{2+}]$ in solution or
>
> $$\text{molar solubility of } CaCO_3 = [Ca^{2+}] = \frac{K_s}{[CO_3^{2-}]} = \frac{4.95 \times 10^{-9}}{0.0010} = 4.95 \times 10^{-6} \text{ mol/L}$$
>
> The added CO_3^{2-} decreases the molar solubility of $CaCO_3$ from 7.04×10^{-5} mol/L to 4.95×10^{-6} mol/L.

K_s: also called the solubility product constant and designated as K_{sp}.

In this experiment you will determine the K_s of $Ca(OH)_2$ and its molar solubility in a saturated $Ca(OH)_2$ solution and its molar solubility in the presence of added Ca^{2+}. A saturated $Ca(OH)_2$ solution is decanted from solid $Ca(OH)_2$; the OH^- is titrated with a standard HCl solution to determine its concentration. According to the equation

$$Ca(OH)_2(s) \rightleftharpoons Ca^{2+}(aq) + 2\,OH^-(aq),$$

for each mole of $Ca(OH)_2$ that dissolves, 1 mol Ca^{2+} and 2 mol OH^- are present in solution. Thus by determining the $[OH^-]$, the $[Ca^{2+}]$, K_s, and the molar solubility of $Ca(OH)_2$ can be calculated.

$$K_s = [Ca^{2+}][OH^-]^2 = \left(\tfrac{1}{2}[OH^-]\right)[OH^-]^2$$

The molar solubility of $Ca(OH)_2 = [Ca^{2+}] = \tfrac{1}{2}[OH^-]$

Likewise we use the same procedure to determine the molar solubility of $Ca(OH)_2$ after the common-ion Ca^{2+} is added; Ca^{2+} is an ion common to the slightly soluble salt, $Ca(OH)_2$, equilibrium.

TECHNIQUES

- Technique 2, page 1   Clean Glassware and Lab Bench
- Technique 8, page 7   Reading a Meniscus
- Technique 9, page 7   Pipetting a Liquid or Solution
- Technique 10, page 8   Titrating a Solution
- Technique 14a, page 12   Separation of a Liquid from a Solid

PROCEDURE

Three trials for Parts A and B are to be completed. To hasten the analysis, clean and label three 125 mL or 250 mL Erlenmeyer flasks. In Part A.3b, pipet 25 mL of the saturated $Ca(OH)_2$ solution into each flask.

A. K_s and Molar Solubility of $Ca(OH)_2$

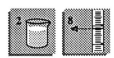

1. Prepare a saturated $Ca(OH)_2$ solution one week[1] before the experiment by adding about 3 g of solid $Ca(OH)_2$ to 100 mL of boiled, distilled (or deionized) water in a 125 mL Erlenmeyer flask. Stir the solution and stopper.

2. Prepare a 50 mL buret for titration. Rinse the clean buret and tip with two 5 mL portions of a standard 0.02 M HCl solution. Fill, read (± 0.02 mL), and record the volume of 0.02 M HCl in the buret. Record the exact molar concentration of the HCl solution.

3. a. Allow the undissolved $Ca(OH)_2$ to remain settled on the bottom of the flask. Without disturbing the insoluble $Ca(OH)_2$, *carefully* decant the saturated solution into a second 125 mL flask.

   b. Rinse a 25 mL pipet with 1 mL or 2 mL of the saturated $Ca(OH)_2$ solution and discard. Pipet 25 mL of the decantate into a clean 125 mL Erlenmeyer flask and add 2 drops of methyl orange indicator. Record the temperature of the solution.

Decantate: the solution that is transferred from above the settled precipitate.

4. Titrate the saturated $Ca(OH)_2$ solution with the standard HCl solution. Record the volume (± 0.02 mL) needed to just turn the yellow color of the methyl orange indicator red.[2] Repeat the titration with two new samples of $Ca(OH)_2$ solution.

B. Solubility of $Ca(OH)_2$ in the Presence of Ca^{2+}

1. Mix 3 g of solid $Ca(OH)_2$ and 1 g of solid $CaCl_2 \bullet 2H_2O$ with 100 mL of boiled, distilled (or deionized) water in a 125 mL Erlenmeyer flask one week[3] before the experiment, stir, and stopper the flask. Decant the solution into a second flask or into a beaker.

2. To complete the analysis, titrate a 25 mL sample of the decantate from Part B.1 to the methyl orange endpoint with the standard 0.02 M HCl solution as was described in Parts A.3 and A.4.

Disposal Information: Dispose of the test solutions in the "Waste Limewater" container.

---

[1]This solution may have been prepared for you. Ask your instructor.

[2]The color change at the endpoint for methyl orange is subtle. Keep the titrated solutions from the successive trials to compare the colors at the endpoint.

[3]This solution may have also been prepared for you. Ask your instructor.

SOLUBILITY CONSTANT AND COMMON ION EFFECT

Date _____Name _____ Lab Sec. _____Desk No. _____

1. How is the molar concentration of OH^- ion in a saturated $Ca(OH)_2$ solution measured in this experiment.

2. a. How many times should a buret and/or pipet be rinsed before it is used to deliver a solution for an analysis?

    b. What criterion is used to determine whether or not a buret, pipet, or any other form of glassware is clean?

    c. Explain how to add less-than-a-drop of titrant from a buret.

3. a. What indicator is used in today's titration?

    b. What color change is observed at the endpoint in the titration?

    c. Does this color change occur as a result of the indicator changing from an acidic to basic form or basic to acidic form?

4. Write the mass action expression for these slightly soluble salt equilibria:

    a. $CuS(s) \rightleftharpoons Cu^{2+}(aq) + S^{2-}(aq)$ $K_s =$

    b. $BaSO_4(s) \rightleftharpoons Ba^{2+}(aq) + SO_4^{2-}(aq)$ $K_s =$

    c. $Ca_3(PO_4)_2(s) \rightleftharpoons 3\,Ca^{2+}(aq) + 2\,PO_4^{3-}(aq)$ $K_s =$

5. a. Calculate the molar solubility of AgI. $K_s = 1.5 \times 10^{-16}$

    b. Calculate the molar solubility of AgI in the presence of 0.020 M KI.

6. The pH of a saturated $Ni(OH)_2$ solution at equilibrium is 8.92.
    a. Calculate the $[OH^-]$ in a saturated $Ni(OH)_2$ solution.

    b. Calculate the $[Ni^{2+}]$ in a saturated $Ni(OH)_2$ solution.

    c. Calculate the K_s of $Ni(OH)_2$.

SOLUBILITY CONSTANT AND COMMON ION EFFECT

Date _____ Name _____ Lab Sec. _____ Desk No. _____

A. K_s and Molar Solubility of $Ca(OH)_2$

Concentration of standard HCl solution (*mol/L*) _____

	Trial 1	Trial 2	Trial 3
1. Buret reading, **final** (*mL*)			
2. Buret reading, **initial** (*mL*)			
3. Volume of standard HCl used (*mL*)			
4. Moles of HCl added (*mol*)			
5. Moles of OH^- in sat'd $Ca(OH)_2$ solution (*mol*)			
6. Volume of sat'd $Ca(OH)_2$ solution (*mL*)	25.0	25.0	25.0
7. Temperature of $Ca(OH)_2$ solution (°C)			
8. $[OH^-]$ at equilibrium (*mol/L*)			
9. $[Ca^{2+}]$ at equilibrium (*mol/L*)			
10. K_s of $Ca(OH)_2$			
11. Average K_s at ___°C			
12. Molar solubility of $Ca(OH)_2$ at ___°C			

B. Solubility of $Ca(OH)_2$ in the Presence of Ca^{2+}

	Trial 1	Trial 2	Trial 3
1. Buret reading, final (*mL*)			
2. Buret reading, initial (*mL*)			
3. Volume of standard HCl used (*mL*)			
4. Moles of HCl added (*mol*)			
5. Moles of OH^- in $Ca(OH)_2/Ca^{2+}$ solution (*mol*)			
6. Volume of $Ca(OH)_2/Ca^{2+}$ solution titrated (*mL*)	25.0	25.0	25.0
7. Temperature of $Ca(OH)_2/Ca^{2+}$ solution (°C)			
8. $[OH^-]$ at equilibrium (*mol/L*)			
9. $[Ca^{2+}]$ at equilibrium (*mol/L*)	*		

10. Molar solubility of $Ca(OH)_2$ in $Ca(OH)_2/Ca^{2+}$ solution at ___°C _____

*Show calculations for Trial 1

============================
QUESTIONS
============================

1. How does the addition of $CaCl_2$ affect the molar solubility of $Ca(OH)_2$? Explain.

2. a. In Part A.3 how would the transfer of some solid $Ca(OH)_2$ into the 125 mL Erlenmeyer flask affect the volume of standard HCl used to reach an endpoint?

   b. How will this affect the reported K_s value of $Ca(OH)_2$? Explain.

3. If the endpoint of the titration is surpassed in Part A.4, will the K_s value for $Ca(OH)_2$ be higher or lower than the accepted value? Explain.

4. Does adding boiled, distilled water to the Erlenmeyer receiving flask, in order to wash the side of the flask and the buret tip, affect the value of K_s for $Ca(OH)_2$? Explain.

5. How would the use of tap water, instead of boiled, distilled water, for preparing the saturated solution in Part A affect the value of K_s for $Ca(OH)_2$.

EXPERIMENT 27

pH, HYDRATION, AND BUFFERS

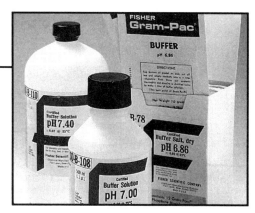

If the solution is aqueous or the substance is damp, then it has a pH. The water in a river, the discharge from a sewage treatment plant, and the fluids in our bodies all have a pH. Most substances dissolved in water affect the pH of the solution; some are obvious—they are called acids and bases—but salts also affect pH. In some aqueous systems it is desirable to change the pH, in others we purposely add substances so that the pH is virtually unaffected by the addition of acids, bases, or salts. The latter are called buffered solutions. Surface water, such as rivers and lakes, and ground water contain dissolved carbonate and bicarbonate salts that serve as buffering agents. The pH of a solution arises from the presence of hydronium and hydroxide ions in an aqueous system. When equal amounts are present then the system is neutral, when there is an imbalance then the solution is acidic or basic.

OBJECTIVES

- To measure the pH of some acids and bases
- To measure the degree of hydrolysis of various ions
- To observe the effectiveness of a buffer system

PRINCIPLES

pH: the negative logarithm of the molar concentration of the hydronium ion in an aqueous solution.

The acidity or basicity of aqueous solutions is due to small amounts of H_3O^+ or OH^- ions. Low concentrations of H_3O^+ ion are often, and more conveniently, expressed as **pH**, the negative logarithm of the molar concentration of H_3O^+ in an aqueous solution.

$$pH = -log\ [H_3O^+]$$

Water dissociates *very* slightly producing equal concentrations of H_3O^+ and OH^-.

$$2\ H_2O(l) \rightleftharpoons H_3O^+(aq) + OH^-(aq)$$

At 25°C, the $[H_3O^+] = [OH^-] = 1.0 \times 10^{-7}$ mol/L, producing a pH = 7.00. Addition of an acid to water increases the $[H_3O^+]$ (and reduces the $[OH^-]$) causing a pH *less than* 7. A base, on the other hand, increases the $[OH^-]$ in solution (and decreases the $[H_3O^+]$) producing a pH *greater than* 7.

Acid-base indicator: a weak organic acid that has a color different from that of its conjugate base.

In this experiment we will measure the pH of some acids and bases using several acid-base indicators. An acid-base indicator, a weak organic acid abbreviated HIn, establishes a dynamic equilibrium with its conjugate base, In^-.

$$HIn(aq) + H_2O(l) \rightleftharpoons H^+(aq) + In^-(aq)$$

When a solution is acidic, the indicator has the color of the HIn form of the organic acid, but, in a basic solution, it has the color of the In^-. Since indicators, like all weak acids, are of varying strengths, the dominant form of the indicator (HIn or In⁻) is pH dependent. The pH ranges over which the color changes from the HIn form to the In⁻ form for a number of indicators are listed in Table 27.1.

Table 27.1 Acid-Base Indicators: pH Range of Color Changes

Indicator	Acid (HIn) Color	pH Range	Base (In⁻) Color
thymol blue	red	1.2–2.8	yellow
methyl orange	red	3.2–4.4	orange/yellow
methyl red	red	4.8–6.0	yellow
litmus	red	4.7–8.3	blue
bromocresol purple	yellow	5.2–6.8	purple
bromothymol blue	yellow	6.0–7.6	blue
m-nitrophenol	colorless	6.8–8.6	yellow
thymol blue	yellow	8.0–9.6	blue
phenolphthalein	colorless	8.2–10.0	pink
alizarin yellow R	yellow	10.1–12.0	red

Hydration: water molecules attracted to ions to varying degrees through ion-dipole interactions.

All salts have a solubility in water producing their respective cations and anions in solution. These ions are hydrated to varying degrees depending upon the size and the charge of the ion. For example, the cations Ba^{2+} and Na^+ are weakly hydrated while the Fe^{3+} and Al^{3+} ions, being smaller and having a larger charge, are more strongly hydrated in solution.

Polar molecule: a molecule that has a nonuniform distribution of charge, often designated by δ⁺ and δ⁻.

A strongly hydrated cation attracts the negative end of the polar H_2O molecule; this weakens the O–H bond of the water molecule and increases the probability of free H^+ being released into the solution. This presence of H^+ produces an acidic solution with a pH less than 7. The Fe^{3+} ion produces an acidic solution, H_3O^+, and $FeOH^{2+}$:

A strongly hydrated anion, on the other hand, attracts the positive end of the polar H_2O molecule; this again weakens the O–H bond and OH⁻ is released into the solution. This produces a basic solution with a pH greater than 7. The PO_4^{3-} ion produces a basic solution with the release of OH⁻ and HPO_4^{2-}.

Cations and anions that are strongly hydrated are characterized by their large charge density, i.e., the ion has a large charge but a small ionic volume. Both the Fe^{3+} and the PO_4^{3-} ions have that property.

Cations and anions with a small charge density have a very weak attraction for the polar water molecules and do not affect the pH of the solution. These ions are called **spectator ions**; examples are:

cations: Group IA (Na^+, K^+, Rb^+, Cs^+), Group IIA (Mg^{2+}, Ca^{2+}, Sr^{2+}, Ba^{2+}) and all other metal cations in the 1^+ oxidation state.

anions: Cl^-, Br^-, I^-, NO_3^-, ClO_4^-, and ClO_3^-

In this experiment data are collected to determine the extent of ion hydration for a number of salt solutions. We'll even use the nose as an indicator for the ammonium salts. The ion of the salt most strongly hydrated in solution is identified and an equation is written to account for the observation.

Buffer solution: a solution that resists large changes in pH when small amounts of acid or base are added to the solution. Buffer solutions contain a weak acid and its conjugate base (or weak base and conjugate acid) as solutes.

For many aqueous solutions, chemists, biologists, and environmentalists do not want a large pH change when H_3O^+ (from a strong acid) or OH^- (from a strong base) is added or produced in the solution. An aqueous solution that adjusts to the H_3O^+ or OH^- addition without encountering a large change in pH is called a **buffer solution.** (Figure 27.1)

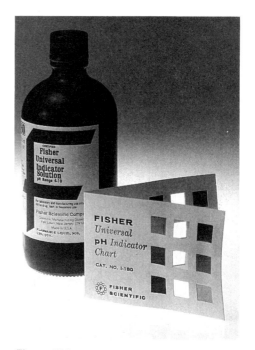

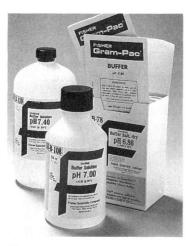

Figure 27.1
Commercial buffer solutions with a desired pH

Figure 27.2
A universal indicator solution and color chart for determining pH

A buffer solution must have two components: one is a substance that will react with H_3O^+ (from a strong acid)—this must be a base, a proton acceptor; the other substance must be capable of reacting with OH^- (from a strong base)—this must be an acid, a proton donor. In either case, the effect of the addition is minimized. The two components of a typical buffer system are a *weak* acid and its conjugate base (or a weak base along with its conjugate acid). A typical buffer system is the $CH_3COOH/CH_3CO_2^-$ system.

$$CH_3COOH(aq) + H_2O(l) \rightleftharpoons H_3O^+(aq) + CH_3CO_2^-(aq) \qquad K_a = 1.8 \times 10^{-5}$$

The addition of OH⁻ to the system, by its reaction with H_3O^+, forms H_2O and more $CH_3CO_2^-$ (a weak base).

$$CH_3COOH(aq) + H_2O(l) \rightleftharpoons H_3O^+(aq) + CH_3CO_2^-(aq)$$
$$\searrow OH^-(aq)$$
$$\hookrightarrow H_2O(l)$$

The equilibrium for the buffer system shifts *right*, the extent of the shift equals the moles of OH⁻ added.

When hydronium ion from a strong acid is added to the $CH_3COOH/CH_3CO_2^-$ system, the $CH_3CO_2^-$ ion (a proton acceptor) reacts with the H_3O^+ forming CH_3COOH; this causes a shift in the buffer system to the *left*, a shift equal to the moles of H_3O^+ added.

$$CH_3COOH(aq) + H_2O(l) \rightleftharpoons H_3O^+(aq) + CH_3CO_2^-(aq)$$
$$\uparrow\!\!\!\_\, H_3O^+(aq)$$

The changes in pH caused by the addition of a strong acid and a strong base to buffered and unbuffered solutions are compared in this experiment.

TECHNIQUES

- Technique 2, page 1   Clean Glassware and Lab Bench
- Technique 3, page 2   Microscale Analyses
- Technique 11, page 10   Testing with Litmus
- Technique 17a, page 16   Testing for Odor

PROCEDURE

A. pH Measurement with Indicators

1. Label a 3 x 6 array of microcells in the 24-cell plate. Make sure that all of the microcells have been cleaned and rinsed with distilled (or deionized) water.

2. The Data Sheet lists 15 solutions that are to be tested for pH. Half-fill each microcell with one of the test solutions (or equivalent substitutes) and add 3–4 drops of universal indicator (Figure 27.2).[1] Swirl the solutions until the colors are uniform. Compare the color of each solution to the "pH Indicator Chart"; estimate the pH of each solution.

3. For the first six solutions write an equation that shows the presence of free H_3O^+ or OH⁻ in solution.

B. Hydration of Ions

1. Set up a 2 x 6 array of microcells. Refer to the Data Sheet; half-fill each microcell with a test solution. Add 3–4 drops of universal indicator and estimate the pH of the solution. Test with litmus paper and other indicators if universal indicator is unavailable. Identify the hydrated ion and write an equation representing the acidity/basicity of the solution.

---

[1]If universal indicator is unavailable, use four 1 x 6 arrays of the microcell plate but test only six solutions at a time. First test each solution with litmus paper. If the red litmus turns blue (indicating that the solutions is basic), further identify the pH of the solution by testing additional samples with several indicators, listed in Table 27.1, that change color in the basic range. If the solution tests acidic (blue litmus turns red), similarly test to further identify the exact pH of the solution.

2. The ammonium ion readily donates a proton to the water vapor in air to release NH_3 gas. Write an equation for this reaction. The extent of this reaction can be detected by the odor of a solid ammonium salt. Cautiously smell the $(NH_4)_2CO_3$, NH_4Cl, and $NH_4CH_3CO_2$ salts. The anion having the strongest affinity (the strongest anionic base) for a proton of the NH_4^+ ion causes the greatest release of NH_3. List the anionic bases (CO_3^{2-}, Cl^-, $CH_3CO_2^-$) in order of increasing strength.

3. A few anions generate H_3O^+ ions in solution. Half-fill three adjacent microcells with 0.1 M $NaHCO_3$, 0.1 M $NaHSO_4$, and 0.1 M NaH_2PO_4. Determine the approximate pH of the solutions. List the three anions in order of increasing acid strength.

4. Baking powders consist of a combination of baking soda, $NaHCO_3$, and a dry acid, a proton donor. When the mixture is added to water, CO_2 is produced and the "dough rises".

$$HCO_3^-(aq) + H_3O^+(aq) \rightarrow 2\ H_2O(l) + CO_2(g)$$

Some common acids in baking powders are cream of tartar, $KHC_4H_4O_6$, calcium dihydrogen phosphate, $Ca(H_2PO_4)_2$, and alum, $NaAl(SO_4)_2 \cdot 12H_2O$.

Add from the tip of a spatula a pinch of each solid acid to separate, clean microcells and dissolve with water. Test each acid to determine their relative acidity.

5. In a clean microcell, place 1 mL of 0.1 M $NaHCO_3$ from Part B.3 with 1 mL of one of the acid solutions in Part B.4. Observe and record.

C. Buffer Solutions

1. Thoroughly clean your 24-cell plate with soap, tap water, and distilled (or deionized) water. Set up a 2 x 2 array of microcells. Mix 15 drops of 0.10 M CH_3COOH with 15 drops of 0.10 M $NaCH_3CO_2$ in microcells 1 and 2. Half-fill microcells 3 and 4 with distilled water. Add 3–4 drops of universal indicator to each microcell and estimate the pH of each solution.[2]

2. Add 10 drops of 0.10 M HCl to microcells 1 and 3, estimate and record the pH, and determine the pH change , ΔpH, of each mixture.[3]

3. Add 10 drops of 0.10 M NaOH to microcells 2 and 4 and estimate the pH.[4] What is the pH change in each test tube?

Disposal Information: Dispose of all of the test solutions in the "Waste Salts" container

---

[2]If universal indicator is unavailable, add 2 drops of methyl orange indicator to microcells 1 and 3 and alizarin yellow R to microcells 2 and 4.

[3]If methyl orange indicator is used, add, count, and record the drops of 1.0 M HCl needed to reach the methyl orange endpoint (when the color change occurs).

[4]If alizarin yellow R indicator is used, then add, count, and record the drops of 1.0 M NaOH need to reach its endpoint.

Notes and Observations

pH, HYDRATION, AND BUFFERS

Date _____ Name _____ Lab Sec. _____ Desk No. _____

1. a. What is an acid-base indicator?

   b. How does it function?

2. a. What color is litmus in an acidic solution?_____

   b. What color is litmus in a basic solution?_____

3. Consider the equilibrium for hydrocyanic acid, $HCN(aq) + H_2O(l) \rightleftharpoons H_3O^+(aq) + CN^-(aq)$:

   a. state the effect of added H_3O^+ from a strong acid, such as hydrochloric acid.

   b. state the effect of added OH^- from a strong base, such as sodium hydroxide.

4. Define and characterize the chemical behavior of a buffer solution.

5. Predict the ion that is more strongly hydrated:

    a. Fe^{3+} or Fe^{2+}. _____ Explain.

    b. Li^+ or Na^+._____ Explain.

6. Write an equation that shows how each of the following can generate OH^- in an aqueous solution.

    N_2H_4 _____

    $CH_3CO_2^-$ _____

7. Write an equation that shows how each of the following can generate H_3O^+ in an aqueous solution.

    $(CH_3)_2NH_2^+$ _____

    Al^{3+} _____

8. In preparing a buffer solution at a desired pH, it is advisable to select a weak acid-conjugate base pair in which the pK_a of the acid equals the desired pH±1. Over what pH range is the $CH_3COOH/CH_3CO_2^-$ buffer most effective? The K_a of CH_3COOH is 1.8×10^{-5}.

9. Briefly describe the technique for testing the pH of a solution using litmus paper.

EXPERIMENT 27

pH, HYDRATION, AND BUFFERS

Date _____ Name _____ Lab Sec. _____ Desk No. _____

A. pH Measurement with Indicators

Solution	Litmus Test (acidic or basic)	Approx pH	Equation
0.10 M HCl			
0.00010 M HCl			
0.10 M CH$_3$COOH			
0.10 M NH$_3$			
0.00010 M NaOH			
0.10 M NaOH			
distilled H$_2$O			
tap water			
vinegar			
lemon juice			
household NH$_3$			
detergent			
409™			
7-UP™			
other			

B. Hydration of Ions

Solution	Litmus test (acidic or basic)	Approx pH	Hydrated Ion	Balanced Equation
0.1 M NaCl				
0.1 M KNO$_3$				
0.1 M NaCH$_3$CO$_2$				
0.1 M NaNO$_2$				
0.1 M Na$_2$CO$_3$				
0.1 M Na$_3$PO$_4$				
0.1 M FeCl$_3$				

0.1 M $ZnCl_2$				
0.1 M NH_4Cl				
0.1 M Na_2SO_4				
0.1 M $Al_2(SO_4)_3$				

2. Equation for the hydration of NH_4^+ _____

List the anionic bases in order of increasing strength according to the increasing strength of NH_3 smell.

_____ < _____ < _____

Write a balanced equation for the hydration of each anion (if hydration occurs).

CO_3^{2-} *(aq)* _____

Cl^- *(aq)* _____

$CH_3CO_2^-$ *(aq)* _____

3. Acid salt solution	Litmus test (acid or base)	Approx pH	Hydrated Ion	Balanced Equation
0.1 M $NaHCO_3$				
0.1 M $NaHSO_4$				
0.1 M NaH_2PO_4				

List the anions in order of increasing acid strength:

_____ < _____ < _____

4. Dry acid	Litmus test (acid or base)	Approx pH	Hydrated Ion	Balanced Equation
$KHC_4H_4O_6$				
$Ca(H_2PO_4)_2$				
$NaAl(SO_4)_2 \cdot 12H_2O$				

5. Observation of reaction of 0.1 M $NaHCO_3$ with an acid from Part B.4

Balanced equation for the reaction _____

C. Buffer Solutions

Observation	CH3COOH/CH3CO2⁻ buffer	Water
initial pH		
pH after 0.10 M HCl		
ΔpH		
(drops of 1.0 M HCl to methyl orange endpoint)		
pH after 0.10 M NaOH		
ΔpH		
(drops of 1.0 M NaOH to alizarin yellow R endpoint)		

Comment on the effectiveness of this buffer solution in resisting large changes in pH.

QUESTIONS

1. Identify those ions in Part B.1 that produce an acidic solution and those that produce a basic solution.

   acidic: _____

   basic: _____

2. a. Which ions in Part B.1 do *not* affect the pH of the solution?

   _____

   b. Which anion hydrates most extensively? _____

   c. Which cation hydrates most extensively? _____

3. At what point does a buffer solution stop resisting a pH change when strong acid is added? Hint: consider the concentration of the components of a buffer solution.

4. Predict whether an aqueous solution of each salt produces an solution with a pH > 7, < 7, or = 7. If the pH > 7 or < 7, write an equation that justifies your prediction.

Salt	pH > 7, < 7, or = 7	Equation justifying your prediction
Na_2SO_3		
KNO_3		
$(CH_3)_2NH_2NO_3$		
$Ca(C_3H_7O)_2$		

5. A student needs 1.00 L of buffer solution with a pH = 7.00. She selects the $H_2PO_4^-/HPO_4^{2-}$ buffer system. The $K_a(H_2PO_4^-) = 6.2 \times 10^{-8}$. What must be the $[HPO_4^{2-}]/[H_2PO_4^-]$ ratio for the buffer system?

EXPERIMENT 28

A STANDARD NaOH SOLUTION

Just how acidic is orange juice or vinegar? How is that determined in the laboratory? The analysis of any commercial or industrial product is extremely important to the consumer. For example, knowing the amount of impurities or by-product that result from a manufacturing process of an over the counter drug is of value to physicians of internal medicine. The symptoms of any ailment could result from the intake of the impurity of what was thought to be a "safe" drug.

For any analysis it is important to start with what is known to determine something that is unknown. In this experiment, the determination of a solution of known concentration is prepared so that analyses can be made on other samples in subsequent experiments.

OBJECTIVE

• To determine the molar concentration of a sodium hydroxide solution.

PRINCIPLES

A titrimetric technique of volumetric analysis can be used to measure the amount of solute in solution. The procedure requires the addition of a solution having a known concentration of solute (a standard solution) to a solution containing the unknown amount of solute until the reaction between the two solutes is complete. The reaction is complete when the mole ratio of the two reacting solutes is the same as that of the balanced equation. This is the **stoichiometric point**[1] in a titration.

Indicator: an acid-base indicator is a weak organic acid that has a color different from that of its conjugate base.

Endpoint: when the volume of titrant dispensed from a buret causes the indicator to change color.

In this experiment a sodium hydroxide solution of unknown concentration is determined in an acid-base titration. The stoichiometric point is detected using the phenolphthalein indicator, which is colorless in an acidic solution but pink (or red) in a basic solution. The point at which the phenolphthalein indicator changes color is the **endpoint** of the titration. Indicators are selected so that the stoichiometric point and the endpoint occur at essentially the same point in the titration.

In this experiment potassium hydrogen phthalate is used as the **primary standard**[2] to determine the molar concentration of a NaOH solution. Potassium hydrogen phthalate, $KHC_8H_4O_4$, is a white, crystalline, nonhygroscopic, acidic substance with a high degree of purity. To determine the molar concentration of the NaOH solution, a measured mass of a dried sample of $KHC_8H_4O_4$ is dissolved in distilled (or deionized) water; to this solution is added a recorded volume (from a buret) of the NaOH solution until the stoichiometric point is reached—when 1 mol NaOH has been added

---

[1]The stoichiometric point is also called the **equivalence point**, indicating the point at which stoichiometrically equivalent quantities of the reacting substances are combined.

[2]A **primary standard** is a substance that has a known purity, is nonhygroscopic, is stable in its pure form and in solution, does not decompose with heat, and has a relatively high molar mass.

for each measured mole of $KHC_8H_4O_4$ for the analysis, according to the equation:

$$K^+HC_8H_4O_4^-(aq) + NaOH(aq) \rightarrow H_2O(l) + Na^+(aq) + K^+(aq) + C_8H_4O_4^{2-}(a_q$$

This is also the point at which the phenolphthalein indicator changes from colorless to pink.

Since the initial measured mass and the molar mass of $KHC_8H_4O_4$ are known, the amount of $KHC_8H_4O_4$ titrated in the analysis is calculated.

$$\text{g } KHC_8H_4O_4 \times \frac{1 \text{ mol } KHC_8H_4O_4}{204.2 \text{ g } KHC_8H_4O_4} = \text{mol } KHC_8H_4O_4$$

At the stoichiometric point equal moles of $KHC_8H_4O_4$ and NaOH have been combined.

$$\text{mol NaOH} = \text{mol } KHC_8H_4O_4$$

The molar concentration of the NaOH solution equals the moles of NaOH needed to react with the known moles of $KHC_8H_4O_4$ divided by the volume of sodium hydroxide solution dispensed from the buret.

$$\text{molar concentration of NaOH, M} = \frac{\text{mol NaOH}}{\text{liter NaOH solution}}$$

TECHNIQUES

- Technique 2, page 1   Clean Glassware and Lab Bench
- Technique 5, page 3   Transferring Liquids
- Technique 8, page 7   Reading a Meniscus
- Technique 10, page 8   Titrating a Solution
- Technique 13, page 11   Using the Laboratory Balance

PROCEDURE

You are to complete at least three trials in the standardization of your NaOH solution. To hasten the analysis, clean, dry, and label three 125 mL or 250 mL Erlenmeyer flasks and measure the mass of three $KHC_8H_4O_4$ samples while occupying the balance. If all the balances are occupied, prepare your NaOH solution (Part A.2) and buret (Part B.2).

In this experiment you are striving for "good" results. The measured molar concentration of the NaOH for the three trials should be within ±1%. If not, a fourth or fifth trial may be necessary. Part A.1 of the experiment may have already been completed for you. Ask your instructor.

A. Preparation of the NaOH Solution

1. One week before the scheduled laboratory period, dissolve 3–4 g of NaOH (pellets or flakes) in 20 mL of distilled water in a 125 mL rubber-stoppered Erlenmeyer flask. Thoroughly mix and allow the solution to stand for the precipitation of any Na_2CO_3.[3] Dry about 2 g of $KHC_8H_4O_4$ at 110°C for several hours in a constant temperature drying oven. Cool the sample in a desiccator (if available).

---

[3] $2 NaOH(aq) + CO_2(aq) \rightarrow Na_2CO_3(s) + H_2O(l)$. Na_2CO_3 has a low solubility in a concentrated NaOH solution.

2. With a graduated cylinder, transfer 10 mL[4] of the concentrated NaOH solution (**Caution:** *a concentrated NaOH solution causes severe skin burns.*) into a 500 mL polyethylene bottle and dilute to 500 mL with previously, boiled, distilled water. The boiled water removes traces of CO_2. Cap the polyethylene bottle to prevent the absorption of CO_2 by the NaOH solution; stir the solution for several minutes and label the bottle. Do *not* shake the bottle (this increases the probability of CO_2 absorption).

B. Preparation of the $KHC_8H_4O_4$ for Titration

1. Measure 0.3–0.5 g (±0.001 g) of the dried $KHC_8H_4O_4$ in a clean, dry, 125 mL or 250 mL Erlenmeyer flask. Add 50 mL of distilled water and 2 drops of phenolphthalein.

2. Clean a buret for titration. Rinse the buret with two 5 mL portions of your NaOH solution, making certain that the NaOH wets its entire inner surface. Have your instructor approve your buret *before* proceeding to Part C. Fill the buret with the NaOH solution and carefully read the meniscus before recording its volume (±0.01 mL). Place white paper beneath the Erlenmeyer flask.

C. Analysis of the NaOH Solution

1. Slowly add the NaOH solution from the buret to the $KHC_8H_4O_4$ solution in the flask, swirling the mixture after each addition. Consult the instructor (or Technique 10) for proper techniques of titrating with the left hand (if right-handed), rinsing the wall of the flask with water from the wash bottle, and adding half-drops of NaOH solution from the buret.

2. As the rate of the phenolphthalein color change (pink, where the NaOH has been added, back to colorless, for the acidic solution) decreases, slow the rate of NaOH addition; proceed with drop and half-drop additions of the NaOH until the phenolphthalein endpoint is reached. This occurs when a single half-drop finally causes the pink color of the phenolphthalein indicator to persist for 30 seconds. Read the meniscus in the buret (±0.01 mL) and record.

3. Repeat the titration at least two more times with varying but accurately known amounts of $KHC_8H_4O_4$ until $\pm1\%$ reproducibility in the molar concentration of the NaOH solution is obtained. **Save** your standardized NaOH solution in the 500 mL polyethylene bottle for later experiments (for example, Experiments 29 and 30).

---

[4]If 1 L of NaOH solution is to be prepared for this experiment, transfer all of the concentrated NaOH solution to a 1 L polyethylene bottle and dilute to 1 L. Do *not* transfer any of the Na_2CO_3 precipitate.

This is a modern titration apparatus. The titrant is automatically added until a desired (programmed) endpoint is reached. The data can be analyzed by preprogramming the computer and can then be printed. The progress of the titration (i.e., a titration curve) can also be printed out.

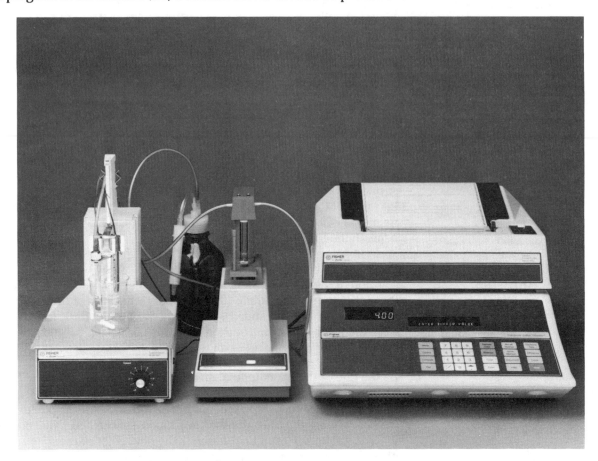

A STANDARD NaOH SOLUTION

Date _____ Name _____ Lab Sec. _____ Desk No. _____

1. Characterize a primary standard.

2. Explain how CO_2 absorbed from the atmosphere affects the molar concentration of a standardized NaOH solution.

3. What is the purpose for placing white paper beneath the Erlenmeyer flask during today's titration? See Procedure, Part B.2.

4. Distinguish between a stoichiometric point and an endpoint.

5. Sulfamic acid can be used as a primary standard for determining the molar concentration of a sodium hydroxide solution. A 0.418 g sample of sulfamic acid, NH_2SO_3H, dissolved in 50.00 mL of water is neutralized by 27.49 mL of NaOH at the phenolphthalein endpoint. What is the molar concentration of the NaOH solution? The molar mass of NH_2SO_3H is 97.1 g/mol.

$$NH_2SO_3H(aq) + NaOH(aq) \rightarrow NH_2SO_3^-Na^+(aq) + H_2O(l)$$

6. A 0.377 g sample of potassium hydrogen phthalate, $KHC_8H_4O_4$, is dissolved in 100 mL of water. If 7.39 mL of a NaOH solution is required to reach the stoichiometric point, what is the molar concentration of the NaOH solution? The molar mass of $KHC_8H_4O_4$ is 204.2 g/mol.

7. a. Where should the meniscus be read when viewing the volume of a solution in a buret (or pipet)?

    b. When rinsing a buret after cleaning it with soap and water, should the rinse be dispensed through the buret tip or at the top of the buret? Explain.

    c. What is the criterion for a clean buret?

    d. In preparing a buret for titration, the final rinse (or two) should be with the solution that is subsequently used in the titration. Why is this solution used for the rinse rather than distilled water?

    e. How is a half-drop dispensed from a buret into the solution?

    f. How long should the color change from an indicator persist to ensure that its endpoint has been reached?

A STANDARD NaOH SOLUTION

Date _____ Name _____ Lab Sec. _____ Desk No. _____

Maintain at least three significant figures when recording data and performing calculations.

	Trial 1	Trial 2	Trial 3
1. Mass of flask + $KHC_8H_4O_4$ (g)			
2. Mass of flask (g)			
3. Mass of $KHC_8H_4O_4$ (g)			
4. Moles of $KHC_8H_4O_4$ (mol)			
Buret approval by instructor			
5. Buret reading of NaOH, **final** (mL)			
6. Buret reading of NaOH, **initial** (mL)			
7. Volume of NaOH used (mL)			
8. Moles of NaOH neutralized (mol)			
9. Molar concentration of NaOH (mol/L)*			

10. Average molar concentration of NaOH (mol/L)   _____

*Show calculation for Trial 1

QUESTIONS

1. Is it quantitatively acceptable to titrate all $KHC_8H_4O_4$ samples with the NaOH solution to the same *dark red* endpoint, just as long as the red intensity is consistent from one sample to the next?

2. A student titrates the $KHC_8H_4O_4$ to a faint pink endpoint which persists for 30 seconds; the buret reading is recorded and the calculations are completed. When he again looks at the receiving flask, the solution is no longer pink. Explain a probable cause of this change. Assume that the $KHC_8H_4O_4$ was pure, completely dissolved, and the stoichiometric point in the titration had been reached.

3. If the endpoint in the titration of the $KHC_8H_4O_4$ with the NaOH solution is mistakenly surpassed (too pink), what effect does this have on the calculated molar concentration of the NaOH solution?

4. Oxalic acid, $H_2C_2O_4$, can also be used as a primary standard in this experiment. What mass of oxalic acid should be measured to neutralize 25.00 mL of your NaOH solution? Oxalic acid is diprotic (Hint: what then is the balanced equation?).

5. If a drop of the NaOH solution adheres to the side of the Erlenmeyer flask during the standardization of the NaOH solution, how does this error affect its reported molar concentration? Explain.

6. If a drop of the NaOH solution adheres to the side of the buret during the standardization of the NaOH solution, how does this error affect its reported molar concentration? Explain.

EXPERIMENT 29

ANALYSIS OF ACIDS

Most of the foods that we eat and the chemicals that we use are acidic. For example, even vegetables such as carrots and peas have a pH less than 7. Some home cleaning agents, such as tub and shower cleaners, are also acidic. Most soils throughout the world are acidic, especially in those areas that have ample rainfall for extensive vegetation. Desert areas tend to be slightly alkaline. Plants most conducive to the acidity of the soil tend to thrive. For example, celery requires a very acidic soil for best results; hence an attempt to grow celery commercially in arid climates would be disastrous. Therefore the analysis of various mixtures and formulations for their acidic content is important to the farmer, the rancher, and the consumer. In this experiment we will find out more about the acidic content of various common formulations.

OBJECTIVES

- To determine the molar mass of an acid
- To determine the percent acid in a commercial product
- To measure the percent acetic acid in vinegar

PRINCIPLES

Molar Mass of an Acid

Stoichiometric point: in a titration when the moles of reactants in the reaction mixture are in the same ratio as that in a balanced equation.

The standard NaOH solution prepared in Experiment 28 is used to titrate an unknown, dry acid to the stoichiometric point. The amount of NaOH used in the titration is determined from the volume of NaOH dispensed from the buret in the titration and its molar concentration.

volume NaOH (*L*) x molar concentration NaOH (*mol/L*) = mol NaOH

The moles of unknown acid that reacts with the NaOH is found from the balanced equation for the reaction. The unknown acid is either monoprotic, HA, diprotic, H_2A, or triprotic, H_3A.

$$HA(aq) + NaOH(aq) \rightarrow NaA(aq) + H_2O(l)$$
$$H_2A(aq) + 2\,NaOH(aq) \rightarrow Na_2A(aq) + 2\,H_2O(l)$$
$$H_3A(aq) + 3\,NaOH(aq) \rightarrow Na_3A(aq) + 3\,H_2O(l)$$

From a measurement of the mass of acid on the laboratory balance and the determination of the moles of acid present in the titration, the molar mass of the acid can be determined.

$$\text{molar mass (acid)} = \frac{\text{mass of acid}}{\text{mol acid}}$$

Percent Acid in Vanish™

Vanish™, a common household cleaning agent, is primarily sodium bisulfate. Sodium bisulfate, $NaHSO_4$, is a safe, easy-to-use acid that is used to remove rust, calcium deposits from shower stalls, bathtubs, and commodes, and to acidify home swimming pools. The solid $NaHSO_4$ dissolves in water to produce the bisulfate ion, HSO_4^-, which acts as a weak acid in aqueous solutions. A measure of the moles of NaOH (OH⁻) that neutralizes the HSO_4^- in the sample (a 1:1 mole ratio) allows us to determine percent composition of $NaHSO_4$ in the cleaning agent.

$$HSO_4^-(aq) + OH^-(aq) \rightarrow H_2O(l) + SO_4^{2-}(aq)$$

$$\text{mol } OH^- = \text{mol } HSO_4^- = \text{mol } NaHSO_4$$

$$\text{mass of } NaHSO_4 = \text{mol } NaHSO_4 \times \frac{120.1 \text{ g } NaHSO_4}{\text{mol } NaHSO_4}$$

$$\% \text{ } NaHSO_4 = \frac{\text{mass of } NaHSO_4}{\text{mass of sample}} \times 100$$

Percent Acetic Acid in Vinegar

Acetic acid, CH_3COOH, is the acid in vinegar. Its concentration varies slightly in vinegars but must be at least 4% (by mass) acetic acid in water to meet the minimum federal standard. Concentrations of acetic acid may even exceed 5% in some vinegars. Generally, caramel flavoring and coloring may also be added to make the product aesthetically pleasing to the consumer.

The percent by mass of CH_3COOH in vinegar is determined by titrating a measured mass of vinegar to a phenolphthalein endpoint with a measured volume of a standard NaOH solution. The amount of CH_3COOH present in the vinegar is calculated from the balanced equation.

$$CH_3COOH(aq) + OH^-(aq) \rightarrow H_2O(l) + CH_3CO_2^-(aq)$$

At the stoichiometric point the moles of OH^- dispensed from the buret equals the moles of CH_3COOH in the vinegar.

$$\text{mol } OH^- = \text{mol } CH_3COOH$$

$$\text{mass of } CH_3COOH = \text{mol } CH_3COOH \times \frac{60.05 \text{ g } CH_3COOH}{\text{mol } CH_3COOH}$$

$$\% \text{ } CH_3COOH = \frac{\text{mass of } CH_3COOH}{\text{mass of vinegar}} \times 100$$

TECHNIQUES

- Technique 2, page 1   Clean Glassware and Lab Bench
- Technique 8, page 7   Reading a Meniscus
- Technique 10, page 8   Titrating a Solution
- Technique 13, page 11   Using the Laboratory Balance

PROCEDURE

For completing Parts A, B, and C of this experiment you will need approximately 250 mL of the standardized NaOH solution prepared in Experiment 28. You may need to standardize additional NaOH, using potassium hydrogen phthalate as the primary standard. Ask your instructor which parts of the experiment you are to complete.

A. Molar Mass of an Acid

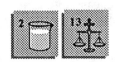

1. Three trials are to be completed in this part of the experiment; successive results should be within ±1%. To hasten the analysis, clean, dry, and measure the mass (±0.001 g) of three 125 mL or 250 mL Erlenmeyer flasks; make all measurements while occupying the same balance. If all balances are occupied, proceed to Part A.4 and prepare the buret for the titrimetric analysis.

2. Ask the instructor whether your unknown acid is monoprotic, diprotic, or triprotic.

3. In one of your Erlenmeyer flasks, measure (±0.001 g) 0.3–0.4 g of your solid, unknown acid. Add 50 mL of distilled (or deionized) water and 2 drops of phenolphthalein.[1]

4. Prepare a buret. Rinse the buret twice with the standardized NaOH solution and drain it through the buret tip. Have your instructor approve the buret before you fill it.

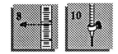

5. Fill the buret with your NaOH solution, remove all air bubbles from the buret tip, and record the initial volume (±0.01 mL). Titrate the sample to the phenolphthalein endpoint. See Experiment 28, Part C of the Procedure. Read the meniscus in the buret after the endpoint has been reached. Record.

6. Successive trials should have a reproducibility of <±1%.

B. Percent Acid in Vanish™[2]

1. Read Part A.1 for increasing the efficiency of your time and technique.

2. Determine the mass (±0.001 g) of a 125 mL (or 250 mL) Erlenmeyer flask. Measure 0.4–0.5 g (±0.001 g) of sample and dissolve it in 100 mL of distilled (or deionized) H_2O. Add 2 drops of phenolphthalein.

3. Prepare the buret as in Part A.4. Obtain your instructor's approval.

4. Titrate the sample to the phenolphthalein endpoint. Follow the procedure in Part A.5.

C. Percent Acetic Acid in Vinegar

1. Read Part A.1 for increasing the efficiency of your time and technique.

2. Determine the mass (±0.01 g) of a 125 mL (or 250 mL) Erlenmeyer flask. Add 5 mL of a selected brand of vinegar from the reagent shelf to the Erlenmeyer flask and remeasure (±0.01 g). Be sure to use the same balance. Add 2 drops of phenolphthalein to the vinegar and wash down the wall of the flask with 20 mL of distilled water.

3. Prepare the buret as in Part A.4. Obtain your instructor's approval.

4. Titrate the sample to the phenolphthalein endpoint. Follow the procedure in Part A.5.

5. Select another brand of vinegar and complete two analyses to determine its percent acetic acid.

6. Compare the percent acetic acid in the two vinegars and then determine which vinegar has the most acetic acid per unit cost.

Disposal Information: **Dispose of the test solutions in the "Waste Salts" container.**

---

[1]The acid may be relatively insoluble, but with the addition of the NaOH solution, it gradually dissolves. The addition of 10 mL of ethanol may be necessary to hasten the acid's dissolution, especially if the phenolphthalein endpoint is reached before it all dissolves.

[2]Other commercial products containing $NaHSO_4$ may be substituted for the Vanish™.

Notes, Observations, and Calculations

ANALYSIS OF ACIDS

Date _____Name _____ Lab Sec. _____Desk No. _____

1. Determine the volume (mL) of 0.150 M NaOH that neutralizes 0.188 g of a *di*protic acid having a molar mass of 152.2 g/mol.

2. a. What mass (in grams) of adipic acid, $C_4H_8(COOH)_2$, will neutralize 28.2 mL of 0.188 M NaOH at the phenolphthalein endpoint (adipic acid is diprotic).

    b. If the mass of the adipic acid sample is 0.485 g, what is the percent purity of the sample.

3. If an air bubble initially trapped in the tip of the buret containing the standardized NaOH disappears during the titration, how will this affect the reported number of moles in the unknown acid in Part A? (Read the Procedure.)

4. How can a half-drop of NaOH solution be added from the buret?

5. A 0.793 g sample of an unknown monoprotic acid requires 31.90 mL of 0.106 M NaOH to reach the phenolphthalein endpoint. What is the molar mass of this acid?

6. a. A 30.84 mL volume of 0.128 M NaOH is required to reach the phenolphthalein endpoint in titrating 5.961 g of vinegar. Calculate the moles of acetic acid in vinegar.

   b. How many grams of acetic acid are in the vinegar?

   c. What is the percent acetic acid in the vinegar?

ANALYSIS OF ACIDS

Date _____ Name _____ Lab Sec. _____ Desk No. _____

Maintain 3 significant figures when recording data and performing calculations.

Data for Part ___ of experiment, entitled _____

	Trial 1	Trial 2	Trial 3	Trial 4
1. Mass of flask + sample (*g*)				
2. Mass of flask (*g*)				
3. Mass of sample (*g*)				
4. Instructor's approval of buret				
5. Buret reading of NaOH, **final** (*mL*)				
6. Buret reading of NaOH, **initial** (*mL*)				
7. Volume of NaOH used (*mL*)				
8. Molar concentration of NaOH (*mol/L*)				
9. Moles of NaOH used (*mol*)				

A. Molar Mass of an Acid

1. Unknown number or name of acid _____; molecular form of acid: HA, H_2A, or H_3A_____

2. Balanced equation for the reaction of the acid with NaOH:

3. Moles of unknown acid (*mol*)

4. Molar mass of acid (*g/mol*)

5. Average molar mass (*g/mol*)

B. Percent Acid in Vanish™

Sample number _____	Trial 1	Trial 2	Trial 3
1. Moles of $NaHSO_4$ in Vanish™ (*mol*)			
2. Mass of $NaHSO_4$ in Vanish™ (*g*)			
3. Percent $NaHSO_4$ in Vanish™ (%)			
4. Average percent $NaHSO_4$ in Vanish™ (%)			

C. Percent Acetic Acid in Vinegar

Brand of Vinegar: _____

	Trial 1	Trial 2	Trial 1	Trial 2
1. Moles of CH$_3$COOH in vinegar sample				
2. Mass of CH$_3$COOH in vinegar sample (g)				
3. Percent CH$_3$COOH in vinegar by mass (%)				
4. Average percent CH$_3$COOH in vinegar by mass (%)				

Comment on the availability of CH$_3$COOH per unit cost in the two vinegars.

QUESTIONS

1. If a drop of NaOH solution had adhered to the side of the Erlenmeyer flask during the standardization of the NaOH solution in Experiment 28, how will this error affect the reported molar mass of the unknown acid in Part A?

2. If the endpoint is surpassed in analyzing for the percent NaHSO$_4$ in Vanish™, will its reported percent concentration be high or low? Explain.

3. A drop of standardized NaOH solution adheres to the side of the Erlenmeyer flask and is not washed down into the vinegar with the wash bottle; how does this error affect the reported percent CH$_3$COOH in vinegar?

4. In determining the percent CH$_3$COOH in vinegar, the mass of each sample was determined rather than its volume. Explain.

EXPERIMENT 30
ANTACID ANALYSIS

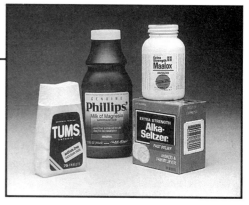

Acid indigestion! What a terrible feeling! Plop, plop, fizz, fizz! Many of the foods we eat are naturally acidic, but many of the foods we eat, in addition, stimulate acidic secretions from the lining of the stomach causing an excess of acid to build, leading to an upset stomach. Various commercial antacids claim to give the "best relief" for acid indigestion, but often they do not substantiate their claims with quantitative data. All antacids, which must therefore be bases, neutralize acid; in this experiment we shall quantitatively determine the effectiveness of various antacids.

How many names of commercial antacids can you name? Which one do you think to be most effective and why?

OBJECTIVE

• To determine the amount of antacid in commercial antacids

PRINCIPLES

Stomach secretions generally have a pH ranging from 1.0 to 2.0; acid indigestion and heartburn occur at a lower pH. Antacids neutralize (or buffer) the excess hydrogen ion in the stomach to relieve this discomfort. The amount of antacid needed for relief is dependent upon its "strength", although there is little danger of consuming too much.

Some of the "old" antacids are still among the most effective, the safest, and the cheapest. Milk of magnesia, an aqueous, milky suspension of magnesium hydroxide, $Mg(OH)_2$, is an antacid that provides hydroxide ions to neutralize hydrogen or hydronium ions, H_3O^+.

$$Mg(OH)_2(s) + 2\,H_3O^+(aq) \rightarrow Mg^{2+}(aq) + 2\,H_2O(l)$$

Maalox™, a "double strength" antacid, contains equal masses of $Mg(OH)_2$ and $Al(OH)_3$.

The advertised antacids that buffer excess acid in the stomach are those containing calcium carbonate, $CaCO_3$, or sodium bicarbonate, $NaHCO_3$. A HCO_3^-/CO_3^{2-} buffer system is established in the stomach with these antacids.

$$CO_3^{2-}(aq) + 2\,H_3O^+(aq) \rightleftharpoons HCO_3^-(aq) + 2\,H_2O(l)$$
$$HCO_3^-(aq) + H_3O^+(aq) \rightleftharpoons 2\,H_2O(l) + CO_2(aq)$$

Baking soda, which is pure $NaHCO_3$, is an inexpensive antacid. The warmth in the stomach converts the $CO_2(aq)$ to $CO_2(g)$ creating "gas on the stomach" with belching as a natural result.

Rolaids™, containing dihydroxyaluminum sodium carbonate, $NaAl(OH)_2CO_3$, is a combination antacid that also reacts with stomach acid.

$$NaAl(OH)_2CO_3(aq) + 3\,H_3O^+(aq) \rightarrow$$
$$Na^+(aq) + Al^{3+}(aq) + 5\,H_2O(l) + HCO_3^-(aq)$$

Buffer system: a solution that resists large changes in acidity.

This experiment determines the effectiveness of several antacids using a strong acid-strong base titration. To avoid the possibility of a buffer system[1] from being established and thus affecting the analysis, an excess of $HCl(aq)$ is added to the dissolved antacid, driving the $HCO_3^-(aq) + H_3O^+(aq) \rightleftharpoons 2\,H_2O(l) + CO_2(aq)$ far to the right. The solution is then heated to expel the $CO_2(aq) \rightarrow CO_2(g)$. The excess $HCl(aq)$ is titrated with a standardized NaOH solution.

Since an antacid has the same neutralizing effect on stomach acid as does NaOH, the amount of antacid in a sample is called its $NaOH_{equivalent}$.[2] To determine the $NaOH_{equivalent}$ for an antacid, we subtract the excess moles of $HCl(aq)$ from the total moles of $HCl(aq)$ added to the antacid.

$$NaOH_{equivalent} = HCl(aq)_{total} - HCl(aq)_{excess}$$

TECHNIQUES

- Technique 2, page 1   Clean Glassware and Lab Bench
- Technique 6b, page 4   Heating Liquids
- Technique 8, page 7   Reading a Meniscus
- Technique 9, page 7   Pipetting a Liquid or Solution
- Technique 10, page 8   Titrating a Solution
- Technique 13, page 11   Using the Laboratory Balance

In addition, the technique for back-titrating a solution is used.

PROCEDURE

A. Dissolving the Antacid

1. a. Determine the mass (± 0.001 g) of a 250 mL Erlenmeyer flask. Accurately measure approximately 0.7 g (± 0.001 g) of a pulverized commercial antacid tablet in the flask.

    b. Pipet 25.0 mL of standardized 0.1 M HCl into the flask and swirl to dissolve the antacid.[3] Record the actual HCl concentration on the Data Sheet.

2. Heat the solution to boiling and continue to heat at a gentle boil for at least 1 minute to expel the dissolved CO_2. Add 4–8 drops of bromophenol blue indicator.[4] If the solution is blue, add an additional 15.0 mL of 0.1 M HCl and boil again.

---

[1]In our analysis, we want to "swamp" the system with the excess HCl to remove this buffering effect and then analyze for the HCl that was *not* neutralized by the antacid.

[2]The $NaOH_{equivalent}$ can also be referred to as the number of "equivalents" of antacid in the sample. An **equivalent** of any substance is merely an expression of its amount in the system, just as a **mole** of a substance indicates an amount of that substance in the system.

[3]The inert ingredients, such as the binder used in manufacturing the tablet, may not dissolve.

[4]Bromophenol blue is yellow in an acidic solution and blue in a basic solution.

B. Analysis of the Antacid Sample

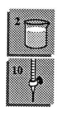

1. Prepare a clean 50 mL buret. Rinse the clean buret with two 5 mL portions of the standard NaOH solution prepared in Experiment 28. Fill the buret with the NaOH solution; read and record its initial volume (±0.02 mL). Remember to use the correct technique for reading the meniscus.

2. Place a white sheet of paper beneath the Erlenmeyer flask. Titrate the excess HCl to the blue endpoint of the bromophenol indicator. Read and record the final volume of NaOH in the buret.

3. Repeat the experiment for a second trial.

4. Select a second antacid for analysis and repeat the procedure. Compare the strengths (amount of antacid per gram of tablet) of the two antacids.

Disposal Information: **Dispose of the test solutions in the "Waste Salts" container.**

Notes, Observations, and Calculations

EXPERIMENT 30

ANTACID ANALYSIS

Date _____ Name _____ Lab Sec. _____ Desk No. _____

1. a. Write the balanced for the reaction of one mole of the active ingredient in Rolaids™ with *excess* H_3O^+ ion.

    b. Write a balanced equation that represents the antacid effect of sodium citrate, $Na_3C_6H_5O_7$, on an excess of stomach acid. In an aqueous solution sodium citrate dissociates into Na^+ ions and citrate, $C_6H_5O_7^{3-}$, ions.

2. Assuming the acidity of the stomach is due to hydrochloric acid, what molar concentration of hydrochloric acid gives a pH of 1.5, the average for that of stomach acid.

3. What acid-base indicator is used in this experiment? _____ Its color in an acidic

    solution is _____ ; in a basic solution it is _____ .

4. a. How much time should be allowed for the titrant to drain from the wall of a buret before a reading is made?

    b. What color should be the background of the receiving flask in today's titration?

5. A volume of 50.0 mL of 0.104 M HCl is added to an unknown base. The HCl *not* neutralized by the base (the excess HCl) is titrated to a bromophenol blue endpoint with 26.7 mL of 0.0841 M NaOH. Calculate the NaOH$_{equivalent}$ for the unknown base.

6. Describe Technique 6b.

ANTACID ANALYSIS

Date _____ Name _____ Lab Sec. _____ Desk No. _____

A. Dissolving the Antacid

Commercial Antacid _____ | _____

	Trial 1	Trial 2	Trial 1	Trial 2
1. Mass of flask + crushed tablet (g)				
2. Mass of flask (g)				
3. Mass of crushed tablet (g)				
4. Volume of HCl added (mL)				
5. Molar conc. of HCl solution (mol/L)				
6. Moles of HCl added (mol)				

B. Analysis of the Antacid Sample

	Trial 1	Trial 2	Trial 1	Trial 2
7. Buret reading, **final** (mL)				
8. Buret reading, **initial** (mL)				
9. Volume of NaOH added (mL)				
10. Molar conc. of NaOH (Exp't 28) (mol/L)				
11. Moles of NaOH added (mol)				
12. Moles of excess HCl (mol)				
13. $NaOH_{equivalent}$ of antacid in tablet				
14. $NaOH_{equivalent}$/g tablet				
15. Cost of antacid/g tablet ($¢/g$)				

Which antacid is the best buy ($¢/NaOH_{equivalent}$)? _____

1. If the CO_2 is *not* removed by boiling after the 0.1 M HCl is added to the antacid, how will this affect the amount of NaOH needed to reach the bromophenol blue endpoint? Explain.

2. If the results from Trials 1 and 2 differ by a substantial amount (>5%), what should you do before presenting your results to your laboratory instructor?

3. If the endpoint in the titration is surpassed, will the reported amount of antacid in the sample be too high or too low? Explain.

EXPERIMENT 31

THE ELECTROLYTIC CELL

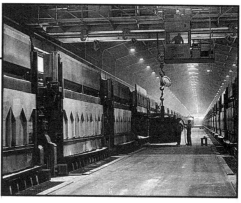

The next time that you purchase a soft drink in a can and before you take that first inviting drink, stop for just a moment and think of all of the chemistry that went into producing that aluminum can. The ore containing the aluminum, called bauxite ore, is mined from the Earth's crust and the ore is shipped to where it is refined to alumina, a white powder with a formula of Al_2O_3. The alumina is then melted in a stainless steel container ("the cell") in the presence of cryolite, Na_3AlF_6, at about 900°C. Electric current (the movement of electrons) is forced through the molten alumina and aluminum metal is produced at the cathode of the cell. Thereafter, the aluminum metal is rolled and shaped into the can. It is subsequently lined with a thin plastic liner before the carbonated soft drink is added and the can is sealed. The passage of an electric current through molten salts or aqueous solutions is very important to the economy of any industrialized nation. Now you can take a drink . . . maybe you can appreciate it a little more!

OBJECTIVES

- To identify the reactions occurring at the anode and cathode in an electrolytic cell
- To determine the amount of charge passed through an electrolytic cell

PRINCIPLES

Electrolysis: the process of using energy to force a nonspontaneous oxidation-reduction reaction to occur.

Cathode: electrode at which reduction occurs in a cell.

Anode: electrode at which oxidation occurs in a cell.

An oxidation-reduction reaction that requires energy occurs in an **electrolytic cell**. Electrical energy is used to direct a flow of electrons through a chemical system causing an otherwise nonspontaneous reaction to occur. For example, most active metals, such as Na and Al, are prepared by the electrolysis of their molten salts—the directed current causes the reduction of the Na^+ or Al^{3+} at the cathode and the oxidation of its anion at the anode. The products are isolated at the electrodes as they form.

A direct current (D.C.) source is used to electrolyze molten NaCl; in the electrolysis Cl^- ions are oxidized at the anode and Na^+ ions are reduced at the cathode.

oxidation, anode: $\qquad 2\,Cl^-(l) \rightarrow Cl_2(g) + 2\,e^-$

reduction, cathode: $\qquad Na^+(l) + e^- \rightarrow Na(l)$

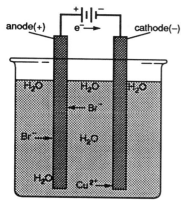

In the electrolysis of an aqueous $CuBr_2$ solution, the Cu^{2+} ions are reduced at the cathode and Br^- ions are oxidized at the anode.

reduction, cathode: $\qquad Cu^{2+}(aq) + 2\,e^- \rightarrow Cu(s)$

oxidation, anode: $\qquad 2\,Br^-(aq) \rightarrow Br_2(aq) + 2\,e^-$

But for electrolysis reactions in aqueous solutions, we must also consider the decomposition of H_2O at the cathode and the anode.

reduction, cathode: $\qquad 2\,H_2O(l) + 2\,e^- \rightarrow H_2(g) + 2\,OH^-(aq)$

oxidation, anode: $\qquad 2\,H_2O(l) \rightarrow O_2(g) + 4\,H^+(aq) + 4\,e^-$

If reduction of H_2O occurs at the cathode, H_2 gas is evolved and the solution near the cathode becomes basic due to the formation of OH^-. When oxidation of H_2O occurs at the anode, O_2 gas is evolved and the solution near the anode becomes acidic because of the formation of H^+. The production of H^+ and/or OH^- can be detected using litmus paper or any acid-base indicator with an endpoint near 7.

When there are two competing reactions at the cathode, the reaction having the greatest natural tendency to undergo reduction[1] is the one that occurs. Conversely, for two competing reactions at the *anode*, the reaction having the greatest natural tendency to undergo oxidation[2] (or the lowest probability to undergo reduction) is the one that occurs. An analysis of these competing reactions is undertaken in Parts A and B of this experiment.

Movement of Charge Through an Electrolytic Cell

Electric current moves by two different modes in an electrolytic cell: ions carry charge in the electrolyte and electrons carry charge in the external wire that connects the electrodes. The quantity of charge passing through the cell can be measured through either mode.

Ampere: unit of electrical current equal to one coulomb per second.

In the external wire, the quantity of charge (measured in **coulombs, C**) is determined by measuring the current (measured in **amperes, C/s**) for a time period (measured in seconds, s): $C = (C/s) \cdot s$.

In the electrolyte, the quantity of charge is determined by measuring a mass loss at the anode (an oxidation of the metal anode) or a mass gain at the cathode (a deposition of a metal ion from solution onto the cathode). For example, when Zn is the anode and it oxidizes, its mass loss is proportional to the charge that passes through the cell—one mole of Zn is oxidized when 2 moles of electrons (or 2 faradays) pass, causing a mass loss of 65.38 g Zn from the anode.

$$Zn(anode) \rightarrow Zn^{2+}(aq) + 2\,e^-$$

Since 1 mole of electrons has a charge of 96 500 C, the charge passing through the cell for the oxidation of 1 mole of Zn is 2(96 500 C) or 1.93×10^5 C.

Half-reaction: the oxidation or reduction reaction that occurs in a redox reaction.

In this experiment you will electrolyze a number of solutions with a D.C. power source. A determination of which of the competing reactions occurs at each electrode is obtained from a test or observation of the products formed in the electrolysis. In addition, the quantity of charge that passes through a cell over a set time period is determined by measuring both the electron movement and the mass change of an electrode.

TECHNIQUES

- Technique 5, page 3
- Technique 11, page 10
- Technique 13, page 11

Transferring Liquid Reagents
Testing with Litmus
Using the Laboratory Balance

---

[1] A reaction that has a high probability to undergo reduction is said to have a high **reduction potential**. Reduction potentials are often used to compare the probability for half-reactions to occur in an electrolysis process.

[2] A reaction with a high probability for oxidation has a high oxidation potential but a low reduction potential.

PROCEDURE

A. Electrolysis of Salt Solutions, Carbon Electrodes

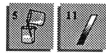

1. Dissolve about 3 g of NaCl in 150 mL of distilled (or deionized) water and transfer it to the U-tube (Figure 31.1; this apparatus can be constructed with a U-shaped bent piece of glass tubing and graphite "lead" from a pencil.). Place two carbon electrodes in the solution and connect them to a D. C. power source.[3] Identify its cathode (the negative terminal) and the anode (the positive terminal).

2. Electrolyze the solution for 5 minutes. During the electrolysis, observe any evidence of a reaction occurring in the anode and cathode chambers:

   • Does the pH of the solution change at the electrodes? Test the solution at each electrode with litmus paper. Compare the color with a similar test on the original salt solution.
   • Is a gas evolved at one of the electrodes? Which one?
   • Does a metal deposit on the cathode? Is there any discoloration of the cathode and/or anode?

3. On the basis of your observations, write balanced half-reactions for the cathode and anode reactions.

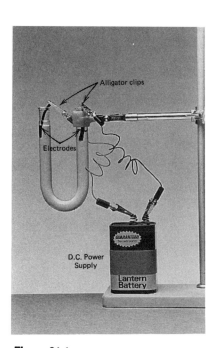

Figure 31.1
An electrolysis apparatus

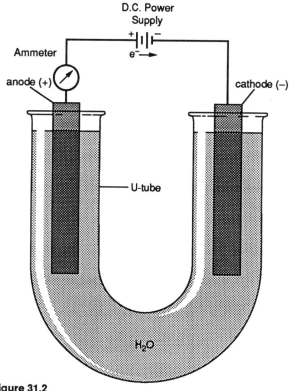

Figure 31.2
Apparatus for the electrolysis of CuSO₄

4. Substitute $CuSO_4$, $Pb(NO_3)_2$, $ZnCl_2$, KI, NaBr, and other salts suggested by your laboratory instructor for NaCl and repeat. Thoroughly clean the U-tube and electrodes after each electrolysis.

---

[3]The most convenient D. C. power source is a 9V transistor battery.

B. Electrolysis of a CuSO₄ Solution, Copper Electrodes

1. Fill the U-tube three-fourths full with 0.1 M $CuSO_4$. Clean, with steel wool, two copper electrodes (copper wire is satisfactory), connect them to the D. C. energy source, insert them into the two arms of the U-tube and electrolyze the solution for 5 minutes. Observe closely the cathode and anode chambers to detect any reactions and test with litmus; write balanced half-reactions that are consistent with your observations.[4]

C. Determination of the Quantity of Charge Passed Through the Cell

1. Set up the apparatus in Figure 31.2. The D. C. power source must provide 3–5V[5] and the ammeter should range should be from 0.2–1.0 amperes.

2. a. Polish a strip of Cu metal with steel wool, momentarily dip it into a 6 M HNO_3 solution to remove any oxide coating, rinse with distilled (or deionized) water and acetone, air-dry, and measure its mass (± 0.0001 g).[6] Use this Cu metal for the anode (the positive terminal). Similarly prepare a Cu metal strip as the cathode. Three-fourths fill the U-tube with 0.1 M $CuSO_4$ to a level just below the top of the Cu electrodes.

   b. Electrolyze the solution for exactly 15 minutes. Record the average current from the ammeter during the 15 minute period

   c. Calculate the coulombs of charge that passed through the wire.

3. *Carefully* remove the Cu anode from the solution, rinse with acetone, air-dry, and again determine its mass. Determine the mass loss and calculate the coulombs of charge that passed through the solution.

Disposal Information: Dispose of all of the test solutions in the "Waste Salts" container.

---

[4]This electrolysis procedure is similar to that used for the industrial electrolytic refining of copper.

[5]A 3-5V direct current power source can be two flashlight batteries, a lantern battery, or a transistor battery.

[6]A balance with 0.1 mg precision is best; however, if only ±0.001 g sensitivity is available, extend the time for electrolysis to 30 min.

THE ELECTROLYTIC CELL

Date _____ Name _____ Lab Sec. _____ Desk No. _____

1. In an electrolytic cell

    a. What is the sign of the anode? _____

    b. What is the sign of the cathode? _____

    c. What type of reaction that occurs at an anode? _____

    d. What type of reaction that occurs at a cathode? _____

    e. Toward which electrode do cations migrate? _____

    f. Toward which electrode do anions migrate? _____

    g. Electrons migrate from the _____ to the _____ .

2. In the electrolysis of an aqueous solution, water can be oxidized and/or reduced.

    a. Write the equation for the electrolysis of water at the anode.

    b. Write the equation for the electrolysis of water at the cathode.

    c. If water is electrolyzed at the cathode, what will be the color of the litmus paper in a test?

3. A current of 0.5 A is passed through an electrolytic cell for 15 min.

   a. How many coulombs of charge passed through the cell?

   b. If this charge deposits zinc metal at the cathode, how many grams of Zn will electroplate (1 mole of electrons = 96 500 C)?

THE ELECTROLYTIC CELL

Date _____ Name _____ Lab Sec. _____ Desk No. _____

A. Electrolysis of Salt Solutions, Carbon Electrodes

Observations, Anode Chamber (+)

Salt	Litmus Test	Evidence of Reaction	Equation for the Reaction
NaCl			
$CuSO_4$			
$Pb(NO_3)_2$			
$ZnCl_2$			
KI			
NaBr			

Observations, Cathode Chamber (-)

Salt	Litmus Test	Evidence of Reaction	Equation for the Reaction
NaCl			
$CuSO_4$			
$Pb(NO_3)_2$			
$ZnCl_2$			
KI			
NaBr			

B. Electrolysis of a $CuSO_4$ Solution, Copper Electrodes

Experimental observation at	Equation
Anode	
Cathode	
Cell reaction	

C. Determination of the Quantity of Charge Passed Through the Cell

Time (*min*)	Current (*A*)	Time (*min*)	Current (*A*)
1		10	
2		11	
3		12	
4		13	
5		14	
6		15	
7		16	
8		17	
9		18	

Electrical Measurement

1. Average current (*amperes* = C/s) ....

2. Time for electrolysis (*s*) ....

3. Coulombs passed (C) ....

Chemical Measurement

1. Mass of polished Cu anode before electrolysis(*g*) ....

2. Mass of Cu anode after electrolysis (*g*) ....

3. Moles of Cu oxidized (*mol*)* ....

4. Moles of e⁻ passed (*faradays*)* ....

5. Coulombs passed (C) ....

*Show calculations of mol Cu and mol e⁻

Compare the chemical and electrical measurements of charge passed through the electrolytic cell. Comment on any differences.

1. a. From the chemical measurement in Part C, how many (total) electrons were generated at the anode in 15 minutes?

   b. From the electrical measurement in Part C, and the answer in **a**, calculate the charge of an electron.

2. a. If nickel electrodes are used instead of carbon electrodes in Part A, could the reaction at the anode have changed? Explain.

   b. If nickel electrodes are used instead of carbon electrodes in Part A, could the reaction at the cathode have changed? Explain.

3. a. In the electrolysis of the KI solution, could a low pH affect the product(s) that form at the anode? Explain.

   b. In the electrolysis of the KI solution, how might a low pH affect the product(s) that form at the cathode? Explain.

EXPERIMENT 32

THE GALVANIC CELL

Over one billion batteries are manufactured in the United States each year, from the "cheap" throw aways used in toys, flashlights, calculators, and automatic-focusing cameras to the rechargeable batteries used in power tools and tooth brushes. The flashlight battery revolutionized the toy industry in the mid 1960s; its technology promises to make even greater changes in the future, especially in the development of rechargeable batteries for personal transportation. All batteries are constructed for the same purpose, that is to provide a convenient and portable source of energy. The energy is provided by a movement of electrons generated from a spontaneous redox reaction.

OBJECTIVES

- To measure the reduction potentials for several redox couples
- To follow the movement of electrons, cations, and anions in a galvanic cell

PRINCIPLES

Redox: abbreviation for oxidation-reduction.

The transfer of electrons from one substance to another occurs in a redox reaction. In a redox reaction, oxidation cannot occur unless reduction also occurs—a substance cannot lose electrons unless another substance accepts the electrons. The substance that loses electrons is oxidized (its oxidation number increases); the substance that gains electrons is reduced (its oxidation number decreases). In the reaction

$$Zn(s) + 2\ Ag^+(aq) \rightarrow 2\ Ag(s) + Zn^{2+}(aq)$$

The oxidation and reduction parts of the overall reaction can be divided into two half-reactions, where each half-reaction is called a **redox couple**.

oxidation half-reaction $Zn \rightarrow Zn^{2+} + 2\ e^-$ Zn^{2+}/Zn redox couple
reduction half-reaction $2\ Ag^+ + 2\ e^- \rightarrow 2\ Ag$ Ag^+/Ag redox couple

Zn metal is oxidized and Ag^+ ions are reduced. When Zn loses electrons, the Ag^+ is reduced; hence, Zn is called the **reducing agent**. Conversely, when Ag^+ gains electrons, the Zn metal is oxidized and Ag^+ is the **oxidizing agent**.

In a galvanic cell aqueous solutions of the oxidation half-reaction (the reducing agent) and reduction half-reaction (the oxidizing agent) are placed in separate compartments called **half-cells**. The two half-cells are connected by a salt bridge which allows for the movement of ions from one half-cell to another.[1] Each half-cell has an electrode; the two electrodes are connected with a wire through a voltmeter. Each electrode must be a part of the half-cell and serve to conduct electrons away from the reducing agent (substance being oxidized) through the wire to the oxidizing agent (substance being reduced). The electrode at which oxidation occurs is called

---

[1]A porous barrier which also prevents the free mixing of the two redox couples can be substituted for a salt bridge. A commonly used porous barrier is an unglazed porcelain cup.

the **anode**, the (–) electrode; reduction occurs at the **cathode**, the (+) electrode.

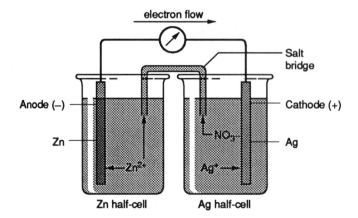

Figure 32.1

Schematic diagram of a galvanic cell

Let's see what happens in a galvanic cell consisting of the Zn^{2+}/Zn and Ag^+/Ag redox couples (Figure 32.1): Zn oxidizes to Zn^{2+}; electrons are deposited on the Zn anode and migrate through the wire to the Ag cathode; the Ag^+ in solution accepts the electrons at the Ag cathode to form Ag metal. In the meantime, the Zn^{2+} migrates away from the Zn anode and through the salt bridge—this occurs because as the positive charge of Ag^+ is being removed from the Ag half-cell the Zn^{2+} diffuses in and replaces the positive charge. Conversely the anion of the Ag^+, such as NO_3^-, can move into the Zn half-cell and out of the Ag half-cell. Both ion movements serve to maintain electrical neutrality in the galvanic cell.

Different metals, such as Zn and Ag, have different tendencies (or potentials) to oxidize; their cations have different tendencies (or potentials) to reduce. We can consider the Zn^{2+} as having a potential (call it $\mathcal{E}_{(Zn^{2+}/Zn)}$) to exist in its reduced state, Zn, and also Ag^+ as having a potential (call it $\mathcal{E}_{(Ag^+/Ag)}$) to exist in its reduced state, Ag.

The ion having the greatest potential to be in the reduced state is the one that undergoes reduction when the two redox couples form a galvanic cell; oxidation must then occur in the other redox couple. The difference in these two reduction potentials is measured as the **cell potential**, $\mathcal{E}_{cell}$, for the galvanic cell. Since Ag^+ has the higher potential for reduction, the positive cell potential is written as

$$\mathcal{E}_{cell} = \mathcal{E}_{(Ag^+/Ag)} - \mathcal{E}_{(Zn^{2+}/Zn)}$$

The measurement of the cell potential is done with an instrument called a potentiometer (or voltmeter). The standard reduction potential (molar concentrations of 1 mol/L for all ions and 1 atm pressure for all gases), $\mathcal{E}^\circ$, for the Zn^{2+}(1 mol/L)/Zn redox couple is -0.76V and that for the Ag^+(1 mol/L)/Ag redox couple is +0.80V. A voltmeter connected to the electrodes would show the difference between these potentials.

$$\mathcal{E}^\circ_{cell} = 0.80V - (-0.76V) = 1.56V$$

Deviations from this potential may be due to surface activity at the electrodes, activity of the ions in solution, and/or current drawn by a voltmeter.

The cell potential, $\mathcal{E}_{cell}$, for a galvanic cell measured at ionic concentrations other than 1 mol/L is related to the standard cell potential, $\mathcal{E}^\circ_{cell}$, by the Nernst equation.

$$\mathcal{E}_{cell} = \mathcal{E}^\circ_{cell} - \frac{0.0592}{n} \log Q$$

Q is the reaction quotient for the cell reaction and n is the moles of electrons transferred in the cell reaction. For the Zn-Ag galvanic cell (where n = 2) the Nernst equation becomes

$$\mathcal{E}_{cell} = 1.56V - \frac{0.0592}{2} \log \frac{[Zn^{2+}]}{[Ag^+]^2}$$

Notice that the concentrations of Zn and Ag are omitted since they, as solids, maintain a constant concentration.

The $\mathcal{E}_{cell}$ of a selected number of redox couples are measured in the experiment; from the data the redox couples will be listed in order of their relative reduction potentials.

In addition the Nernst equation is used to determine an unknown Cu^{2+} concentration using the Zn^{2+}/Zn redox couple as a reference potential. The cell reaction is

$$Cu^{2+}(aq) + Zn(s) \rightarrow Cu(s) + Zn^{2+}(aq)$$

The corresponding Nernst equation is

$$\mathcal{E}_{cell} = 1.10V - \frac{0.0592}{2} \log \frac{[Zn^{2+}]}{[Cu^{2+}]}$$

If 0.1 M Zn^{2+} is used for the measurement the Nernst equation becomes,

$$\mathcal{E}_{cell} = 1.10V - \frac{0.0592}{2} \log [0.1] + \frac{0.0592}{2} \log [Cu^{2+}] \text{ or}$$
$$\mathcal{E}_{cell} = 1.13V + 0.0296 \log [Cu^{2+}]$$

Therefore, $\mathcal{E}_{cell}$ is directly proportional to log $[Cu^{2+}]$; cell potentials can therefore be used to determine the concentrations of ions in a solution. This is the basic principle used for pH determinations using a pH meter.

TECHNIQUES

• Technique 3, page 2	Microscale Analyses
• Technique 8, page 7	Reading a Meniscus
• Technique 9, page 7	Pipetting a Liquid or Solution

PROCEDURE

A. Reduction Potentials of Selected Half-Reactions

1. Set up a 2 x 2 array of microcells in a 24-cell plate and transfer about 2 mL of 0.1 M $Zn(NO_3)_2$, 0.1 M $Pb(NO_3)_2$, 0.1 M $Cu(NO_3)_2$, and 0.1 M $FeSO_4$ to the microcells. Place a polished[2] strip of each metal, Zn, Pb, Cu, and Fe, in the respective cells. Prepare several strips of filter paper, cut to about 1" x 1", fold lengthwise, and wet the strips with a sodium sulfate solution from a dropper bottle.

2. Measure the potential of the Zn-Cu cell.
   a. Connect the two microcells with a wet filter paper strip; connect one electrode to the positive terminal of the voltmeter and the other electrode to the negative terminal. If the needle shows a positive deflection, the connections are correct; if not, reverse the connections to the electrodes.

   b. Read and record the E_{cell}.[3] Which electrode is connected to the positive (+) terminal (cathode)? Does oxidation occur at the Zn or Cu electrode? Write an equation for the reaction at each electrode.

3. Repeat the measurements for all possible combinations of cells from the remaining half-cells, using a freshly prepared filter strip for the salt bridge each time in connecting the half-cells.

4. Assuming the reduction potential of the Zn^{2+}(0.1 M)/Zn redox couple is -0.79V, calculate the reduction potentials of all other redox couples.[4]

5. Other redox couples: Determine the reduction potential for the following redox couples designated by your instructor. Polish each electrode.

Ag^+/Ag	(0.1 M $AgNO_3$ solution)
Ni^{2+}/Ni	(0.1 M $Ni(NO_3)_2$ solution)
Mg^{2+}/Mg	(0.1 M $MgSO_4$ solution)
M^{n+}/M (unknown)	(0.1 M solution of M^{n+})

Disposal Information:	Rinse the microcells with deionized water and discard in the "Waste Salts" container.

B. Measuring an Unknown Concentration

"The mark": the etched line that calibrates the volume of a volumetric flask.

1. Obtain a 1 mL pipet and three 100 mL volumetric flasks. To prepare solution A, pipet 1 mL of 1 M $CuSO_4$ solution into a 100 mL volumetric flask and dilute to the mark with distilled (or deionized) water. Prepare solution B by pipetting 1 mL of A into a second 100 mL volumetric flask and diluting to the mark. Similarly, prepare solution C. Calculate the $[Cu^{2+}]$ in solutions A, B, and C. Use the equation, $E_{cell} = 1.13V + 0.0296 \log [Cu^{2+}]$, to calculate the theoretical E_{cell} for the 1 M $CuSO_4$ and solutions A, B, and C.

---

[2]Polish each metal strip with steel wool.

[3]A positive E_{cell} indicates that the redox couple at the cathode, the one connected to the (+) terminal of the voltmeter, has the higher reduction potential by the voltage read on the voltmeter.

[4]These are not standard reduction potentials because 1 M solutions at 25°C were not used for the measurements.

Share this set of solutions with other students in the laboratory.

1 M CuSO$_4$	–1 mL→	dilute to 100 mL A	–1 mL→	dilute to 100 mL B	–1 mL→	dilute to 100 mL C

2. Set up a 2 x 2 array of microcells. Place a Zn electrode in a clean microcell that is three-fourths filled with 0.1 M Zn(NO$_3$)$_2$. Connect a polished Zn electrode to the negative (–) terminal of the voltmeter. Similarly fill an adjacent microcell with solution A; join the microcells with a freshly prepared (wetted with a sodium sulfate solution) strip of filter paper. Connect a polished Cu electrode to the positive (+) terminal and record the $\mathcal{E}_{cell}$.

3. Repeat Part B.2 to record the $\mathcal{E}_{cell}$ for solutions B and C relative to the Zn^{2+}(0.1 M)/Zn redox couple. Be sure to use a new filter paper strip for each measurement.

4. Obtain a solution with an unknown [Cu^{2+}] from your instructor. As before, determine the $\mathcal{E}_{cell}$. From an analysis of your data in Part B, estimate the [Cu^{2+}] in your unknown.

C. Concentration Cell, Cu^{2+}(0.01 mol/L)/Cu and Cu^{2+} (1 x 10^{-6} mol/L)/Cu Redox Couples

1. Join the adjacent microcells of solution A and C with a new wet strip of filter paper. Connect the electrodes to the voltmeter. Determine the $\mathcal{E}_{cell}$, the anode reaction, and the cathode reaction. Explain *why* a potential is recorded. Write an equation for the reaction at each electrode.

2. Add about 20 drops of 6 M NH$_3$ (**Caution:** *avoid inhalation*) to solution C. Observe any changes in $\mathcal{E}_{cell}$. Explain.

Disposal Information: Rinse the microcells with deionized water and discard in the "Waste Salts" container.

Activity Series for Hydrogen and Some Metals

Element	Reduced State	Oxidized State
Cesium	Cs	Cs^+
Rubidium	Rb	Rb^+
Potassium	K	K^+
Barium	Ba	Ba^{2+}
Strontium	Sr	Sr^{2+}
Calcium	Ca	Ca^{2+}
Sodium	Na	Na^+
Magnesium	Mg	Mg^{2+}
Aluminum	Al	Al^{3+}
Manganese	Mn	Mn^{2+}
Zinc	Zn	Zn^{2+}
Chromium	Cr	Cr^{3+}
Iron	Fe	Fe^{2+}
Cadmium	Cd	Cd^{2+}
Cobalt	Co	Co^{2+}
Tin	Sn	Sn^{2+}
Lead	Pb	Pb^{2+}
Hydrogen	H_2	H^+
Copper	Cu	Cu^{2+}
Silver	Ag	Ag^+
Mercury	Hg	Hg^{2+}
Gold	Au	Au^{3+}

THE GALVANIC CELL-LAB PREVIEW

Date _____ Name _____ Lab Sec. _____ Desk No. _____

1. State the purpose of the wet filter paper strip in today's experiment.

2. What is the sign of the anode in a galvanic cell?_____

   What electrochemical process occurs at the anode?_____

Redox Couples		$\mathcal{E}^\circ$
$Au^{3+} + 3\,e^- \rightleftharpoons$	Au	+1.42 V
$Fe^{3+} + e^- \rightleftharpoons$	Fe^{2+}	+0.77 V
$Sn^{4+} + 2\,e^- \rightleftharpoons$	Sn^{2+}	+0.15 V
$Cr^{3+} + 3\,e^- \rightleftharpoons$	Cr	-0.74 V

3. Consider a galvanic cell consisting of the Au^{3+}(1 mol/L)/Au and Sn^{4+}(1 mol/L)/Sn^{2+}(1 mol/L) redox couples.

   a. Write the reduction half-reaction and corresponding $\mathcal{E}^\circ$ for each redox couple.

   b. Which redox couple undergoes reduction?_____

   c. Write an equation for the reaction that occurs at the

      anode_____

      cathode_____

   d. Write the equation for the cell reaction.

   e. What is the cell potential, $\mathcal{E}^\circ_{cell}$, for the cell?

4. Determine the cell potential, $\mathcal{E}^{\circ}_{cell}$, of a galvanic cell consisting of the Sn^{4+}(1 mol/L)/Sn^{2+}(1 mol/L) and Cr^{3+}(1 mol/L)/Cr redox couples.

5. Use the Nernst equation to determine the cell potential, $\mathcal{E}_{cell}$, of the galvanic cell consisting of the following redox couples.

$$Sn^{4+}(10^{-3}\ mol/L) + 2\ e^{-} \rightleftharpoons Sn^{2+}(10^{-5}\ mol/L)$$
$$Fe^{3+}(1\ mol/L) + e^{-} \rightleftharpoons Fe^{2+}(10^{-5}\ mol/L)$$

THE GALVANIC CELL-LAB PREVIEW

Date _____ Name _____ Lab Sec. _____ Desk No. _____

A. Reduction Potentials of Selected Half-Reactions

Cell	$\mathcal{E}_{cell}$	anode	anode reaction	cathode	cathode reaction
Zn-Cu					
Cu-Pb					
Zn-Pb					
Fe-Pb					
Zn-Fe					
other					
unknown					

1. Compare the sum of the Zn-Pb + Cu-Pb cell potentials with the Zn-Cu cell potential. Explain.

2. Compare the sum of the Zn-Fe + Fe-Pb cell potentials with the Zn-Pb cell potential. Explain.

3. Arrange the redox couples in order of *decreasing* reduction potentials, the most positive first. List the reduction potential for each redox couple relative to that of the Zn^{2+}(0.1 mol/L)/Zn couple which is -0.79 V.

Redox Couple	Reduction Potential

B. Measuring an Unknown Concentration

Solution	Molar Concentration	log [Cu^{2+}]	$\mathcal{E}_{cell}$(calc)	$\mathcal{E}_{cell}$(exp'tl)
1 M CuSO$_4$	1.0	0		
A				
B				
C				

2. Compare $\mathcal{E}_{cell}$ (calc) to $\mathcal{E}_{cell}$ (exp'tl). Are differences consistent? Is the trend of cell potentials as predicted? Explain.

3. $\mathcal{E}_{cell}$ for the unknown galvanic cell: Zn^{2+}(0.1 mol/L)/Zn–Cu^{2+}(? mol/L)/Cu._____

    [Cu^{2+}] in solution of unknown concentration._____

C. Concentration Cell, Cu^{2+}(0.01 mol/L)/Cu and Cu^{2+} (1 x 10^{-6} mol/L)/Cu Redox Couples

1. $\mathcal{E}_{cell}$ _____

    anode reaction (oxidation)_____

    cathode reaction (reduction)_____

    Why is a potential recorded?_____

2. $\mathcal{E}_{cell}$ _____
    Account for the $\Delta\mathcal{E}_{cell}$ from Part B.1

1. Identify the oxidizing and reducing agent of each cell in Part A.

Cell	Oxidizing Agent	Reducing Agent
Zn-Cu		
Cu-Pb		
Zn-Pb		
Fe-Pb		
Zn-Fe		

2. List two factors accounting for the experimental cell potential *not* being equal to the theoretical cell potential.

3. If in Part B, the $Zn(NO_3)_2$ solutions were successively diluted instead of the $CuSO_4$ solutions, how would the relative cell potentials have differed?

EXPERIMENT 33
REACTION RATES

What is the secret to quickly building a roaring fire in a fireplace on a cold, romantic evening? Lets see, we need some wood, preferably some small pieces (kindling) and a few larger pieces, a match, some starter fluid (nice but not really necessary), and some soft music (you decide if that is really necessary!). The match head has a few chemicals that quickly become thermally activated with friction causing the ignition of the match (paper or wood). The kindling (old newspaper also works) has a large surface area exposed to the oxygen in air, thereby making it easy to reach its ignition temperature from the heat of the match. Fanning the burning kindling causes it to burn faster and therefore generate more heat; a larger piece of wood, now at a higher temperature and with a plentiful supply of air also ignites. Opening the draft now causes the larger and larger pieces to catch fire and before you know it, BINGO, you have a nice roaring fire! Now, the rest of the evening is not a part of this course or manual!

OBJECTIVE

- To identify factors that affect the rate of a chemical reaction

PRINCIPLES

A number of conditions can alter the rate of a chemical reaction. Experimental parameters that we, as chemists, can quickly change include temperature, pressure (if gases), and reactant concentrations. In addition the substitution of a more reactive reactant, the subdividing of a reactant (if a solid or liquid), or the inclusion of a catalyst can increase a reaction rate. Each of these factors are observed and studied in this experiment.

Temperature Effects

A temperature increase generally increases reaction rate—a rule of thumb is that a 10°C temperature rise doubles the rate. Because of a temperature increase, the average kinetic energy of reactant molecules (or ions) also increases[1]; when the molecules (or ions) collide, this added kinetic energy becomes a part of the collision system. The distribution of the additional energy among the atoms in the colliding molecules increases the probability of bonds being broken; subsequent recombination of the fragments can result in the formation of product.

Concentration Effects

Quantitative: measurements that are as exact as the chemical apparatus will allow.

An increase in the concentration of a reactant increases the probability of collision between the reactant molecules (or ions). As a result, most reaction rates increase, although there are systems in which a decrease or no change in reaction rate occurs. Experiment 34 focuses on a *quantitative* investigation of the effect of concentration changes on reaction rate.

---

[1]The kinetic energy-temperature relationship is expressed by the equation
$$K.E. = (1/2) mv^2 = (3/2) kT$$
k is the called the Boltzman constant and T is the temperature in kelvins.

Nature of Reactants

Some substances are naturally more reactive than others, and, therefore undergo more rapid chemical changes. To manufacture an automobile that does not rust, alternative construction materials, such as stainless steel or fiberglass can be used. The reaction between water and sodium is very rapid and exothermic, whereas with copper, no reaction occurs. Hydrogen and fluorine react explosively at room temperature; hydrogen and iodine react very slowly, even at elevated temperatures. Therefore, the nature of the reactants chosen for a particular process can affect the reaction rate.

Particle Size of Reactants

The smaller particle size of the reactant, the greater the reaction rate. A large piece of wood in the fireplace burns much more slowly than the kindling. Coal mine explosions and grain elevator explosions result from the very finely divided (large surface area) dust particles being exposed to the oxygen of the air—a spark ignites the very rapid reaction.

Presence of a Catalyst

A catalyst increases the reaction rate without undergoing any *net* change in its properties for the reaction. Catalysts generally reroute the pathway (or mechanism) of the reaction in such a way that the reaction requires less energy and therefore proceeds more rapidly. The many biochemical reactions that occur in the body are catalyzed by specific proteins called enzymes.

TECHNIQUES

- Technique 3, page 2   Microscale Analyses
- Technique 6d, page 4   Heating Liquids
- Technique 8, page 7   Reading a Meniscus
- Technique 9, page 7   Pipetting a Liquid or Solution
- Technique 13, page 11   Using the Laboratory Balance
- Technique 17b, page 16   Handling Gases
- Technique 18, page 18   Graphing Techniques

PROCEDURE

Each portion of the experiment requires that you time the reaction. The time lapse should be recorded in seconds; therefore, before you begin this experiment, make sure you have available a clock or watch that can be read to the nearest second.

A. Temperature Effects

The oxidation-reduction reaction that occurs between HCl and $Na_2S_2O_3$ produces insoluble sulfur as a product. The time required for the cloudiness from the formation of sulfur is a measure of the reaction rate.

$$2\,HCl(aq) + Na_2S_2O_3(aq) \rightarrow S(s) + SO_2(g) + 2\,NaCl(aq) + H_2O(l)$$

Measure each volume with a pipet. Work in pairs to best complete this part of the experiment.

1. Pipet 5.0 mL of 0.1 M $Na_2S_2O_3$ into one set of three (clean) 150 mm test tubes. Into a second set of three 150 mm test tubes, pipet 5.0 mL of 0.1 M HCl. Do not intermix the pipets. Place a $Na_2S_2O_3$/HCl pair of test tubes in a salt/ice-water bath until thermal equilibrium is established (about 5–6 minutes). You may now want to proceed to Part A.2.

   O.K., get ready to time for the appearance of the sulfur cloudiness with a watch that reads to the nearest second. As one student pours the solutions together, the other notes the time . . . transfer one solution to the other and agitate the mixture vigorously for several seconds. Return the reaction mixture to the ice bath and record the time for the sulfur to appear. Record the temperature (±0.1°C) of the mixture.

2. Place a second pair of $Na_2S_2O_3$/HCl test tubes in a hot water bath adjusted to a temperature of about 70°C. Repeat the mixing of the two solutions as in Part A.1 and record the time for the sulfur cloudiness to appear. You can predict the relative time even before the mixing. Record the actual temperature of the water bath.

3. Combine and time the remaining pair of $Na_2S_2O_3$/HCl solutions at room temperature.

4. Plot the temperature (°C) *vs* time (seconds) for the appearance of sulfur on linear graph paper. Have your instructor approve your graph.

B. Concentration Effects

1. Set up a 1 x 3 array of microcells in a 24-cell plate. Pipet 2.0 mL of 1 M HCl, 3 M HCl, and 6 M HCl (**Caution:** HCl *causes a skin irritation*) into the respective cells. Polish a 30 mm strip of Mg metal, tear into three equal lengths, and determine the mass (±0.001 g) of each.

2. Add a Mg strip (be sure you have recorded its mass) to the 1 M HCl and start timing (in seconds). When all traces of Mg ribbon disappear, stop timing. Record the elapsed time on the Data Sheet. Repeat the experiment with the 3 M HCl and 6 M HCl solutions.

3. Plot (mol HCl/mol Mg) *vs* time (seconds) on linear graph paper. Have the instructor approve your graph.

C. Nature of Reactants

1. In a 1 x 4 array of microcells place about 2 mL of 3 M H_2SO_4, 6 M HCl, 6 M HNO_3, and 6 M H_3PO_4 (**Caution:** *carefully handle acids*). Place a strip of polished Mg ribbon into each microcell and note the relative reaction rates. Record your observations.

2. Place about 1 mL of 6 M HCl into a 1 x 3 array of microcells. Add a polished 10 mm strip of Zn to the first, one of Fe to the second and another of Cu to the third. Note the relative reaction rates and record your observations.

D. Particle Size of Reactants

1. Place a small piece of chalk, $CaCO_3$, in a microcell and add several drops of 6 M HCl. Place some chalk dust in an adjacent microcell, and repeat the addition. Account for your observations.

Disposal Information:	Dispose of the test chemicals from Parts A through D in the "Waste Acids" container.

E. Presence of a Catalyst

The thermal decomposition of potassium chlorate generates small quantities of oxygen gas. Several tests of the chemical reactivity of oxygen are included in this experiment.

Caution: *This experiment can be dangerous if not performed properly. Before you begin, read the procedure carefully and adhere to the following guidelines:*

• Wear safety glasses
• Use a clean 200 mm Pyrex test tube—there should be no evidence of a black residue. Do *not* dry the inside of the test tube with a paper towel.
• Your instructor must approve your apparatus before you begin.

1. Set up the apparatus in Figure 33.1. Fill a 50 mL graduated cylinder with water in the pneumatic trough and invert it over the gas collecting port. No air bubbles should be present in the cylinder. Into a clean, dry 200 mm Pyrex test tube, place about 0.5 g of $KClO_3$. Gently heat the $KClO_3$ and record the time (in seconds) required to collect 20 mL of O_2 gas.

$$2\ KClO_3(s)\ \rightarrow\ 2\ KCl(s)\ +\ 3\ O_2(g)$$

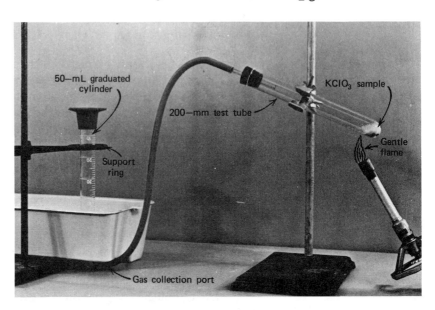

Figure 33.1
Apparatus for the collection of $O_2(g)$

2. Before removing the heat from the test tube, first disconnect the gas delivery tube from the test tube. **Caution:** *it is very important that the delivery tube be disconnected while the tube is hot; if it isn't, cool water will be drawn from the trough into the test tube and may cause it to break.* Then, allow the test tube and its contents to cool.

3. Repeat Parts E.1 and E.2, but in addition, add a pinch of MnO_2 to the 0.5 g $KClO_3$. Record the time required to collect 20 mL of O_2 gas.

4. Replace the inverted graduated cylinder with a gas collecting bottle and continue heating. Use additional gas collecting bottles until no further oxygen is evolved; then follow the cool-down procedure described in Part E.2.

5. Ignite a wood splint, put out the flame so that only a glow remains. Place the glowing splint in one of the flasks.

Disposal Information: **Dispose of the chemicals from Part E in the "Waste Oxidants" container.**

EXPERIMENT 33

REACTION RATES

Date _____ Name _____ Lab Sec. _____ Desk No. _____

1. If the rate of a chemical reaction doubles for every 10°C temperature increase, by what factor will a chemical reaction increase if the temperature is increased over a 30°C range?

2. Explain why coal dust burns more rapidly that larger pieces of coal?

3. What volume (in mL) of oxygen gas (at STP) can be produced from the thermal decomposition of 0.50 g $KClO_3$ See Part E.

4. When hydrogen peroxide, H_2O_2, as an antiseptic, is placed on an open wound, bubbles form because of the reaction

$$2\,H_2O_2(aq) \rightarrow 2\,H_2O(l) \,+\, O_2(g)$$

However, H_2O_2 only slowly decomposes in the bottle. How does the wound accelerate its decomposition?

5. Wood burns more rapidly in a fireplace that has a good draft of air. Explain.

6. List factors that can affect the rate of ammonia production by the Haber process.

$$N_2(g) \ + \ 3\,H_2(g) \rightarrow 2\,NH_3(g) \ + \ 46.1\;kJ$$

7. What is Technique 6d? How will you need to modify the technique to adapt it to this experiment?

8. Describe how Technique 17b is to be used in this experiment.

REACTION RATES

Date _____ Name _____ Lab Sec. _____ Desk No. _____

A. Temperature Effects

Na$_2$S$_2$O$_3$-HCl Test Pair	Time Elapsed (*seconds*)	Temperature (°C)
1.		
2.		
3.		

Instructor's approval of graph _____

Based upon your data, what can you conclude about the effect of temperature on the rate of this reaction?

From your graph of the data and assuming the same set of solutions, estimate the time for the appearance

of sulfur when the reaction occurs at 40°C _____; at 95°C _____.

B. Concentration Effects

Molar conc. of HCl	mol HCl	mass Mg (±0.01 g)	mol Mg	$\dfrac{\text{mol HCl}}{\text{mol Mg}}$	time (*seconds*)
1.0					
3.0					
6.0					

Instructor's approval of graph _____

How does a change in the molar concentration of HCl affect the time for a known amount of Mg to react?

From your graph of the data, estimate the time it would take for 0.2 g Mg to react in 3 mL of 0.40 M HCl.

C. Nature of Reactants

1. Record the relative reaction rates of the four acids with Mg in order of *increasing* activity.

    _____, _____, _____, _____

    What can you conclude about the relative chemical reactivity of the four acids?

2. List the three metals in order of the increasing reaction rate with 6 M HCl.

    _____, _____, _____

    What can you conclude about the chemical reactivity of the three metals?

D. Particle Size of Reactants

1. Explain how the physical state of calcium carbonate affects the rate of the chemical reaction.

E. Presence of a Catalyst

1. Instructor's approval of apparatus_____

2. Time to collect 20 mL of O_2 without a catalyst _____

    Time to collect 20 mL of O_2 with a catalyst _____

3. Describe the affect that a catalyst has on the rate of evolution of O_2 gas.

4. Describe the reaction of the glowing splint with oxygen. How can the observation be explained in terms of factors affecting reaction rates?

RATE LAW DETERMINATION

A recent mishap at a fertilizer plant released large quantities of nitrogen dioxide, NO_2, into the atmosphere. A brown haze has formed over the area. What eventually happens to the brown haze? Physically the haze dissipates because of air currents and dilution; chemically, nitrogen dioxide reacts with water vapor to form nitrous acid, with ammonia gas (that occurs naturally in the atmosphere) and with water vapor to form ammonium nitrite, and with limestone from buildings and the soil to form calcium nitrite.

Initially the chemical reactions of NO_2 are rapid because of its high concentration and the immediate presence of the other reactants, but as it and the other reactants combine chemically and as it dilutes in the atmosphere, the reactions slow; therefore the quantity of NO_2 in the atmosphere decreases, finally reaching low, undetectable levels. At these low levels we say that its concentration has reached an equilibrium with the concentration of NO_2 that existed prior to the mishap. The levels of NO_2 as a function of time can be plotted as shown in the figure.

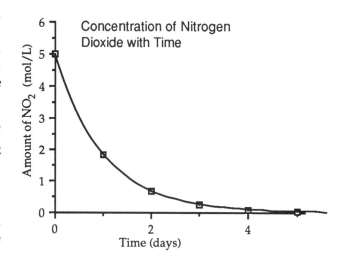

The progress of a chemical reaction can be measured quantitatively by any one of a number of methods described in Experiment 33. For example, NO_2 can be measured spectrophotometrically since NO_2 is a red-brown gas.

OBJECTIVES

- To determine the rate law for a chemical reaction
- To use a graphical analysis of experimental data

PRINCIPLES

The rate of disappearance of a reactant or the rate of appearance of a product are indicative of the rate of a chemical reaction. The technique used to monitor and measure a reaction rate is dependent upon a unique property of one of the substances in the reaction. For example, a reactant may lose its color, a product may produce an odor, the reaction may generate heat or become more acidic, a product may be a gas or precipitate, etc., all of which can be monitored as a function of time. For our purposes, we will monitor the time required for a color to appear, the color appears after a known amount of reactant has been consumed in the reaction.

The rate of a chemical reaction is often affected by the amount of reactant initially placed in the reaction system. Quantitatively, the rate is proportional to the molar concentration of each reactant raised to some power, called the **order** of the reactant. For the reaction $A_2 + 2\,B_2 \rightarrow 2\,AB_2$, this relationship between rate and concentration can be expressed as:

α: *means "proportional to".*

$$\text{rate } \alpha \ [A_2]^p \ \textit{and} \ \text{rate } \alpha \ [B_2]^q \ \textit{or} \ \text{rate } \alpha \ [A_2]^p[B_2]^q$$

The superscripts, p and q are the orders for reactants A_2 and B_2 respectively. When the proportionality sign, α, is replaced with a proportionality constant, k, the **rate law** for the reaction is

$$rate = k[A_2]^p[B_2]^q$$

The value of **k**, called the **specific rate constant**, is determined experimentally and varies with temperature and the presence of a catalyst, but is independent of reactant concentrations.

The orders, **p** and **q**, are also determined experimentally. The effect that a change in a reactant concentration has on a reaction rate is expressed by the order of that reactant. For example, suppose that in doubling the molar concentration of A_2 (while holding the $[B_2]$ constant) the reaction rate increases by a factor of four. To maintain the proportionality between rate and $[A_2]$ (rate α $[A_2]^p$) **p** must equal 2. Experimentally the molar concentration of B_2 can remain virtually unchanged if a large excess of B_2 (compared to the amount of A_2) is initially placed in the reaction system. For the reaction the $\Delta[A_2]$ is therefore much larger than the $\Delta[B_2]$ and therefore is more pronounced in affecting reaction rate.

In this experiment, the values of **k** and **p** are determined for the reaction of the iodate anion, IO_3^-, with the sulfite anion, SO_3^{2-}, in an acidic solution. The order of the IO_3^- in the reaction will be designated as **p**. The reaction occurs in a series of steps.

(Step #1) $IO_3^-(aq) + 3\,SO_3^{2-}(aq) \rightarrow I^-(aq) + 3\,SO_4^{2-}(aq)$

(Step #2) $5\,I^-(aq) + 6\,H^+(aq) + IO_3^-(aq) \rightarrow 3\,H_2O(l) + 3\,I_2(aq)$

(Step #3) $3\,I_2(aq) + 3\,SO_3^{2-}(aq) + 3\,H_2O(l) \rightarrow$
$$6\,I^-(aq) + 3\,SO_4^{2-}(aq) + 6\,H^+(aq)$$

(Net equation) $2\,IO_3^-(aq) + 6\,SO_3^{2-}(aq) \rightarrow 2\,I^-(aq) + 6\,SO_4^{2-}(aq)$

How do we detect the reaction rate? Lets note the function and progress of the SO_3^{2-} anion in the reaction: we see that SO_3^{2-} reacts with not only the IO_3^- anion in step #1, but also with I_2 in step #3. What happens when all of the SO_3^{2-} anion has been consumed in the reaction? The I_2 that is produced in step #2 is "free" in solution; we detect its presence with starch, which together form a deep-blue solution due to the $I_2 \bullet$ starch complex.

$$I_2(aq) + starch\ (aq) \rightarrow I_2 \bullet starch\ (aq, \text{deep blue color in solution})$$

This color appearance is the signal for detecting the consumption of the SO_3^{2-} anion and therefore becomes our method for monitoring the reaction rate. If the amount of IO_3^- is increased in the reaction system, step #1 proceeds faster and, accordingly, the SO_3^{2-} is consumed more rapidly. Therefore, we have a method for observing the effect that a change in the molar concentration of IO_3^- has on the reaction rate.

The rate law for the reaction is

$$rate = k[IO_3^-]^p[SO_3^{2-}]^q$$

In the experiment, the $[SO_3^{2-}]$ will remain constant in each trial while the $[IO_3^-]$ will change. This modifies the rate law to

$$\text{rate} = k[IO_3^-]^p \cdot \text{constant} \quad \text{or} \quad \text{rate} = k'[IO_3^-]^p$$

The rate of the reaction can be expressed as the amount of I_2 generated per unit of time. Because the same amount of SO_3^{2-} is used in repeated experiments, the amount of I_2 produced will also be the same; the change in time of the $I_2 \cdot$ starch blue color will be dependent upon the molar concentration of IO_3^-. Therefore for this reaction the rate can be expressed in reciprocal time, gauged by the appearance of the deep-blue $I_2 \cdot$ starch per time.

$$\text{rate (sec}^{-1}) = \text{appearance of } I_2 \cdot \text{starch/time}$$

Determination of the Reaction Order, p, and Rate Constant, k'

The order of $[IO_3^-]$ in the reaction rate is determined by changing its concentration in several trials while maintaining a constant $[SO_3^{2-}]$ and $[H^+]$. The data generated in the experiment is graphed to determine p and k'. The plot of the logarithmic form of the rate law, $\text{rate} = k'[IO_3^-]^p$, yields an equation with a straight line.

$$\log (\text{rate}) = \log k' + p \log[IO_3^-] \quad \text{or}$$
$$y = b + m x$$

Therefore, a plot of log (rate) *vs* log $[IO_3^-]$ produces a straight line with a slope of **p**, the order of the reaction with respect to $[IO_3^-]$. The order, **q**, of $[SO_3^{2-}]$ in the reaction, could be similarly determined.

The reaction rate constant, **k'**, is calculated by substituting the rate, p, and $[IO_3^-]$ for a given trial into the rate law. It can also be determined from the y-intercept of the graphed data, where $b = \log k'$.

Activation Energy

Reaction rates are temperature dependent. Higher temperatures increase the kinetic energy of the (reactant) molecules, such that when two reacting molecules collide, they do so with a much greater force (more energy is dispersed within the collision system) causing bonds to rupture, atoms to rearrange, and new bonds (products) to form. The sum of these kinetic energies required for a reaction to occur is called the **activation energy** for the reaction.

The relationship between the reaction rate constant, **k'**, at a measured temperature, **T(K)**, and the activation energy, E_a, is expressed in the Arrhenius equation.

$$k' = Ae^{-E_a/RT}$$

A is a collision parameter for the reaction and **R** is the gas constant (=8.314 J/mol K). When a reaction is performed at two temperatures, $T_{(1)}$ and $T_{(2)}$, the ratio of the reaction rate constants can be determined.

$$\frac{k_1'}{k_2'} = \frac{Ae^{-E_a/RT_{(1)}}}{Ae^{-E_a/RT_{(2)}}}$$

We can simplify the ratio of these two equations by using natural logarithms.

$$\ln \frac{k_1'}{k_2'} = - \frac{E_a}{R} \left[\frac{1}{T_{(1)}} - \frac{1}{T_{(2)}} \right]$$

Therefore, the activation energy for a reaction can be calculated from the determination of rate constants at two temperatures.

In this experiment you will determine the activation energy, E_a, for the $[IO_3^-]/[SO_3^{2-}]$ system by determining the k' at room temperature and again at 50°C.

TECHNIQUES

- Technique 2, page 1   Clean Glassware and Lab Bench
- Technique 6d, page 6   Heating Liquids
- Technique 8, page 7   Reading a Meniscus
- Technique 9, page 7   Pipetting a Liquid or Solution
- Technique 18, page 18   Graphing Techniques

In addition, some techniques for timing reactions will be used.

PROCEDURE

Read the entire experimental procedure before beginning. You should work with a partner—one of you will need to determine the time lapse for the reaction and the other will need to mix the solutions.

A. Test Reactions

The test reactions are prepared according to Table 34.1.

Table 34.1. Composition of Test Reactions

Test Reaction	Solution A			Solution B
	0.02 M HIO₃	Starch	Distilled H₂O	0.005 M H₂SO₃
1	5.0 mL	2 drops	0 mL	5.0 mL
2	4.0 mL	2 drops	1 mL	5.0 mL
3	3.0 mL	2 drops	2 mL	5.0 mL
4	2.0 mL	2 drops	3 mL	5.0 mL
5	1.0 mL	2 drops	4 mL	5.0 mL

1. **Test Reaction #1.** Prepare Solution A for Test Reaction #1 in a 200 mm test tube and Solution B in a 150 mm test tube. The volumes of water and the HIO₃ and H₂SO₃ solutions should be measured with pipets.[1] Agitate each solution. Place a white sheet of paper behind Solution A so that the appearance of the color change is more evident.

2. The reaction begins when the H₂SO₃ from Solution B is added to the HIO₃ from Solution A. Start timing, in **seconds**, the reaction when the two solutions are combined—be as exact as possible in starting and stopping the timing of the reaction.[2]

    O.K., get ready; quickly pour Solution B into Solution A (B → A), **start** timing, and swirl. The blue color appears suddenly; watch and **stop**

---

[1]Do not intermix pipets from one trial to the next. Cleanliness is important, especially in this experiment.

[2]The time (seconds) elapsed from the initial mixing of the solutions (A and B) until the deep-blue color *first* appears is to be recorded.

timing at its *first* appearance. Record the time lapse to the nearest second. Record the temperature of the solution at $T_{(1)}$ on the Data Sheet; $T_{(2)}$ is the temperature for the reactions in Part C.

3. Thoroughly clean the 200 mm test tube. Prepare Solutions A and B for Test Reaction #2. Repeat the reaction procedure as described in Part A.2.

4. Continue with the data collection for the other test reactions in Table 34.1. If your instructor approves, repeat the tests or prepare and collect data for other volumes of HIO_3 in Solution A.

Disposal Information: **Dispose of the test solutions in the "Waste Salts" container.**

B. Data Analysis and Calculations

1. Determine the reciprocal of the time (seconds) elapsed for the appearance of the blue color and record it as the reaction rate. Calculate the logarithm of this rate and record on the Data Sheet.

2. Calculate and record the initial molarity of the IO_3^- ion, $[IO_3^-]_i$, and the logarithm of $[IO_3^-]_i$ for each test reaction. Remember the total volume of the test reaction is 200 mL.

3. Plot on linear graph paper log (rate) *vs* log $[IO_3^-]_i$. Draw a straight line that passes as close as possible to the 5 data points. Calculate the slope of the line; its value should approximate a whole number, but record its actual value on the Report Sheet. This is the order, **p**, of IO_3^- in the reaction. Ask your instructor to approve your graph.

4. Use the values of the rates of the test reactions (from Part B.1), $[IO_3^-]_i$ (from Part B.2), **p** (from Part B.3), and the rate law, **rate = k'$[IO_3^-]^p$**, to determine **k'** for the five test reactions. Calculate the average value of **k'** and express it with its proper units.

5. From the plot, extrapolate the straight line until it intersects the y-axis at x = 0 (or log $[IO_3^-]$ = 0). This y-value equals log k'; calculate **k'**.

C. Activation Energy

1. Repeat all of Parts A and B but increase the temperature of solutions A and B to no more than 50°C. How are you going to do this? Prepare a water bath using a 600 mL beaker half-filled with water and heat to about 50°C. Place Solutions A and B in the hot water bath for several minutes until thermal equilibrium has been reached. At that point, quickly pour solution B into A, and record the time lapse as before. Record the temperature (±0.1°C) of the water bath[3] and use this value for your calculations.

2. Perform the same data analysis as before. Using the values of k' (at $T_{(1)}$ and $T_{(2)}$), calculate the activation energy for the reaction. Remember to use kelvins and R = 8.314 J/mol K.

3. Repeat the calculation of the activation energy using the k' values (at $T_{(1)}$ and $T_{(2)}$) determined from the two graphs at x = 0.

---

[3]The temperature does not have to be exactly at 50°C, but the *actual* temperature should be recorded to ±0.1°C.

Notes, Observations, and Calculations

EXPERIMENT 34

RATE LAW DETERMINATION

Date _____ Name _____ Lab Sec. _____ Desk No. _____

1. a. If 3.5 hours are required to travel 165 miles, what is the *rate* of travel?

   b. If 4 minutes and 10 seconds (250 s) elapse in completing four laps of a 400 m track, what is the pace (running rate), in m/s, of the track star?

   c. A total of 27.3 seconds elapses before a reaction is complete. What is the reaction rate?

2. A reaction between the gaseous substances C and D proceeds at a measurable rate to produce a sudden color change. The following set of data was collected for the reaction

$$3\ C(g) + 2\ D(g) \rightarrow 2\ F(g) + G(g)$$

$[C]_i$	$[D]_i$	Time for Color Change (*seconds*)	Rate (sec^{-1})
1.0×10^{-2}	1.0	30	
1.0×10^{-2}	3.0	10	
2.0×10^{-2}	3.0	1.3	
2.0×10^{-2}	1.0	3.7	
3.0×10^{-2}	3.0	0.37	

   a. What is the (whole number) order of C in the reaction? _____; of D in the reaction? _____

   b. What is the rate law?

c. What is the specific rate constant, k?

d. At any given instant during the reaction,

the rate of appearance of F is _____ times the rate of disappearance of C.

the rate of disappearance of D is _____ times the rate of appearance of G.

the rate of appearance of G is _____ times the rate of appearance of F.

3. A set of rate data is plotted for the reaction of A at several concentrations while the B_2 concentration remains constant for $2 A + B_2 \rightarrow A_2B + B$

   a. From the graph, determine the order of A in the reaction.

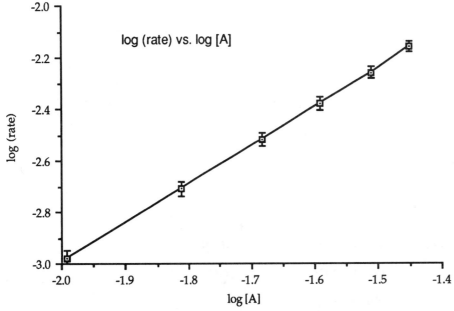

b. From the equation, log (rate) = p log [A] + log k, the graphed data, and the order of A in the reaction, calculate k, the specific rate constant for the reaction.

4. A 1:1 mixture of 0.040 M KIO_3 and 0.040 M H_2SO_4 is used to prepare the 0.020 M HIO_3 solution for today's experiment.

   a. Describe the procedure for the preparation of 150 mL of a 0.040 M KIO_3 solution, starting with solid KIO_3?

   b. How do you prepare 150 mL of a 0.040 M H_2SO_4 solution, starting with 6.0 M H_2SO_4?

RATE LAW DETERMINATION

Date _____ Name _____ Lab Sec. _____ Desk No. _____

A. Test Reactions

Test Reactions		1	2	3	4	5	6
1. Time elapsed (*sec*)	at $T_{(1)}$						
	at $T_{(2)}$						
2. Reaction temperature	$T_{(1)}$						
	$T_{(2)}$						

B. Data Analysis and Calculations

		1	2	3	4	5	6
1. rate (sec^{-1})	at $T_{(1)}$						
	at $T_{(2)}$						
2. log (rate)	at $T_{(1)}$						
	at $T_{(2)}$						
3. $[IO_3^-]_i$ (*mol/L*)	at $T_{(1)}$						
	at $T_{(2)}$						
4. log $[IO_3^-]_i$	at $T_{(1)}$						
	at $T_{(2)}$						

5. Plot log (rate) *vs* log $[IO_3^-]_i$

Instructor's approval of graph   at $T_{(1)}$ _____   at $T_{(2)}$ _____

6. Value of **p** from graph _____

7. Value of **k'** for each trial	at $T_{(1)}$						
	at $T_{(2)}$						

8. Average value of **k'** (calculated)   at $T_{(1)}$ _____   at $T_{(2)}$ _____

9. Value of **k'** from graph   at $T_{(1)}$ _____   at $T_{(2)}$ _____

C. Activation Energy

E_a from calculated k' values*　　　　　　　　~~~~~~~~~~~~~~~~~~~~~~~~~~~~~~

E_a from k' values from graph　　　　　　　~~~~~~~~~~~~~~~~~~~~~~~~~~~~~~

* Show calculation of E_a here.

QUESTIONS

1. What would each of of the following changes in the experimental procedure have on the reaction rate?

   a. a slight increase in the concentration of starch? Assume no volume change. Explain.

   b. a slight increase in the concentration of the HIO_3 in solution A. Explain.

   c. a slight increase in the amount of water. Explain.

2. Two test reactions are the *minimum* needed to obtain the values of p and k' in this experiment. Explain the advantage of using additional test reactions, as were performed in this experiment.

3. What was happening in today's reaction between the times Solutions A and B were mixed and the appearance of the blue color?

PREFACE

CATION IDENTIFICATION

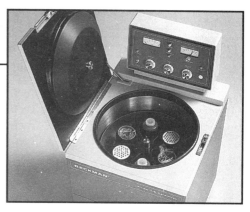

Quick and simple procedures for the identification of a substance are often necessary in forensic science. For example rapid and easy tests have been developed for the detection of confiscated drugs and for the detection of drugs in a urinanalysis. Home pregnancy test kits and tests for diabetes are currently available at the local pharmacy. A quick and easy identification of a cation in a salt mixture or a drinking water sample is necessary in water laboratories when metal contamination is suspect and immediate action for public safety is necessary. Tests kits for determining the levels of chlorine, the alkalinity, and the pH in home swimming pools are available at pool supply stores. All of these quick and ready tests must be reproducible, reliable, and must use stable testing chemicals. These criteria are upheld, only if a very carefully outlined and researched procedure is followed.

PRINCIPLES

Ammoniacal solution: a solution that contains ammonia as the principal solute.

The characteristic physical and chemical properties of a cation allow us, in the next series of experiments, to separate it from a mixture of cations and to characteristically identify its presence. For example, the Ag^+ ion is identified as being present by its precipitation as the chloride, i.e., $AgCl(s)$. While other cations precipitate as the chloride, $AgCl$ is the only one that is soluble in an ammoniacal solution.[1] Similarly, from the solubility rules that you have learned earlier in the course, Ba^{2+} can be separated from a number of other cations with the addition of SO_4^{2-}; $BaSO_4$ forms a white precipitate, while other cations (except Pb^{2+}, Sr^{2+}, and Hg_2^{2+}) generally are soluble as sulfates.

Therefore, with enough knowledge of the chemistry of the various cations, a unique separation and identification procedure can be developed. Some procedures are one step, quick tests, while others are more exhaustive. *The procedure, however, must systematically eliminate all other ions that may interfere with the test.* Cations can be classified according to groups that have similar chemical properties; cations within each group can then be further separated and identified. A procedure that follows this pattern of analysis is called **Qualitative Analysis**, one that we will follow in the next several experiments.

The separation and identification of the cations use many chemical principles, many of which we will cite as we proceed. These principles will include precipitation, ionic equilibrium, acids and base properties, pH, oxidation and reduction, and complex ion formation. To help you to understand these test procedures, each experiment presents some pertinent chemical equations, and you are asked to write equations for other reactions that occur in the separation and identification of the cations. This usually appears in the Lab Preview.

---

[1]This was one test procedure that prospectors for silver used in the early prospecting days.

TECHNIQUES

In order to complete these procedures, you will need to practice the following laboratory techniques and, in addition, develop several new techniques. The techniques that you should review are

- Technique 5, page 3   Transferring a Liquid
- Technique 6a, c, d, page 4   Heating Liquids
- Technique 7a, b, page 6   Evaporation of Liquids
- Technique 11, page 10   Testing with Litmus
- Technique 14a, d page 12   Separation of a Liquid from a Solid
- Technique 17a, c, page 16   Handling Gases

PROCEDURES

Most of the tests for the cations will be performed in 3 mL (75 mm) test tubes and reagents will be added with medicine droppers ($\approx$20 drops/mL) or Beral pipets. Do not mix the medicine droppers (or Beral pipets) with the various reagents you will be using. When mixing solutions in a test tube, agitate by tapping the side of the test tube. Break up a precipitate with a stirring rod or cork the test tube and invert, but *never* use your thumb! (Figure QA.1).

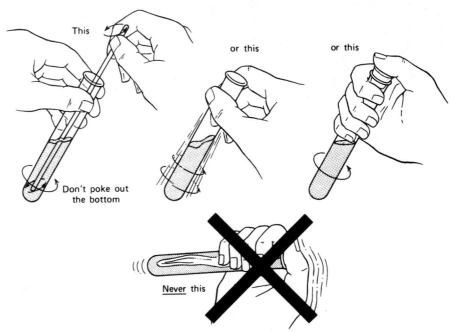

Figure QA.1
Technique for mixing solutions in a test tube

Washing a precipitate is often necessary to remove occluded impurities in your analysis. Add the water or wash liquid to the precipitate, mix thoroughly with a stirring rod, centrifuge, and decant. Failure to properly wash precipitates often leads to errors in the analysis (and arguments with your laboratory instructor!).

Since you will be frequently using the centrifuge in the next several experiments, *be sure* to read Technique 14d very closely in the Techniques section of this manual.

Flow Diagrams

To organize the qualitative analysis test procedure for the various cations, a flow diagram is often used. A flow diagram utilizes several standard notations:

• Brackets, [ ], indicate the use of a test reagent, written in molecular form.

• A single horizontal line, _____, indicates that separation is made with a centrifuge.

• A double horizontal line, _____ , indicates soluble cations present

• Two short vertical lines, ‖ , indicate that a precipitate has formed; these lines are drawn to the *left* of the single horizontal line.

• One short vertical line, | , indicates a supernatant and is drawn to the *right* of the single horizontal line.

• A square box, ☐ , placed around the compound or test reagent, confirms the presence of the cation.

Flow Diagram for Cation Identification

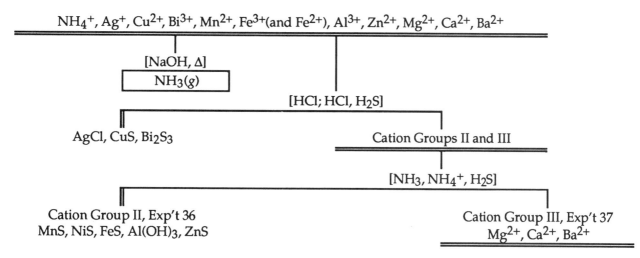

An unknown consisting of any number of the 12 cations analyzed in the next three experiments may be assigned at the conclusion of Experiment 37. Ask your instructor about that assignment. The preceding flow diagram is for these cations. A flow diagram for each group is presented in the corresponding experiment.

The following suggestions are offered while performing these "qual" experiments:

• *Always* read the procedure in detail and with an understanding of the principles before lab. Is extra equipment necessary? Is a hot water bath needed? What precautions are to be taken? Why is this reagent added at this time?
• Mark with a file, 1, 2, and 3 mL intervals on a 75 mm test tube in order to quickly estimate volumes.
• Keep a wash bottle filled with distilled (or deionized) water at all times.
• Keep a number of *clean* medicine droppers (or Beral pipets) and 75 mm test tubes available; always rinse several times with distilled (or

deionized) water immediately after their use.

•Maintain a "file" of confirmatory tests in the test tubes from the analysis of the *known* solution for each Group of ions, while you are conducting the unknown analysis; in that way, color and precipitate comparisons can be quickly done.

•Closely follow simultaneously the principles used in each test, the flow diagram, the Procedure, and the Data Sheet, during the analysis.

 Caution: *In the next several experiments you will be handling a large number of chemicals (acids, bases, oxidizing and reducing agents, and, perhaps, even some toxic chemicals), some of which are more concentrated than others.*

*Carefully, handle all chemicals. Do not intentionally inhale the vapors of any chemical unless you are specifically told to do so; **avoid** skin contact with any chemicals—wash the skin immediately in the laboratory sink, eye wash fountain, or safety shower; **clean up** any spilled chemical. If you are uncertain of the proper cleanup procedure, flood with water, and consult your laboratory instructor; **be** aware of the techniques and procedures of neighboring chemists—discuss potential hazards with them.*

*And finally, **dispose of the waste chemicals** in the appropriately labelled waste containers. Consult your laboratory instructor to ensure proper disposal.*

QUALITATIVE ANALYSIS

Date _____ Name _____ Lab Sec. _____ Desk No. _____

1. a. The approximate volume of a standard 75 mm test tube is _____.

    b. Small volumes of reagents are usually added using _____.

    c. A _____ is commonly used to break up a precipitate.

    d. A _____ is an instrument used to separate and compact a precipitate in a test tube.

    e. The clear solution above a precipitate is called the _____.

    f. The number of drops of water equivalent to 1 mL is about _____.

    g. A solution should be centrifuged for (how long?) _____.

    h. On a flow diagram ‖ means _____.

    i. On a flow diagram, a single horizontal line means _____.

    j. On a flow diagram, ☐ means _____.

2. Explain the procedure for balancing a centrifuge.

3. Explain the procedure for washing a precipitate.

EXPERIMENT 35

CATION IDENTIFICATION, I. NH_4^+, Ag^+, Cu^{2+}, Bi^{3+}

The world's largest active open pit copper mine is located near Salt Lake City, Utah. The copper is chemically combined in various compounds but its percent abundance in the ore is only 0.6%! To determine the presence of copper in an ore that is that low in natural abundance requires some basic chemical skills to first detect its presence, a qualitative analysis, and then to determine its amount, a quantitative analysis. The copper ore is processed through a series of metallurgical processes until 98% pure copper is obtained prior to a final electrolytic refining process.

Similarly, most silver found in the Earth's crust is not found in the elemental state, but rather combined chemically with various components of its ore. Prospectors performed qualitative tests for the presence of silver by a procedure that is analogous to the one you will use in this experiment.

OBJECTIVE

• To identify the presence of one or more cations from a mixture of the cations, NH_4^+, Ag^+, Cu^{2+}, and Bi^{3+}

PRINCIPLES

This experiment is a first in a series in which a set of reagents and a number of separation techniques are used to identify the presence of a particular cation among a group of cations that have similar chemical properties. We approach this study from that of an experimental chemist—we will perform some tests, write down observations, and then write a balanced equation that agrees with the data.[1]

The identification of a particular cation present in a mixture of cations is most often accomplished through a scheme of selective precipitation steps. The optimal conditions for the precipitation of a cation are very selective; the position of equilibrium (LeChatelier's Principle, see Experiment 24) is controlled by pH adjustment, occasionally with a buffer, by the presence of a limited concentration range of the precipitating anion, and by the presence of oxidizing or reducing conditions. For example, the Ag^+ is "insoluble" at low concentrations of chloride ion, but at higher concentrations of chloride ion the soluble complex anion, $AgCl_2^-$ forms.

The sulfide salts of the cations in this group (exclusive of the ammonium ion) are insoluble in solutions that are at least 0.3 M H^+ (pH = 0.5). If the pH is greater than 0.5, sulfide precipitation of cations in the second group (Experiment 36) may also precipitate. Therefore, pH control is extremely important for both controlling the amount of sulfide ion present in solution and in separating these two groups of cations—interference with the cation identification between these two groups can be problematic.

[1]For a more complete description and procedure for the qualitative analysis of cations, see Beran & Brady, *Laboratory Manual for General Chemistry, 4th Ed.*, John Wiley & Sons, Inc., 1990.

The sulfide ion is produced *in situ* from $H_2S(aq)$. The thermal degradation of thioacetamide, CH_3CSNH_2, in an acidic or basic solution generates small quantities of hydrogen sulfide.

$$CH_3CSNH_2(aq) + 2\,H_2O(l) \xrightarrow{\Delta} CH_3CO_2^-(aq) + NH_4^+(aq) + H_2S(aq)$$

The slow generation of H_2S in the solution minimizes the amount of this foul-smelling, highly toxic gas in the laboratory. When H_2S is produced slowly, more compact sulfide precipitates form, making them easier to separate by centrifugation.

$H_2S(aq)$ is a weak, diprotic acid ionizing slightly to produce the sulfide ion that is necessary for cation precipitation.

$$H_2S(aq) + 2\,H_2O(l) \rightleftharpoons 2\,H_3O^+(aq) + S^{2-}(aq)$$

Applying LeChatelier's Principle, we can infer that a high H_3O^+ concentration (low pH) shifts this equilibrium to the left, leaving a low sulfide concentration; a low H_3O^+ concentration (high pH) shifts the equilibrium to the right, increasing the sulfide concentration. Therefore, if cations precipitate as the sulfide at a low pH, then only small amounts of sulfide ion are needed for precipitation; this means that these sulfide salts have a very low solubility; the sulfide salts also have a small K_s value. A high pH provides more sulfide ion for precipitation of the cation; these sulfides salts are more soluble and have larger K_s values.

The chemical principles and laboratory techniques that are used to separate and identify the four cations in this experiment require you to be a careful, conscientious chemist. Carefully read the Procedure, review your laboratory techniques, and complete the Lab Preview before beginning the analysis—it will save you time and minimize frustration.

The flow diagram outlines the procedure for the separation and the identification of the NH_4^+, Ag^+, Cu^{2+}, and Bi^{3+} cations. Review the Preface to Qualitative Analysis for an understanding of the symbolism in the flow diagram.

Ammonium Ion

The NH_4^+ ion is stable in an acidic solution, but when its solution is made basic, NH_3 gas is evolved. With gentle heat, the NH_3 gas is driven from the solution and can be detected by its odor or by a litmus test.

$$NH_4^+(aq) + OH^-(aq) \xrightarrow{\Delta} NH_3(g) + H_2O(l)$$

Silver Ion

The silver ion precipitates with low concentrations of chloride ion. In an ammoniacal solution, AgCl dissolves to form the diamminesilver ion, $Ag(NH_3)_2^+$. The $Ag(NH_3)_2^+$ ion is unstable in an acidic solution; the NH_3 forms NH_4^+ and the Ag^+ combines with the Cl^- that is still in solution to reform white insoluble AgCl.

$$Ag(NH_3)_2^+(aq) + Cl^-(aq) + 2H^+(aq) \rightarrow AgCl(s) + 2NH_4^+(aq)$$
$$\text{(white)}$$

Flow Diagram for NH_4^+, Ag^+, Cu^{2+}, Bi^{3+}

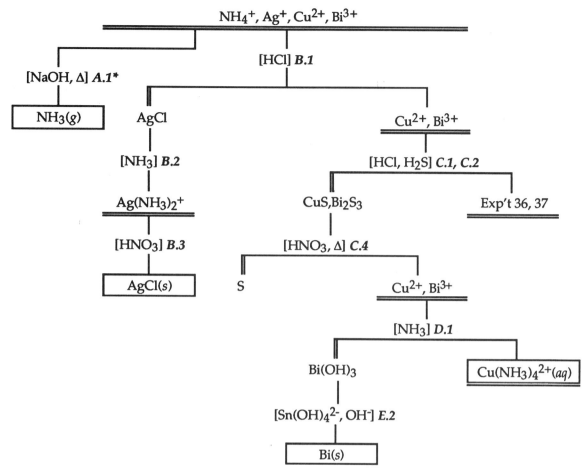

* Refers to Part A.1 of the Procedure

Copper(II) Ion

After removal of the silver ion, the solution is made acidic with hydrochloric acid. Thioacetamide is added, generating the sulfide ion that precipitates the copper and bismuth cations. Aqueous NH_3 precipitates Bi^{3+} as a white hydroxide, but complexes the Cu^{2+} as a deep-blue $Cu(NH_3)_4^{2+}$ complex ion, a confirmation of the presence of Cu^{2+}.

$$Cu^{2+}(aq) + 4\,NH_3(aq) \rightarrow Cu(NH_3)_4^{2+}(aq)$$
$$\text{(blue)}$$

Bismuth(III) Ion

When a freshly prepared sodium stannite, $Na_2Sn(OH)_4$, solution is added to the bismuth hydroxide precipitate, the Bi^{3+} ion is reduced to black bismuth metal, confirming the presence of bismuth in the sample.

$$2\,Bi(OH)_3(s) + 3\,Sn(OH)_4^{2-}(aq) \rightarrow 2\,Bi(s) + 3\,Sn(OH)_6^{2-}(aq)$$
$$\text{(black)}$$

PROCEDURE

To become familiar with the identification of these cations, take a sample that contains the four cations and analyze it according to the procedure. At each numbered superscript (i.e., [1]), STOP, and record on the Data Sheet. After the presence of each cation has been confirmed, SAVE the test tube so that its appearance can be compared to that for your unknown sample.

Caution: *A number of concentrated and 6 molar acids and bases are used in the analysis of these cations. Handle each of these solutions with care. Read the Lab Safety section in Experiment 1 for instructions in handling acids and bases.*

A. Ammonium Ion Test

1. Place 1 mL of the test solution in a 75 mm test tube. Moisten a piece of red litmus paper. Add 5 drops of 6 M NaOH (**Caution!!**) (Note: be careful not to let NaOH contact the litmus paper) and place red litmus paper over the mouth of the test tube. Warming with a *gentle* flame may be necessary.[#1]

2. Using the proper technique, check the odor of the gas evolved.

B. Silver Ion Test

1. Place 1 mL of the sample solution in a 75 mm test tube, add 2 drops of 6 M HCl (**Caution:** *handle acid with care.*) (Figure 35.1), and centrifuge (see Technique 14d).[#2] Ask the instructor about the safe operation of the centrifuge. Test the supernatant with drops of 6 M HCl for the additional formation of precipitate. Again centrifuge, if necessary. Decant the supernatant liquid and save for Part C.

2. To the precipitate add 5 drops of 6 M NH₃ (**Caution:** *avoid inhalation and skin contact*)[#3].

3. Re-acidify to litmus the solution with drops of 6 M HNO₃. (**Caution:** *avoid skin contact. Clean up spills immediately.*)[#4]

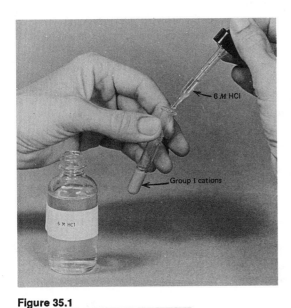

Figure 35.1

Precipitation of Ag⁺(aq)

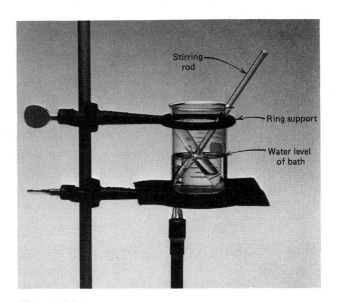

Figure 35.2

The setup of a hot water bath

C. Sulfide Precipitation of the Cations

Supernatant: the liquid/solution that remains above the precipitate after centrifuging.

1. To the supernatant from Part B.1 add several more drops of 6 M HCl. This should adjust the pH of the solution to about 0.5.

2. Add 10 to 15 drops of 1 M CH_3CSNH_2, heat in a hot water bath ($\approx 95°C$, Figure 35.2) for several minutes (see Technique 6d), cool, and centrifuge. Test for complete precipitation by repeating the thioacetamide addition to the supernatant. The precipitate contains the CuS and Bi_2S_3 salts;[5] the supernatant contains cations that do not precipitate under these conditions (Experiments 36 and 37).[2]

3. Wash the precipitate twice with 2 M NH_4NO_3, stir, and discard each washing.

4. Add 15 drops of 6 M HNO_3 to the precipitate and heat to boiling until the precipitate dissolves (a direct flame may be necessary. **Caution:** *read Technique 6a for heating test tubes with a direct flame*). Cool and centrifuge. Transfer the supernatant[6] to a 75 mm test tube; discard any free sulfur.

D. Copper(II) Ion Test

1. Add drops of conc NH_3 (**Caution:** *avoid inhalation or skin contact!*) to the supernatant from Part C.4. The deep-blue solution confirms the presence of Cu^{2+} in the sample.[7] Centrifuge, decant the solution, and save the precipitate[8] for Part E.

E. Bismuth(III) ion Test

1. Prepare a fresh $Na_2Sn(OH)_4$ solution by placing 2 drops of 1 M $SnCl_2$ in a 75 mm test tube, followed by drops of 6 M NaOH until the $Sn(OH)_2$ precipitate just dissolves. Shake or stir the solution.

2. Add several drops of the $Na_2Sn(OH)_4$ solution to the precipitate from Part D. The immediate formation of a black precipitate[9] confirms the presence of bismuth in the sample.

F. Unknown Test

1. Obtain an unknown sample from your instructor. Determine which cation(s) is(are) present.

Disposal Information: **Dispose of the test solutions in the "Waste Salts" container**

---

[2]If your sample is to be analyzed for additional cations in Experiments 36 and 37, save this supernatant for that analysis; otherwise discard it.

Notes and Observations

EXPERIMENT 35 LAB PREVIEW

CATION IDENTIFICATION, I. NH_4^+, Ag^+, Cu^{2+}, Bi^{3+}

Date ____________Name ______________________________ Lab Sec. ______Desk No. ________

1. a. Find the K_s values of AgCl and CuS in your text book.

   $K_s(AgCl) =$

   $K_s(CuS) =$

   b. Calculate the molar solubility of each salt. Which salt is the least soluble?

2. Write balanced equations for:

   a. the reaction of chloride ion with silver ion.

   b. the reaction of NH_3 with AgCl.

   c. the reaction of $Ag(NH_3)_2^+$ with acid and chloride ion.

3. What reagent separates Cu^{2+} from Bi^{3+} in solution?

4. a. Explain why HNO₃ is added to the precipitate in Part C.4.

    b. What is the origin of the elemental sulfur?

5. What is the color of
    a. $Cu(NH_3)_4^{2+}$

    b. bismuth metal

6. Write balanced equations for the following reactions cited in this experiment. Read the Principles and Procedure sections to assist you in writing the equations. Also use your textbook as resource information.

    a. The formation of ammonia gas from ammonium ion.

    b. The precipitation of Cu^{2+} and Bi^{3+} as the sulfide salts.

    c. The reactions of CuS and Bi_2S_3 with hot nitric acid.

    d. The reaction of Bi^{3+} with aqueous ammonia.

EXPERIMENT 35

DATA SHEET

CATION IDENTIFICATION, I. NH_4^+, Ag^+, Cu^{2+}, Bi^{3+}

Date ____________ Name ________________________________ Lab Sec. ______ Desk No. ________

Procedure Number and Ion	Test Reagent or Technique		Observation (Color or General Appearance)	Chemicals Responsible for Observation	Check (√) if Observed in Unknown
#1 NH_4^+		gas			
#2 Ag^+		ppt			
#3		spnt			
#4		ppt			
#5		ppt			
#6		spnt			
#7 Cu^{2+}		spnt			
#8		ppt			
#9 Bi^{3+}		ppt			

Cations present in unknown: ________________________________

Instructor's approval: ________________________________

QUESTIONS

1. Name and give the color for the following.

   a. AgCl ________________________ ____________

   b. $Ag(NH_3)_2^+$ ________________________ ____________

   c. CuS ________________________ ____________

   d. $NaSn(OH)_4$ ________________________ ____________

2. What happens in Part B.1 if 6 M HNO_3 is substituted for 6 M HCl?

3. What happens in Part B.2 if 6 M NaOH is substituted for 6 M NH_3?

4. What happens in Part D.1 if NaOH is substituted for NH_3?

5. What happens if conc HCl is added to the precipitate in Part D.1?

6. Identify a single reagent that separates

   a. Ag^+ from Cu^{2+} _____

   b. Cu^{2+} from Bi^{3+} _____

EXPERIMENT 36

CATION IDENTI-FICATION, II. Mn^{2+}, Ni^{2+}, Fe^{3+}(and Fe^{2+}), Al^{3+}, Zn^{2+}

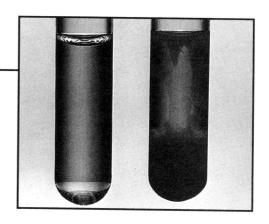

Trace amounts of metals are found throughout our bodies, the most abundant being iron, zinc, copper, and manganese. The daily requirement of metals designated as trace metals (or trace elements) is less than 100 mg/day. The trace metals are chemically bound to large molecules, such as proteins, to form enzymes, which function as catalysts for specific biochemical reactions, hormones, and vitamins. Iron, for example, is a part of the hemoglobin molecule that carries oxygen throughout the body and returns to the lungs with the carbon dioxide waste. Zinc is important to growth, copper deficiencies leads to anemia, and manganese is important to bone development. Excesses of aluminum have been linked to Alzheimer's disease. Over-the-counter mineral supplements are readily available containing various amounts of these elements.

The presence of trace amounts of metal ions in the body can be determined through procedures that are only slightly more complicated than what appears in this experiment.

OBJECTIVE

• To identify one or more cations from a mixture of the cations, Mn^{2+}, Ni^{2+}, Fe^{3+}(and Fe^{2+}), Al^{3+}, and Zn^{2+}

PRINCIPLES

Systematic separation: a series of sequential steps in a chemical analysis that are necessary for the successful separation of a substance.

In a systematic separation and identification of cations, the cations studied in this experiment form insoluble salts in a solution containing sulfide ion at a pH near neutrality. The cations of this group do *not* precipitate as the chloride salts or as the sulfide salts in a solution with an approximate pH of 0.5 (Experiment 35). Therefore, a separation of the cations in this experiment from those cations in the previous experiment is rather easily accomplished.

Preparation of the Cations for Analysis

The Mn^{2+}, Ni^{2+}, Fe^{3+}(and Fe^{2+}), and Zn^{2+} cations precipitate as sulfide salts but Al^{3+} precipitates as the hydrated hydroxide in a basic solution containing sulfide ion. The sulfide ion is generated in the same manner as was done in Experiment 35; that is, from the decomposition of thioacetamide, CH_3CSNH_2, producing H_2S. The H_2S ionizes to produce low concentrations of sulfide ion.

$$H_2S(aq) + 2H_2O(l) \rightleftharpoons 2H_3O^+(aq) + S^{2-}(aq)$$

The addition of base, such as NH_3, causes the equilibrium to shift *right*, increasing the sulfide concentration to a high enough level to where the product of the molar concentrations of the cation and the sulfide ion exceeds the salt's K_s value and the salt precipitates. In this basic solution, the Al^{3+} precipitates as $Al(OH)_3$.

The H_2S also serves as a reducing agent, reducing any Fe^{3+} ion to Fe^{2+} ion, and forming elemental sulfur as its oxidized product. Therefore, when a sample containing all the test cations in this experiment is treated with thioacetamide in a solution made basic with NH_3, a mixed precipitate of MnS, NiS, FeS, ZnS, and $Al(OH)_3$ forms.

The procedure for the separation and identification of these five cations is summarized in the following flow diagram. Follow it as you read through the Principles and Procedure.

Flow Diagram for Mn^{2+}, Ni^{2+}, Fe^{3+}(and Fe^{2+}), Al^{3+}, Zn^{2+} Identification

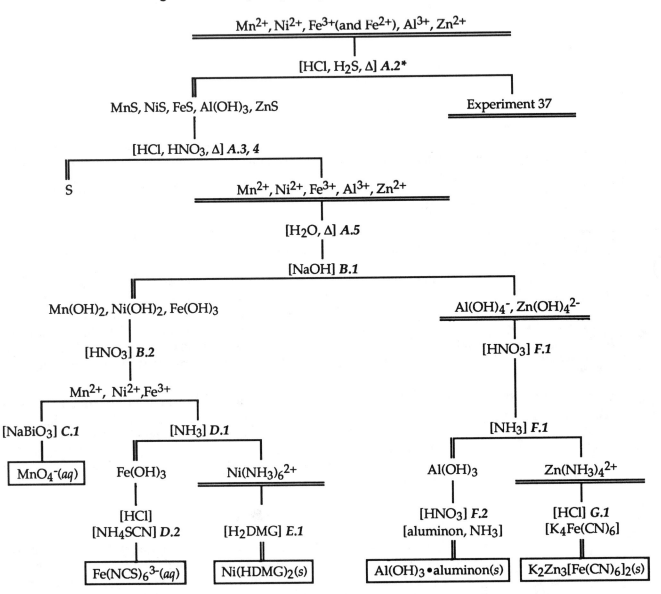

*Refers to Part A.2 in the Procedure

The MnS, FeS, ZnS, and $Al(OH)_3$ precipitates can be dissolved with a strong acid; however hot, conc HNO_3 is necessary to dissolve the NiS. The HNO_3 oxidizes the sulfide ion of the NiS, forming free, elemental sulfur and NO as products; in addition, the HNO_3 oxidizes Fe^{2+} to Fe^{3+} in the solution.

Therefore, HNO_3 dissolves the precipitates producing the soluble cations, sulfur, and NO gas as products.

Separation of Mn^{2+}, Ni^{2+}, and Fe^{3+} from Zn^{2+} and Al^{3+}

Addition of strong base to an aqueous solution of the five cations precipitates Mn^{2+}, Ni^{2+}, and Fe^{3+} as gelatinous hydroxides; the Zn^{2+} and Al^{3+} hydroxides initially precipitate, but, being amphoteric hydroxides, dissolve in the strong base, forming the aluminate, $Al(OH)_4^-$, and zincate, $Zn(OH)_4^{2-}$, ions.

The gelatinous hydroxides of Mn^{2+}, Ni^{2+}, and Fe^{3+} dissolve in nitric acid.

Manganese(II) Ion

A portion of the solution containing the dissolved hydroxides is treated with sodium bismuthate, $NaBiO_3$, a strong oxidizing agent that oxidizes Mn^{2+} to the characteristic purple permanganate ion, MnO_4^-, confirming the presence of Mn^{2+} in the sample.

$$14\,H^+(aq)\ +\ 2\,Mn^{2+}(aq)\ +\ 5\,BiO_3^-(aq)\ \rightarrow$$
$$2\,MnO_4^-(aq)\ +\ 5\,Bi^{3+}(aq)\ +7\,H_2O(l)$$
$$\text{(purple)}$$

Iron(III) Ion

A second portion of the same solution is treated with an excess of NH_3; this precipitates brown $Fe(OH)_3$ but forms the blue hexaammine complex of Ni^{2+}, $Ni(NH_3)_6^{2+}$. Acid dissolves the $Fe(OH)_3$ precipitate, whereupon the solution is treated with thiocyanate ion, SCN^-, forming a blood-red complex ion with Fe^{3+}, $Fe(NCS)_6^{3-}$, a confirmation of the presence of Fe^{3+} in the sample.

$$Fe^{3+}(aq)\ +\ 6\,SCN^-(aq) \rightarrow Fe(NCS)_6^{3-}(aq)$$
$$\text{(blood-red)}$$

Nickel(II) Ion

The confirmation of Ni^{2+} is the appearance of the bright pink-red precipitate formed when dimethylglyoxime, H_2DMG,[1] is added to a solution of the hexamminenickel(II) complex ion.

$$Ni(NH_3)_6^{2+}(aq)\ +\ 2\,H_2DMG(aq) \rightarrow$$
$$Ni(HDMG)_2(s)\ +\ 2\,NH_4^+(aq)\ +\ 4\,NH_3(aq)$$
$$\text{(pink-red)}$$

Aluminum Ion

An excess of acid added to a solution containing the aluminate and zincate ions restores the formation of the Al^{3+} and Zn^{2+} ions. With NH_3, the Al^{3+} reprecipitates as the gelatinous hydroxide, but the Zn^{2+} forms the soluble tetraammine complex ion, $Zn(NH_3)_4^{2+}$.

The $Al(OH)_3$ precipitate is dissolved with HNO_3; aluminon reagent[2] and ammonia are added and $Al(OH)_3$ reprecipitates. The aluminon reagent, a red dye, adsorbs onto the surface of the gelatinous $Al(OH)_3$ precipitate giving it a pink/red appearance. This confirms the presence of Al^{3+} in the sample.

---

[1]Dimethylglyoxime, abbreviated as H_2DMG for convenience in this experiment, is an organic complexing agent, specific for the precipitation of the nickel ion.

[2]The aluminon reagent is the ammonium salt of aurin tricarboxylic acid, a red dye.

$$Al^{3+}(aq) + 3\,NH_3(aq) + 3\,H_2O(l) + aluminon(aq) \rightarrow$$
$$Al(OH)_3 \cdot aluminon(s) + 3\,NH_4^+(aq)$$
$$(pink/red)$$

Zinc Ion

When potassium hexacyanoferrate(II), $K_4Fe(CN)_6$, is added to an acidified solution of the $Zn(NH_3)_4^{2+}$ ion, a light green precipitate of $K_2Zn_3[Fe(CN)_6]_2$ forms, confirming the presence of Zn^{2+} in the sample.

$$3\,Zn(NH_3)_4^{2+}(aq) + 2\,K_4Fe(CN)_6(aq) + 12\,H^+(aq) \rightarrow$$
$$K_2Zn_3[Fe(CN)_6]_2(s) + 12\,NH_4^+(aq) + 6\,K^+(aq)$$
$$(light\ green)$$

PROCEDURE

To become familiar with the identification of these cations, take a sample that contains the five cations and analyze it according to the procedure. At each numbered superscript (i.e.,[1]), STOP, and record on the Data Sheet. After the presence of each cation has been confirmed, SAVE the test tube so that its appearance can be compared to that for your unknown sample.

Caution: *A number of concentrated and 6 molar acids and bases are used in the analysis of these cations. Handle each of these solutions with care. Read the Lab Safety section in Experiment 1 for instructions in handling acids and bases.*

A. Preparation of the Cations for Analysis

1. If your sample contains only the cations of this experiment, you will *not* need to precipitate the cations in a sulfide solution at the relatively high pH, but instead you may proceed directly to Part B.

2. To 2 mL of a test solution that may contain cations for analysis in Experiment 37, add 1 mL of 2 M NH_4Cl. Add 6 M NH_3 until the solution is basic to litmus and then add 5 additional drops. Saturate the solution with H_2S by adding 6 drops of 1 M CH_3CSNH_2. Heat the solution in a hot water bath ($\approx$ 95°C, see Figure 35.2) for several minutes, cool, and centrifuge. Save the precipitate[1] for Part A.3 and the supernatant for analysis in Experiment 37.

3. Wash twice the precipitate[3] with 2 mL of water and discard the washings. Add 5 drops of 6 M HCl and 5 drops of 6 M HNO_3 (**Caution:** *be careful in handling acids*) and again heat in the hot water bath until the precipitates dissolve. Cool and centrifuge; save the supernatant[2] and discard any free, elemental sulfur.

4. Transfer the supernatant to an evaporating dish and heat gently until a moist residue remains (*not* to dryness). Add 1–2 mL of conc HNO_3 (**Caution:** *conc HNO_3 is a strong oxidizing agent—do not allow contact with the skin or clothing!*) and reheat with a "cool" flame (Figure 36.1) until a moist residue again forms.

5. Dissolve the residue in 1–2 mL of water and transfer the solution to a 75 mm test tube.

---

[3]See the Preface to Qualitative Analysis (page 340) for the proper technique in washing precipitates.

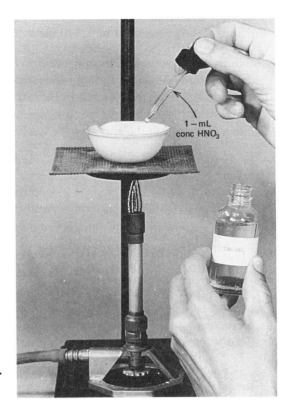

Figure 36.1
Addition of conc HNO_3 (Caution!!) to moist residue

B. Separation of Ni^{2+}, Fe^{3+}, and Mn^{2+} from Zn^{2+} and Al^{3+}

1. To the solution obtained from Part A.5 *or* from an original test solution (in a 75 mm test tube) containing the five test cations for this experiment only, add 15 drops of 6 M NaOH (**Caution:** *do not permit skin contact*). Centrifuge and save the precipitate.[#3] Decant the supernatant[#4] into a 75 mm test tube for Part F. Wash the precipitate with 5 drops of 6 M NaOH, centrifuge, and combine the wash with the supernatant.

2. Dissolve the precipitate[#5] with 1–2 mL of conc HNO_3 (**Caution:** *be careful!!*). A heating of the solution in the hot water bath may be necessary.

C. Manganese(II) Ion Test

1. Decant about 0.5 mL of the solution from Part B.2 into a 75 mm test tube. Add a pinch of solid $NaBiO_3$ to the solution, agitate, and centrifuge.[#6] The deep-purple MnO_4^- confirms the presence of manganese in the sample.[4]

D. Iron(III) Ion Test

1. To the remaining solution from Part B.2, add 10 drops of 2 M NH_4Cl and then drops of conc NH_3 (**Caution:** *do not inhale—use a fume hood if available*) until the solution is basic to litmus and then a 1 drop excess to ensure the complexing of the Ni^{2+}. Centrifuge, save the precipitate,[#7] and transfer the supernatant[#8] to a 75 mm test tube for testing in Part E.

2. Dissolve the precipitate with 6 M HCl and add 2 drops of 0.1 M NH_4SCN.[#9] The blood-red color confirms the presence of iron in the sample.

---

[4] If the deep-purple color forms and then fades, Cl^- is present and is being oxidized by the MnO_4^-; add more $NaBiO_3$.

E. Nickel(II) Ion Test

1. To the supernatant solution from Part D.1, add 3 drops of dimethylglyoxime solution.[10] The appearance of a pink-red precipitate confirms the presence of nickel in the sample.

F. Aluminum Ion Test

1. Acidify the supernatant from Part B.1 to litmus with 6 M HNO_3. Add drops of 6 M NH_3 until the solution is now basic to litmus; then add 5 more drops. Heat the solution in the hot water bath for several minutes to digest the precipitate.[11] Centrifuge and decant the supernatant[12] into a 75 mm test tube and save for the Zn^{2+} analysis in Part G.

2. Wash the precipitate twice with 2–3 mL of hot water and discard each washing. Centrifugation may be necessary. Dissolve the precipitate with drops of 6 M HNO_3. Add 2 drops of the aluminon reagent, stir, and add drops of 6 M NH_3 until the solution is again basic to litmus and a precipitate reforms.[13] Centrifuge the solution; if the $Al(OH)_3$ precipitate is now pink or red and the solution is colorless, Al^{3+} is present in the sample.

G. Zinc Ion Test

1. To the supernatant from Part F.1, add 6 M HCl until the solution is acid to litmus; then add 3 drops of 0.2 M $K_4Fe(CN)_6$ and stir. A *very light green* precipitate[14] confirms the presence of Zn^{2+} in the sample. Centrifugation may be necessary.

H. Unknown

1. Ask your laboratory instructor for a sample containing one or more of the cations from this experiment and determine those present.

| Disposal Information: | Dispose of the test solutions in the "Waste Salts" container. |

EXPERIMENT 36 LAB PREVIEW

CATION IDENTIFICATION, II. Mn^{2+}, Ni^{2+}, Fe^{3+}(and Fe^{2+}), Al^{3+}, Zn^{2+}

Date ____________ Name ________________________________ Lab Sec. ______ Desk No. ________

1. Give the formula of a reagent that precipitates

   a. Ag^+ but not Ni^{2+} ____________________

   b. Mn^{2+} but not Zn^{2+} ____________________

   c. Bi^{3+} but not Al^{3+} ____________________

2. Give the formula of a reagent that dissolves

   a. $NiCl_2$ but not NiS ____________________

   b. FeS but not NiS ____________________

   c. $Al(OH)_3$ but not $Mn(OH)_2$ ____________________

3. $Al(OH)_3$ and $Zn(OH)_2$ are amphoteric. What does the word amphoteric mean?

4. What is the color of

   a. $Ni(NH_3)_6^{2+}$ ____________________

   b. $Fe(NCS)_6^{3-}$ ____________________

   c. $Ni(HDMG)_2$ ____________________

   d. MnO_4^- ____________________

5. Write balanced equations for the following reactions cited in this experiment. Read the Principles and Procedure to assist you in writing the equations. Also read your textbook as a resource for information.

a. The precipitation of the five cations in a sulfide solution with a pH near neutrality.

b. The reduction of Fe^{3+} with H_2S.

c. The dissolving of FeS, ZnS, MnS, and $Al(OH)_3$ with a strong acid.

d. The dissolution of NiS with HNO_3. Elemental sulfur and NO are products of the reaction.

e. The reaction of the five cations with an excess of strong base.

f. The reactions of Fe^{3+} and Ni^{2+} with NH_3.

g. The reactions of Al^{3+} and Zn^{2+} with NH_3.

h. The oxidation of S^{2-} with HNO_3; NO is the reduction product of HNO_3.

CATION IDENTIFICATION, II.
Mn^{2+}, Ni^{2+}, Fe^{3+} (and Fe^{2+}), Al^{3+}, Zn^{2+}

Date _____ Name _____ Lab Sec. _____ Desk No. _____

Procedure Number and Ion	Test Reagent or Technique		Observation (Color or General Appearance)	Chemicals Responsible for Observation	Check (√) if Observed in Unknown
#1		ppt			
#2		spnt			
#3		ppt			
#4		spnt			
#5		spnt			
#6 Mn^{2+}		spnt			
#7 $Fe^{3+(2+)}$		ppt			
#8		spnt			
#9		spnt			
#10 Ni^{2+}		ppt			
#11 Al^{3+}		ppt			
#12		spnt			
#13		ppt			
#14 Zn^{2+}		ppt			

Cations present in unknown: _____

Instructor's approval: _____

QUESTIONS

1. What is the reducing agent that reduces Fe^{3+} to Fe^{2+} in Part A.2?

2. Nitric acid functions as an oxidizing agent and as an acid in Parts A.3 and A.4.

a. Identify the two substances that HNO_3 oxidizes.

b. How does it function only as an acid?

3. What ions precipitate from solution in Part B.1 if NH_3 is substituted for NaOH? Explain.

4. What happens in Part D.1 if NaOH is substituted for NH_3?

5. What happens in Part F.1 if NaOH is substituted for NH_3?

6. Identify a reagent(s) that will

a. precipitate Cu^{2+} but not Ni^{2+} _____

b. dissolve $Al(OH)_3$ but not $Fe(OH)_3$ _____

c. precipitate Fe^{3+} but not Ni^{2+} _____

d. dissolve ZnS but not NiS _____

EXPERIMENT 37

CATION IDENTI-FICATION, III.
Mg^{2+}, Ca^{2+}, Ba^{2+}

Calcium is the most abundant metallic element in the human body, present for the most part in bones and teeth generally as the phosphate known as an apatite. Magnesium is also a major component of bones and teeth, but it also assists in the transfer of electrical pulses between the cells. Magnesium is the trace metal in chlorophyll that is responsible for photosynthesis. Calcium carbonate is the most common compound of calcium, being the principal component of limestone, stalactites, stalagmites, marble, and a few antacids. Calcium and magnesium ions are the major components of "hard" water. Barium sulfate is used as a drilling mud in oil exploration and for assisting in the x-ray analysis of the gastrointestinal tract.

OBJECTIVE

• To separate and identify the presence of Mg^{2+}, Ca^{2+}, and Ba^{2+} ions in a composite sample

PRINCIPLES

The chloride and sulfide salts (Experiments 35 and 36) of Mg^{2+}, Ca^{2+}, and Ba^{2+} ions are soluble. Many cations emit characteristic colors after being excited in a Bunsen flame. A "flame test" will be made on each of the metallic cations to help in our identification process. When more than one cation that exhibits a positive flame test is present in the mixture, conflicting flame data renders flame tests inconclusive.

To separate and identify the alkaline-earth metal ions, Mg^{2+}, Ca^{2+}, and Ba^{2+}, you must practice good laboratory techniques. We will use the anions $NH_4PO_4^{2-}$, $C_2O_4^{2-}$, and SO_4^{2-} to respectively identify the cations. The $NH_4PO_4^{2-}$ ion forms from the HPO_4^{2-} ion in an ammoniacal solution. The tests are reasonably straight-forward, but some care in pH adjustments is necessary.

A flow diagram for the analysis is to be completed for the Lab Preview. Be sure to do this prior to entering the laboratory.

PROCEDURE

To become familiar with the identification of these cations, take a sample that contains the three cations and analyze it according to the procedure. At each numbered superscript (i.e., [1]), STOP, and record on the Data Sheet. After the presence of each cation has been confirmed, SAVE the test tube so that its appearance can be compared to that for your unknown sample.

Caution: *A number of concentrated and 6 molar acids and bases are used in the analysis of these cations. Handle each of these solutions with care. Read the Lab Safety section in Experiment 1 for instructions in handling acids and bases.*

A. Flame Tests

1. Concentrate 2 mL of a sample solution or the supernatant solution from Experiment 36, Part A.2 to a moist residue in an evaporating dish.

2. Clean a coiled platinum wire, sealed in a nonheat conducting handle, by heating it with the nonluminous flame of a Bunsen burner until the flame is colorless (Figure 37.1). Dip the platinum wire into the moist residue and return it to the flame. Note the color.[1] Note that if your unknown contains several of these cations, a mixture of colors will result and therefore may prove inconclusive.

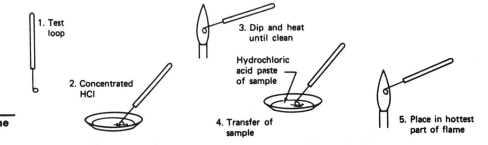

Figure 37.1
Technique for performing a flame test

1. Test loop
2. Concentrated HCl
3. Dip and heat until clean
Hydrochloric acid paste of sample
4. Transfer of sample
5. Place in hottest part of flame

B. Barium Ion Test

1. Dissolve the moist residue from Part A.1 with 1 mL of water. Acidify the solution to litmus with 6 M HCl (**Caution!**) and transfer to a 75 mm test tube. Add drops of 1 M K_2CrO_4 until any precipitation appears complete.[2] Centrifuge and save the supernatant for Part C.

2. Dissolve any precipitate with conc HCl (**Caution:** *do allow skin contact. Avoid inhalation. Clean up any spills.*) and perform a flame test.[3]

C. Calcium Ion Test

1. Adjust the supernatant from Part B.1 to be slightly basic to litmus with 6 M NH_3 (**Caution!**). Add 2–3 drops of 1 M $K_2C_2O_4$.[4] Centrifuge and save the supernatant for Part D.

2. Dissolve any precipitate with 6 M HCl and perform a flame test.[5]

D. Magnesium Ion Test

1. Add 1–2 drops of 6 M NH_3 to the supernatant from Part C.1. Add 2–3 drops of 1 M Na_2HPO_4, heat in a hot water (≈90°C) bath (see Technique 6d), and allow to stand. Any precipitate[6] may be slow in forming; be patient.

2. Dissolve any precipitate with 6 M HCl and perform a flame test.[7]

E. Unknown

1. Obtain a sample containing an unknown number of cations from this experiment. Analyze it according to this procedure and report your findings to your laboratory instructor.

Disposal Information: Dispose of the test solutions in the "Waste Salts" container.

CATION IDENTIFICATION, III.
Mg^{2+}, Ca^{2+}, Ba^{2+}

Date _____ Name _____ Lab Sec. _____ Desk No. _____

1. Record the K_s values (if they are considered insoluble salts) of the following salts. Use your textbook or other reference book, such as the CRC, *Handbook of Chemistry and Physics*.

K_s	Mg^{2+}	Ca^{2+}	Ba^{2+}
SO_4^{2-}			
$C_2O_4^{2-}$			
$NH_4PO_4^{2-}$			

2. Write balanced equations for the precipitation reactions of

   a. Mg^{2+} and $NH_4PO_4^{2-}$

   b. Ca^{2+} and $C_2O_4^{2-}$

   c. Ba^{2+} and SO_4^{2-}

3. Explain the difficulty that may arise if a flame test were used to identify a single cation in a mixture containing several cations.

4. Construct a flow diagram for the four cations in this experiment. Read the Principles and Procedure to help you in its construction.

CATION IDENTIFICATION, III.
Mg^{2+}, Ca^{2+}, Ba^{2+}

Date _____ Name _____ Lab Sec. _____ Desk No. _____

Procedure Number and Ion	Test Reagent or Technique		Observation (Color or General Appearance)	Chemicals Responsible for Observation	Check (√) if Observed in Unknown
#1	flame				
#2 Ba^{2+}		ppt			
#3	flame				
#4 Ca^{2+}		ppt			
#5	flame				
#6 Mg^{2+}		ppt			
#7	flame				

Cations present in unknown: _____

Instructor's approval: _____

QUESTIONS

1. What is the color of the flame test for

   a. Ba^{2+} _____

   b. Ca^{2+} _____

   c. Mg^{2+} _____

2. What is the color of each salt?

   a. $MgNH_4PO_4$ _____

   b. CaC_2O_4 _____

   c. $BaSO_4$ _____

3. Write the formula of a reagent that separates

    a. Cu^{2+} from Ba^{2+} _____

    b. Ba^{2+} from Ca^{2+} _____

    c. Ba^{2+} from Mg^{2+} _____

    d. Mg^{2+} from Ca^{2+} _____

4. What happens if H_2SO_4 is used instead of HCl in Part B.2? Refer to the solubility rules in your textbook.

5. What happens if NaOH is used instead of NH_3 in Part D.1? (Use the solubility rules in your textbook to support your statement.)

EXPERIMENT 38

COORDINATION COMPOUNDS

The Ragu Food Company of Rochester, NY produces a number of "ready-to-use" sauces for the quick preparation of chicken dinners. The sauces must have a relatively long shelf life; from the time the sauce is prepared until it is used may span several months without refrigeration. Various components of the sauce tend to promote the degradation (spoilage) of the food products; among those are the various trace metal ions that enter the food during the various stages of the preparation of the sauce, from the pans, ovens, kettles, etc. To guard against spoilage, Ragu adds a trace of calcium disodium ethylenediaminetetraacetate, $CaNa_2EDTA$. The EDTA is an anion that binds many of the trace metal ions, such as zinc, aluminum, and iron, into a coordination compound; the metal is bound so that it cannot act as a catalyst in the air oxidation of various food components leading to spoilage. Consequently, EDTA is also used in the processing of meats, fish products, dairy products, and vegetables.

The EDTA anion has the structure shown at right. It has six *sites* (designated with asterisks) which can donate an electron pair to a metal ion. Consequently, it has the effect of enveloping the metal ion so that it can offer no other chance of a reaction. EDTA is also used as an antidote for metal poisoning, to bind hardening ions from water, and, in analytical chemistry, can be used for the analysis of nearly every metal.

$$\text{*}^-\text{O-C-CH}_2 \quad \text{*} \qquad \text{*} \quad \text{CH}_2\text{-C-O}^- \text{*}$$
$$\text{N-CH}_2\text{-CH}_2\text{-N}$$
$$\text{*}^-\text{O-C-CH}_2 \qquad \qquad \text{CH}_2\text{-C-O}^- \text{*}$$

OBJECTIVES

- To prepare several complex ions and compare their stabilities
- To synthesize an inorganic coordination compound
- To determine the purity and stability of the prepared compound

PRINCIPLES

A coordination compound consists of at least one complex ion; a complex ion consists of a metal ion, usually a transition metal ion, bonded to one or more Lewis bases. These bases, called **ligands**, generally form 2, 4, or 6 bonds to the metal ion; the number of bonds is the **coordination number** of the metal ion. Most ligands donate only a single electron-pair to the metal ion, but others may donate as many as 2, 3, or, more rarely, 4 or 6.

In the complex ion, $Ag(NH_3)_2{}^+$, the two ammonia molecules are the ligands, and the lone electron pair on each nitrogen bonds to the Ag(I) ion. The structure is linear.

$$[H_3N : Ag : NH_3]^+$$

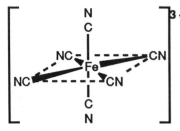

Figure 38.1
The structure of the $[Fe(CN)_6]^{3-}$ complex ion

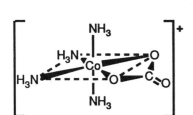

Figure 38.2
The structure of the $[Co(NH_3)_4CO_3]^+$ complex ion

The iron(III) ion in the complex ion, $[Fe(CN)_6]^{3-}$, has a coordination number of 6; the complex ion forms an octahedral structure, with the electron-pair donor site on the C atom of the $:C≡N:^-$ ligand. The $[Fe(CN)_6]^{3-}$ complex ion (Figure 38.1) carries a 3^- charge and each CN^- ligand has a 1^- charge; the Fe in the complex ion is therefore Fe(III). Since each $:C≡N:^-$ is a monodentate ligand, the coordination number of the iron(III) ion is six.

In this experiment, you will prepare a number of Cu(II), Ni(II), and Co(II) complex ions and compare their stability using a number of different ligands. You will then synthesize the coordination compound $[Co(NH_3)_4CO_3]NO_3$ (Figure 38.2). Four $:NH_3$ molecules and one CO_3^{2-} ion form bonds to Co(III). Two oxygen atoms on the carbonate ion serve as Lewis bases to the Co(III) ion; note again that the coordination number of the cobalt(III) ion is six.

The complex ion has a 1^+ charge and the NO_3^- serves as the neutralizing anion to the complex cation.

The coordination compound, $[Co(NH_3)_4CO_3]NO_3$, dissolves in water to produce two ions, $[Co(NH_3)_4CO_3]^+$ and NO_3^-, like that of $NaNO_3$.

$$[Co(NH_3)_4CO_3]NO_3(aq) \xrightarrow{H_2O} [Co(NH_3)_4CO_3]^+(aq) + NO_3^-(aq)$$
$$NaNO_3(aq) \xrightarrow{H_2O} Na^+(aq) + NO_3^-(aq)$$

Addition of Ca^{2+} to the $[Co(NH_3)_4CO_3]NO_3$ solution yields *no* $CaCO_3$ precipitate because the CO_3^{2-} ion remains bound to the cobalt(III) ion in the complex ion. How do you suppose the presence of NH_3 ligands affects the pH of the solution? We'll find out.

TECHNIQUES

• Technique 3, page 2	Microscale Analyses
• Technique 5, page 3	Transferring Liquids
• Technique 7a, b, page 7	Evaporation of Liquids
• Technique 11, page 10	Testing with Litmus
• Technique 13, page 11	Using the Laboratory Balance
• Technique 14c, d, page 12	Separation of a Liquid from a Solid

PROCEDURE

A. A Look at Some Complex Ions

1. **Chloro Complexes.** Set up a 2 x 3 array of microcells in a 24-cell plate. Place 10 drops of 0.1 M $CuSO_4$ in paired microcells. Repeat with 0.1 M $Ni(NO_3)_2$, and 0.1 M $CoCl_2$. Add 1 mL of conc HCl (**Caution:** *handle carefully, do not allow conc HCl to contact skin or clothing. Flush immediately with water.*) to each microcell of a 1 x 3 array. Stir the solutions. Compare the color of the two 1 x 3 arrays, the original and the one with the added HCl. Record your observations on the Data Sheet.

   Slowly add 1–2 mL of water to the 1 x 3 array of microcells having the added HCl. Compare the solution colors to those of the original solution. Does the original color return? What does this tell you about the stability of the chloro complex ions?

2. **Copper and Cobalt Complex Ions.** Set up two 1 x 5 arrays of microcells on the 24-cell plate. In the first labeled set place 10 drops of 0.1 M $CuSO_4$ in each microcell. In the second labeled set place 10 drops of 0.1 M $Co(NO_3)_2$ in each microcell.

3. Place 5 drops of conc NH₃ (**Caution:** *avoid breathing its vapors.*) in the first microcell of each solution, 5 drops of ethylenediamine—abbreviated, en (**Caution:** *avoid breathing its vapors.*)—in the second, of 0.2 M ammonium tartrate, $(NH_4)_2$tar in the third, and of 0.1 M KSCN in the fourth. If a precipitate forms in any one of the solutions, add an excess of the ligand-containing solution. Compare the solutions with that in the fifth micocell.

4. Add 3 drops of 1 M NaOH to each of the four test solutions containing ligands for the copper ion array and the cobalt ion array. Explain your observations.

B. Preparation of [Co(NH₃)₄CO₃]NO₃

1. Dissolve about 10 g (±0.01 g) of $(NH_4)_2CO_3$ in 30 mL of distilled (or deionized), boiled H_2O. Cautiously add 30 mL of conc NH_3[1] (**Caution !**)

2. Dissolve 8 g (±0.01 g) of $Co(NO_3)_2 \cdot 6H_2O$ in 15 mL of H_2O contained in a 250 mL Erlenmeyer flask. While stirring, add the $(NH_4)_2CO_3$ from Part B.1. Cool the mixture to 10°C and slowly add 5 mL of 30% H_2O_2 (**Caution:** 30% H_2O_2 *causes severe skin burns in the form of white blotches; wash the affected skin immediately with water*) to this solution.

3. Transfer the solution to an evaporating dish; using a steam bath (Figure 38.3) reduce the total volume by one-half. During the evaporation periodically add a total of 2.5 g (±0.01 g) of $(NH_4)_2CO_3$ in small amounts. Clean a vacuum filter flask and then vacuum filter the *hot* solution. Cool the filtrate in an ice bath.

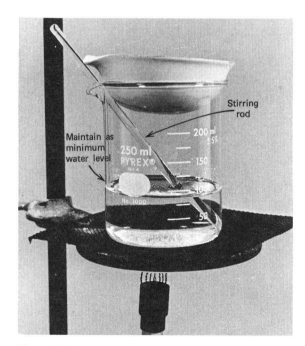

Figure 38.3
The steam bath is used for reducing the volume of the reaction mixture

Figure 38.4
The apparatus for testing the stability of the complex ion

---

[1] If a fume hood is available, use it while transferring the conc NH_3.

4. Red crystals should now be present in the filtrate. Vacuum filter the crystals, wash the crystals twice with 3–5 mL of methanol. Air-dry the crystals. Determine the mass of product and calculate the percent yield.

C. Testing the $[Co(NH_3)_4CO_3]^+$ Ion

Convex: the outward curvature of a piece of glassware.

1. **Litmus test and $CaCl_2$ test.** Dissolve a small pinch of the coordination compound in H_2O in a microcell. Test the solution with red and blue litmus paper. Record your observation. Add several drops of 0.1 M $CaCl_2$ solution. Record your observation.

2. **Na_2CO_3 test.** To an equal-size sample dissolved in H_2O in a microcell, add several drops of 0.1 M Na_2CO_3 solution. Centrifuge and record your observations. What is the formula of the precipitate?

3. **Stability test.** Place a pinch of the compound in an evaporating dish and add 5 mL of water. Add 20 drops (1 mL) of 1 M HCl. Adhere a piece of red litmus paper to the convex side of a watchglass (Figure 38.4). Now add 2 mL of 0.05 M NaOH to the solution and cover with the watchglass, convex side down. Gently heat. What do you observe? Comment on the stability of the $[Co(NH_3)_4CO_3]^+$ ion.

Disposal Information: Dispose of the test solutions in the "Waste Salts" container.

COORDINATION COMPOUNDS

Date _____ Name _____ Lab Sec. _____ Desk No. _____

1. a. What is a complex ion?

    b. What is a ligand?

    c. What is meant by the statement, "the coordination number of the cobalt(II) ion is usually six" in a complex ion?

2. Write the formula of the complex ion formed between

    a. the ligand, NH_3, and copper(II) ion, assuming a coordination number of four.

    b. the ligand, CN^-, and iron(II) ion, assuming a coordination number of six.

    c. the ligand, H_2O, and nickel(II) ion, assuming a coordination number of six.

3. Identify the ligand(s) in the following coordination compounds.

   a. $[Pt(NH_3)_2Cl_2]$ _____

   b. $[Cu(NH_3)_4]SO_4$ _____

   c. $[Cr(NH_3)_4(H_2O)_2]Cl_3$ _____

4. What is the coordination number of the metal ion in each complex ion in Question 3?

   a. _____ ; b. _____ ; c. _____

5. The CO_3^{2-} that is a ligand in the $[Co(NH_3)_4CO_3]^+$ complex ion is called a bidentate. Refer to your text and define bidentate.

COORDINATION COMPOUNDS

Date _____ Name _____ Lab Sec. _____ Desk No. _____

A. A Look at Some Complex Ions

Solution	Color/H_2O	Color/HCl	Formula of Complex Ion	Effect of H_2O
0.1 M $CuSO_4$				
0.1 M $Ni(NO_3)_2$				
0.1 M $CoCl_2$				

Cu^{2+} and ligand:	NH_3	en	tar^{2-}	SCN^-	H_2O
color					
formula					
effect of OH^-					

Co^{2+} and ligand:	NH_3	en	tar^{2-}	SCN^-	H_2O
color					
formula					
effect of OH^-					

B. Preparation of $[Co(NH_3)_4CO_3]NO_3$

1. Mass of $Co(NO_3)_2 \bullet 6H_2O$ (g) _____

2. Mass of $[Co(NH_3)_4CO_3]NO_3$ (g) _____

3. Theoretical yield of $[Co(NH_3)_4CO_3]NO_3$
   based on $Co(NO_3)_2 \bullet 6H_2O$ (g)* _____

4. Percent Yield (%) _____

C. Testing the $[Co(NH_3)_4CO_3]^+$ ion

Test	Observation	Conclusion
Litmus test		
$CaCl_2$ test		
Na_2CO_3 test		
Stability		

1. a. Compare the relative stability of the chloro complex ions of Cu^{2+}, Ni^{2+}, and Co^{2+}.

    b. Compare the relative stability of the NH_3 and ethylenediamine complex ions of Co^{2+}.

2. a. What is the purpose of the 30% H_2O_2 in Part B.2 of the experimental procedure?

    b. Write an equation for the reaction. Ask your instructor for assistance.

3. In the stability test (Part C.3), why was it first necessary to add 1 M HCl?

4. For the same concentration as $[Co(NH_3)_4CO_3]NO_3$, which of the following salts show the same conductivity when dissolved in water? Explain your answer.

    a. $Co(NO_3)_3$

    b. KCl

    c. Na_2CO_3

    d. $NaNO_3$

EXPERIMENT 39
ORGANIC COMPOUNDS

Enzymes, leaves on trees, sharks, humans, pesticides, carpeting, plastics and synthetic fibers . . . it seems as though most matter with which we are familiar is constructed of compounds we call **organic**. The shoes we wear, the clothes we buy, the furniture in our homes, the sex pheromones of insects, the smell of an orange blossom, and the gourmet hamburger are made of organic compounds. The breadth, depth, and versatility of the element carbon, an element of all organic compounds, extend far beyond that of any other element in the Periodic Table. Organic chemistry alone is a specialized field of study that also includes biochemistry and many subdisciplines of biology.

But what is organic gardening? Virtually all of the compounds in plants and vegetables are organic, but it is the nonuse of synthetic organic pesticides in the growth of the garden vegetables that label the gardening practice "organic".

OBJECTIVE

• To study the chemical behavior of hydrocarbons, alcohols, organic acids and bases, and esters

PRINCIPLES

Carbon has the unusual property of bonding to itself. Although other atoms of the elements do this, carbon does it much more extensively. Because of this unique property, over 3 million **organic compounds** have been reported in the literature. As a result, a complete knowledge of the chemical characteristics of each organic compound would indeed be very difficult to acquire. The complexity is lessened somewhat by the characteristic chemical behavior of a large segment of compounds. Organic compounds group into natural classes because members of each class possess similar chemical properties. We will study five of these classes: the hydrocarbons, alcohols, the organic acids and bases, and the esters.

Hydrocarbons

Hydrocarbons are compounds that contain only the elements carbon and hydrogen. According to their chemical properties, the hydrocarbons are subdivided into three subgroups: the saturated hydrocarbons, the unsaturated hydrocarbons, and the aromatic hydrocarbons.

Inert: substances that react chemically only under extreme or specified conditions.

The **saturated hydrocarbons** are commonly referred to as alkanes. All C–C and C–H bonds are single bonds and are relatively inert to chemical attack. By far, their most significant reaction is combustion (e.g., propane C_3H_8) with O_2 forming CO_2 and H_2O as products.

$$C_3H_8 + 5\,O_2 \rightarrow 3\,CO_2 + 4\,H_2O$$

The alkanes slowly react with chlorine and bromine through a substitution of the halogen atom for a hydrogen atom.

$$C_3H_8 + Br_2 \rightarrow C_3H_7Br + HBr$$

The **unsaturated hydrocarbons** have two subgroups: the alkenes and the alkynes. The alkenes have *at least* one C=C bond and the alkynes have at least one C≡C bond. These double and triple bonds, called unsaturated bonds, are chemically quite reactive. For example, bromine quickly reacts across the C=C bond in propene, C_3H_6, to form 1,2-dibromopropane.

$$CH_3-CH=CH_2 + Br_2 \rightarrow CH_3-CHBr-CHBr$$

The unsaturated hydrocarbons also react with oxidizing agents, such as $KMnO_4$, to produce alcohols (Baeyer's test). The $KMnO_4$, a purple reagent, is reduced to MnO_2, a brown precipitate, in the reaction. The brown precipitate may appear as a red-brown solution.

$$3\,CH_2=CH_2(l) + 2\,MnO_4^-(aq) + 4\,H_2O(l) \rightarrow$$
(colorless)   (purple)

$$3\,CHOH-CHOH(aq) + 2\,MnO_2(s) + 2\,OH^-(aq)$$
         (colorless)   (brown)

Aromatic hydrocarbons are generally characterized by the presence of the six-membered carbon ring called benzene, C_6H_6. Although somewhat resistant to chemical attack, aromatic compounds are usually more reactive than the alkanes. For example, bromine reacts very slowly with the alkanes but readily displaces hydrogen from the benzene ring with a catalyst.

Many organic compounds are substituted hydrocarbons where an atom or group of atoms have displaced a hydrogen or carbon atom. The atom or group of atoms is commonly referred to as a **functional group**, which imparts characteristic properties to the compound.

Alcohols

An **alcohol** is a hydrocarbon in which an –OH group (called a **hydroxyl** group) replaces a hydrogen atom. An alcohol is also like water in that an alkyl group[1] replaces a hydrogen atom in the water molecule. Therefore, alcohols have physical properties that are intermediate between those of hydrocarbons and water.

Primary alcohol: the –OH group is attached to a carbon atom that is bonded only to a carbon atom or hydrogen atom.

Secondary alcohol: the –OH group is attached to a carbon atom that is bonded to two other carbon atoms.

Tertiary alcohol: the –OH group is attached to a carbon atom that is bonded to three other carbon atoms.

Depending upon where the –OH group is attached to the hydrocarbon, alcohols are classified as 1°, 2°, or 3°.

CH_3-CH_2-OH, is a 1° alcohol, called ethanol (grain alcohol)
$CH_3-CHOH-CH_3$, is a 2° alcohol, called isopropanol (rubbing alcohol)
$CH_3-COH(CH_3)-CH_3$, is a 3° alcohol, called tertiary butanol

Each alcohol reacts differently with a mild oxidizing agent, such as $K_2Cr_2O_7$. Only 1° and 2° alcohols react to form aldehydes and ketones respectively, as the primary product. A color change from a brilliant orange, due to $Cr_2O_7^{2-}$, to green, due to Cr^{3+}, and/or a change in odor signifies the chemical reaction.

---

[1] An alkyl group is simply a hydrocarbon from which a hydrogen atom has been removed; for example, the ethyl group, C_2H_5-, results when a hydrogen atom is removed from ethane, C_2H_6.

The iodoform test distinguishes between the 1°, 2°, and 3° alcohols. The test is positive when iodine in the presence of a base, such as NaOH, oxidizes the alcohol, producing the acid (with one less carbon atom) and iodoform, CHI_3. A very characteristic odor and a light yellow precipitate confirms the reaction.

Acids and Bases

Organic **acids** are hydrocarbons in which a –COOH group (called a **carboxyl** group) substitutes for a –CH_3 group of an alkane. The acids readily react with $NaHCO_3$ releasing CO_2 gas.

R–: represents a bonded H atom or any organic group.

$$R–COOH(aq) + Na^+HCO_3^-(aq) \rightarrow R–COO^-,Na^+(aq) + H_2O(l) + CO_2(g)$$

Organic acids can easily be prepared by the oxidation of an alcohol with a strong oxidizing agent, such as $KMnO_4$.

Ligand: a Lewis base that bonds to a metal ion for the formation of a complex ion.

Organic **bases** are hydrocarbons in which an –NH_2 group (called an **amine** group) substitutes for a hydrogen atom of an alkane. The lone electron pair on the nitrogen atom serves as a Lewis base. As a Lewis base, an amine can be a ligand in a complex ion. A pH test of an aqueous solution of many organic compounds shows that most are neutral; however, the acids have a low pH and the bases have a high pH. Organic bases tend to have a "fishy" odor as well.

Esters

An organic **ester** is the product of the reaction between an organic acid and an alcohol. Esters tend to be volatile and have a pleasant odor. The natural scents of many flowers and flavors of many fruits are due to one or more esters. The presence of the R–COO–R bond structure is present in all esters. Table 39.1 list some common esters along with their flavor or aroma.

Important natural esters are fats (e.g., butter, lard, and tallow) and oils (e.g., linseed, cottonseed, peanut, and olive) used in the synthesis of oleomargarine, peanut butter, and vegetable shortening.

Table 39.1 Chemical Formulas for Natural Flavors and Aromas

Formula	Flavor/Aroma
$C_3H_7–COO–C_2H_5$	pineapple
$C_2H_5–COO–C_5H_{11}$	apricot
$CH_3–COO–C_8H_{15}$	orange
$H_2N–C_6H_4–COO–CH_3$	grape
$CH_3–COO–C_5H_{11}$	banana
$H–COO–C_2H_5$	rum

TECHNIQUES

- Technique 6d, page 4   Heating Liquids
- Technique 7a, page 6   Evaporation of Liquids
- Technique 11, page 10   Testing with Litmus
- Technique 13, page 11   Using the Laboratory Balance
- Technique 17a, page 16   Handling Gases

PROCEDURE

Caution: *Many organic compounds are volatile and flammable. Therefore, do not have any open flames away from the hood or in designated "flame" areas.*

Obtain an unknown compound from your instructor. Perform the following tests on the unknown at the same time you are testing the sample that is known to have the functional group being studied. Use this procedure to identify the class of organic compounds in your unknown.

The sequence of Parts A through E can be in any desired order.

A. Chemical Reactivity of Hydrocarbons

A number of suggested organic compounds, listed on the Data Sheet, are to be tested for unsaturation. Unsaturated compounds show a positive test for both Br_2 and $KMnO_4$. On occasion however, a test is positive even though the compound has no double or triple bonds. Refer to your text for the structural formulas and note the exceptions on the Data Sheet.

1. **Br_2 test.** In 150 mm test tube, add 4–5 drops of liquid or dissolve 0.1 g of solid organic compound in 2 mL of CH_2Cl_2 (**Caution:** *avoid breathing* CH_2Cl_2 *vapors—use a fume hood if available*). Add drops of 2% Br_2/CH_2Cl_2 (**Caution:** *avoid contact with the* Br_2/CH_2Cl_2. Br_2 *can cause skin burns.*) and agitate after each drop.

   An immediate disappearance of the bromine color indicates the presence of an alkene. If the color of bromine slowly fades, add drops (5 drops maximum) and set the reaction mixture aside for the remainder of the laboratory period and make an occasional observation. The slow color fade results from a substitution reaction rather than an addition reaction. Observe and record.

2. **$KMnO_4$ test.** In a 150 mm test tube, add 4–5 drops of liquid or dissolve 0.1 g of solid organic compound in 2 mL of H_2O or acetone. Add drops of 1% $KMnO_4$ (**Caution:** $KMnO_4$ *is a strong oxidizing agent; avoid skin contact*) and agitate. If a brown precipitate forms within 3 minutes, an alkene is likely present.

B. Chemical Reactivity of Alcohols

Four alcohols and your unknown are listed on the Data Sheet. Use these while completing the following procedure.

1. **Iodoform test.** Dissolve 10 drops of each alcohol in 5 mL of H_2O in separate 150 mm test tubes. Add 5 mL of 10% NaOH. While shaking, add drops of 10% KI/I_2 until the definite dark brown of I_2 persists. Look for CHI_3, a yellow crystalline precipitate. Note any characteristic odor.[2]

   If the iodoform does not form, then warm gently in a hot water bath, but do not exceed 60°C. If the brown color disappears add more KI/I_2 at the elevated temperature until the dark brown color persists for 2 minutes. Allow the test tube and contents to cool.

   While shaking, add drops of 10% NaOH (**Caution:** *avoid skin contact, a concentrated base solution removes layers of skin*) to expel excess I_2. Nearly fill the test tube with distilled (or deionized) water and allow to stand for 10 minutes. Summarize the comparative results for the alcohols on the Data Sheet.

---

[2]Whenever the amount of alcohol is small, iodoform may not separate; however, the characteristic odor establishes its formation.

2. **Alcohol Oxidation.** In a 150 mm test tube, place 2 mL of 0.1 M $K_2Cr_2O_7$ and *slowly* add 1 mL of conc H_2SO_4 (**Caution:** *conc H_2SO_4 is a severe skin irritant and causes clothes to disappear; wash the affected area with large amounts of water*). Swirl to dissolve the $K_2Cr_2O_7$ and cool with tap water. Slowly add 2 mL of the test alcohol. Note any color change and odor. Compare its odor with the test alcohol.

C. Chemical Reactivity of Acids and Bases

1. Place 4–5 drops of each organic acid test solution listed on the Data Sheet in a 75 mm test tube. Test the pH of each sample with litmus paper.

   Repeat the litmus test with an *n*-butylamine. Test the odor of the *n*-butylamine.

2. Place 4–5 drops of each acid listed on the Data Sheet in a 75 mm test tube. Add 1 mL of 10% $NaHCO_3$. A distinct "fizzing" sound is detectable even if the visual evolution of CO_2 is questionable. Record.

3. Place 2 mL of 0.1 M $KMnO_4$ in a 150 mm test tube. Slowly add 1 mL of ethanol. Watch for a color change. Compare its odor with that of acetic acid. What is your conclusion?

4. Add several drops of 1 M $CuSO_4$ to *n*-butylamine. Compare the color of the $CuSO_4$ solution with the Cu^{2+}/*n*-butylamine mixture.

D. Preparation of an Ester

1. **Banana Oil.** Place 2 mL of glacial CH_3COOH (**Caution:** *do not inhale*) and 3 mL of *n*-$C_5H_{11}OH$ (called pentanol) in a 150 mm test tube. Slowly and cautiously add 1 mL of conc H_2SO_4 (**Caution:** *conc H_2SO_4 causes severe skin burns*). Heat the mixture gently on a hot plate or very carefully over a low flame in the hood. If the odor changes, the product is an ester. Have the instructor assist you in writing the balanced equation for the reaction.

2. **Oil of Wintergreen.** In a 75 mm test tube, place a pinch of salicylic acid, $HO–C_6H_4–COOH$. Add 1 drop of 3 M H_2SO_4 and 3 drops of water. After about 30 s, add 3–4 drops of methanol, CH_3OH. Place a loose cotton plug into the mouth of the test tube; place the test tube in a hot water bath (about 60°C) for 20–30 minutes. Note the odor. Have the instructor assist you in writing the balanced equation for the reaction.

Disposal Information: Dispose of all organic test solutions in the "Waste Organics" container

Functional Groups in Organic Compounds

Family	Characteristic Structural Features of Molecules[a]	Examples
Hydrocarbons	Only C and H Alkanes: only single bonds Alkenes: $C=C$ Alkynes: $C\equiv C$	CH_3CH_3, ethane $CH_3CH=CH_2$, propene $HC\equiv CH$, acetylene
	Aromatic: benzene ring	$CH_3-\bigcirc$, toluene
Alcohols	R—OH	CH_3CH_2OH, ethyl alcohol
Aldehydes	$\overset{\displaystyle O}{\overset{\|}{(H)R-C}}-H$	$\overset{\displaystyle O}{\overset{\|}{CH_3-C}}-H$, acetaldehyde
Ketones	$\overset{\displaystyle O}{\overset{\|}{R-C}}-R'$	$\overset{\displaystyle O}{\overset{\|}{CH_3-C}}-CH_3$, acetone
Carboxylic acids	$\overset{\displaystyle O}{\overset{\|}{(H)R-C}}-OH$	$\overset{\displaystyle O}{\overset{\|}{H-C}}-OH$, formic acid
Esters	$\overset{\displaystyle O}{\overset{\|}{(H)R-C}}-O-R'$	$\overset{\displaystyle O}{\overset{\|}{CH_3-C}}-O-CH_3$, methyl acetate
Amines	$R-NH_2$, $R-NH-R'$, $R-\overset{\displaystyle R''}{\overset{\|}{N}}-R'$	CH_3NH_2, methylamine CH_3NHCH_3, dimethylamine $CH_3\overset{\displaystyle CH_3}{\overset{\|}{N}}CH_2CH_3$, dimethylethylamine
Amides	$R-\overset{\displaystyle O}{\overset{\|}{C}}-\overset{\displaystyle R''(H)}{\overset{\|}{N}}-R'(H)$	$CH_3\overset{\displaystyle O}{\overset{\|}{C}}-NH_2$, acetamide

[a] The feature (H)R— means that the group can be either H or a hydrocarbon group. A prime (') or a double prime (") means that the hydrocarbon groups may be alike or different.

ORGANIC COMPOUNDS

Date _____ Name _____ Lab Sec. _____ Desk No. _____

1. Distinguish between

    a. a saturated and an unsaturated hydrocarbon.

    b. an alkene and an alkyne.

    c. an alcohol and a hydrocarbon.

2. a. Write the chemical equation for the combustion of heptane, C_7H_{16}.

    b. Write the chemical equation for the reaction of Br_2 with $CH_3–CH=CH–CH_2Br$.

3. Describe a general procedure for preparing organic acids in the laboratory?

4. What is the principal product in the oxidation of ethanol with $KMnO_4$?

5. How does a chemist quickly distinguish an organic acid from an organic base?

ORGANIC COMPOUNDS

Date _____ Name _____ Lab Sec. _____ Desk No. _____

A. Chemical Reactivity of Hydrocarbons

Identify the tests as positive (+) or negative (-).

Compound	Structural Formula[3]	Br_2/CH_2Cl_2 (+ or -)	$KMnO_4$ (+ or -)
n-hexane or cyclohexane			
1-pentene or 2-pentene			
cyclohexene			
1-hexyne			
ethanol			
naphthalene			
p-xylene			
methane (see Figure 39.1)			
unknown	to be determined		

Figure 39.1
Apparatus for testing natural gas for unsaturation

Natural gas

Test solution

Below level of test solution

_____

[3]Use your text to assist you in writing the structural formulas.

B. Chemical Reactivity of Alcohols

1. Iodoform test Alcohol	Subclass 1°, 2°, 3°	Observation	Oxidation Product
methanol			
ethanol			
i-propanol			
t-butanol			
unknown			

Conclusion of observations on alcohols.

2. Oxidation of Alcohol	Subclass 1°, 2°, 3°	Observation	Oxidation Product
methanol			
ethanol			
i-propanol			
t-butanol			
unknown			

C. Chemical Reactivity of Acids and Bases

1. Name of Acid/Base	Results of Litmus Test	Check (√) for Unknown
acetic acid		
propionic acid		
n-butylamine		

2. Name of Acid	Results of NaHCO$_3$ Test	Check (√) for Unknown
acetic acid		
propionic acid		
butyric acid		

3. Oxidation of Ethanol

Observation   ~~~~~~

Conclusion   ~~~~~~

4. $CuSO_4$ with *n*-butyl amine

Observation   ~~~~~~

E. Preparation of an Ester

1. **Banana Oil.** Write an equation for the reaction.

2. **Oil of Wintergreen.** Write the formula for the compound.

QUESTIONS

1. How can you distinguish between the following compounds?

   a. CH_3NH_2 and CH_3OH

   b. C_2H_5OH and C_4H_9OH, a 3° alcohol

   c. C_3H_8 and C_3H_7OH

   d. $CH_3CH_2CH_3$ and $CH_3CH_2CH_2OH$

2. What reactants make the ester characteristic of the following flavors? (See Table 39.1)

a. pineapple

b. orange

c. grape

3. Acetic acid is prepared by the oxidation of ethanol with permanganate ion in an acidic solution. Use the half-reaction method for writing a balanced (redox) equation for the reaction. The reduction product of the permanganate ion in an acidic solution is Mn(II).

EXPERIMENT 40

CARBOHYDRATES AND PROTEINS

Steel and concrete are used in the construction of many tall buildings. But what is used to build the tall, natural structures such as the redwood trees of California? The natural construction of simple molecules bonding together to build larger and longer molecules forms a framework for the stems of plants and the trunks of trees. Cellulose is this structural **biopolymer** in plants. Cellulose is built from a large number (over 3000) of simpler glucose molecules bonded together. Glucose belongs to a class of compounds called **carbohydrates**.

Are hair and wool also constructed of cellulose? While hair and wool have a definite structure like plants, these biopolymers instead are constructed from a large number (20–60) of amino acids and are called **proteins**. Proteins are also the building blocks for silk, hemoglobin, myoglobin, enzymes that catalyze biochemical reactions, hormones (such as insulin), and DNA.

OBJECTIVES

- To identify monosaccharides, disaccharides, and polysaccharides
- To test samples for various carbohydrates and proteins
- To observe some characteristic reactions of proteins

PRINCIPLES

Carbohydrates

Hydrolysis: the reaction of a substance with water.

Carbohydrates are a class of organic compounds that consist of only carbon, hydrogen, and oxygen in an approximate 1:2:1 mole ratio. Those having a sweet taste are called **sugars**. The carbohydrates are classified according to their complex molecular structure, which is also according to different stages of their hydrolysis. The stages of hydrolysis are

$$\text{polysaccharides} \xrightarrow{\text{H}_2\text{O/H}^+} \text{disaccharides} \xrightarrow{\text{H}_2\text{O/H}^+} \text{monosaccharides}$$

- **Monosaccharides**, such as glucose, fructose, and galactose, are the simplest sugars. Each molecule consists of a single aldehyde (–CHO) or ketone (–CO–) group with hydroxyl groups on the remaining carbon atoms. All monosaccharides have the same molecular formula, $C_6H_{12}O_6$.

Glucose is also known as dextrose or blood sugar and fructose is commonly known as levulose or fruit sugar.

- **Disaccharides**, such as sucrose (table sugar), maltose (malt sugar), or lactose (milk sugar), are chemical combinations of monosaccharides and readily hydrolyze in aqueous acidic solutions forming two monosaccharide molecules.

$$\begin{array}{llll} C_{12}H_{22}O_{11} +H_2O & \rightarrow & C_6H_{12}O_6+ & C_6H_{12}O_6 \\ \text{sucrose} & & \text{glucose} & \text{fructose} \\ \text{or maltose} & & \text{glucose} & \text{glucose} \\ \text{or lactose} & & \text{glucose} & \text{galactose} \end{array}$$

Sucrose, maltose, and lactose (as do all disaccharides) have the same molecular formula, $C_{12}H_{22}O_{11}$, but different structural formulas.

   • **Polysaccharides**, such as starch and cellulose, consist of many monosaccharide units bonded together. The monosaccharide units that make up a polysaccharide molecule are generally the same; for example, only glucose molecules form from the complete hydrolysis of starch and cellulose molecules.

Reducing sugars are oxidized by weak oxidizing agents. An aldehyde group of the reducing sugar is oxidized to a carboxylic acid group. Cupric hydroxide, $Cu(OH)_2$, a weak oxidizing agent in a specially prepared solution called Fehling's solution, produces a red-orange insoluble salt, Cu_2O, in its reaction with a reducing sugar. For the reducing sugar having the formula ...–CHO the equation for the reaction is

$$\text{...–CHO} + 2\,Cu(OH)_2(aq) \rightarrow Cu_2O(s) + 2\,H_2O(l) + \text{...–COOH}$$
$$\text{(blue)} \qquad \text{(red-orange)}$$

The controlled oxidation of monosaccharides in the body is our heat and energy source, about 17 kJ per gram of monosaccharide. The final products of its oxidation are CO_2 and H_2O. If the body cannot effectively oxidize the monosaccharides, they are excreted in the urine indicating diabetes (*diabetus mellitus*).

This experiment uses a general test, called the Molisch test, for identifying carbohydrates followed with a test that identifies the presence of reducing sugars. Sucrose, starch, and cellulose are hydrolyzed and tested for the presence of reducing sugars.

Proteins

α-amino acid: a carboxylic acid that has an amine group bonded to the #2 carbon atom, called the α-carbon atom.

Proteins are formed by a chemical combination of approximately 20 to possibly thousands of α-amino acid molecules resulting in molar masses from about 6000 to several million! The peptide linkage, –CO–NH–, is the connector of the many α-amino acids into a long polymeric chain; the carboxyl group on one α-amino acid condenses with another's amino group, eliminating water in the formation of the peptide link.

R' and R" can be any alkyl group.

In the body, proteins serve many functions: they regulate metabolites (hormones, such as insulin), defend against disease (antibodies), catalyze biochemical reactions (enzymes), and transport oxygen (hemoglobin). Proteins are either *fibrous* (present in hair, wool, muscles, connecting tissue) or *globular* (such as egg or blood albumin, casein in milk, and most enzymes).

Proteins that undergo a change in physical and biochemical properties and, therefore, a change in physiological activity are *denatured*. Heat, alcohol, acids, bases, heavy metal ions, and alkaloids weaken and even break the intermolecular bonding to denature proteins. Coagulation of the protein generally results.

This experiment performs several characteristic tests on egg albumin, a water soluble protein with a molar mass of 34,500. In addition, several color tests are used to detect proteins.

Biuret Test. This is the most general test for proteins. One of the following groups must be present in the protein for a positive test:

$$-\overset{\overset{\displaystyle O}{\|}}{C}-NH_2 \qquad -\overset{\overset{\displaystyle O}{\|}}{C}-NH- \qquad -CH_2-NH_2 \qquad -CHOH-CH_2-NH_2$$

Biuret is a compound with the formula,

$$H_2N-\overset{\overset{\displaystyle O}{\|}}{C}-NH-\overset{\overset{\displaystyle O}{\|}}{C}-NH_2$$

Xanthoproteic Test. This test detects benzene rings with an attached $-NH_2$ or $-OH$ group.

Millon's Test. This test detects the phenolic ring, ⟨benzene ring⟩-OH, in proteins.

Since Cl^- and NH_4^+ interfere with the test, it is unreliable for a urinanalysis.

Hopkins-Cole Test. This test is specific for identifying the trytophan system:

$$\text{(indole ring structure) } CH_2-CH-\overset{\overset{\displaystyle O}{\|}}{C}-OH \atop \underset{NH_2}{|}$$

TECHNIQUES

- Technique 3, page 2 — Microscale Analyses
- Technique 6a, d, page 4 — Heating Liquids
- Technique 11, page 10 — Testing with Litmus

PROCEDURE

A. Carbohydrates. Molisch Test

1. Pour 5 mL of a 1% glucose solution into a 150 mm test tube and add 2 drops of Molisch Reagent. Mix the solution.

2. Incline the test tube; *slowly* and *carefully* add 1–2 mL of conc H_2SO_4 (**Caution:** *conc H_2SO_4 is a severe skin irritant.*) down its side (Figure 40.1). Since conc H_2SO_4 is more dense than the aqueous solution, it forms the lower layer. The purple color at the interface of the aqueous and sulfuric acid layers indicates the presence of carbohydrates. Record your observations.

3. Repeat the test with 1% solutions of fructose, sucrose, maltose, lactose, and starch.

B. Carbohydrates. Fehling's Test, A Test for Reducing Sugars

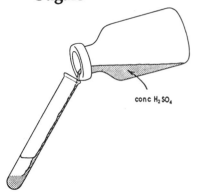

Figure 40.1
Carefully pour conc H₂SO₄ down the side of the test tube

1. Clean six 75 mm test tubes and number them in the test tube rack. Fill them as follows.

Test Tube No.	Fehling's Solution	Sugar Solution
1	1 mL	3–4 drops of 1% glucose
2	1 mL	3–4 drops of 1% fructose
3	1 mL	3–4 drops of 1% sucrose
4	1 mL	3–4 drops of 1% maltose
5	1 mL	3–4 drops of 1% lactose
6	1 mL	3–4 drops of 1% starch

2. Place the test tubes in a boiling water bath for 5–7 minutes. Record your observations.

C. Carbohydrates. Hydrolysis of Sucrose and Starch

1. Place 5 mL of a 1% sucrose solution in one 150 mm test tube and 5 mL of a 1% starch solution in another; add 5 drops of conc HCl (**Caution:** *avoid skin contact*) and place in a boiling water bath for about 10 minutes. Add water to maintain a constant volume in the hot water bath.

2. Cool the solutions and add 2 M NaOH until the solutions are just alkaline to litmus.

3. Test for the presence of reducing sugars in each solution, using the Fehling's solution. Record your observations.

4. A separate test for the presence of starch is to add iodine. A deep-blue I₂•starch complex signifies the presence of starch. To 5 mL of a 1% starch solution, add 1–2 drops of a KI/I₂ solution.

5. Repeat Part C.1 and 2 with 5 mL of a 1% starch solution and test for the presence/absence of starch.

D. Proteins. Solubility and pH

1. Place 10 drops of a 2% aqueous albumin solution[1] into five separate microcells of a 24-cell tray. Add 10 drops of 2 M NaOH to the first, 10 drops of 1 M HCl to the second, 10 drops of 1 M HNO₃ to the third, 10 drops of 0.05 M Na₂CO₃ to the fourth, and 10 drops of distilled (or deionized) H₂O to the fifth.

2. Record your observations on the Data Sheet.

E. Proteins. Denaturation of a Protein

Place the results for each of the following tests in a test tube rack. At the end of Part E, compare the results and record on the Data Sheet as (a) no precipitate, (b) slight precipitate, or (c) heavy precipitate.

---

[1]Alternatively, separate an egg white from its yolk. Add the white of the egg to 150 mL of water and mix until the solution appears to be homogeneous. A mechanical stirrer works well. Filter the solution and use the filtrate for the various tests.

1. Place 5 mL of the 2% albumin solution in a 150 mm test tube and *carefully* heat the upper portion to boiling. Compare the upper portion to the solution in the lower part of the test tube.

2. Place 3 mL of 2% albumin in a 150 mm test tube. Hold it at an angle, and very *slowly* and *carefully* pour 3 mL of conc HNO_3 (**Caution:** *conc HNO_3 is very corrosive to the skin.*) down its side (as is shown with the conc H_2SO_4 in Figure 40.1). The conc HNO_3 underlays the aqueous layer. The appearance of coagulated protein at the aqueous/HNO_3 interface confirms the presence of protein. Describe its appearance.

3. Set up a 1 x 6 array of microcells in the 24-cell plate and place 1 mL of a 2% albumin solution in each.

4. To the respective cells add 5 drops of the following: denatured (95%) ethanol, 10% tannic acid solution, 5% $AgNO_3$, 5% $Pb(CH_3CO_2)_2$, 5% NaCl, and distilled water (as a blank).

5. Compare and record your observations as described above.

F. Proteins. Color Tests

Complete the following tests on 2% solutions of albumin, gelatin, casein, phenol, and glycine.

1. **Biuret Test.** Mix 2 mL of 3 M NaOH with 2 mL of test solution. Add 1 drop of 0.1% $CuSO_4$. Mix thoroughly and note any color change. If a color does not develop, add more drops (up to 10) of 0.1% $CuSO_4$ and mix.

2. **Xanthoproteic Test.** Add 1 mL of conc HNO_3 (**Caution!!**) to 2 mL of test solution. Mix and note any change. Warm carefully; note the color change. Cool and add drops of 3 M NaOH. Note the color change.

3. **Millon's Test.** Add 5 drops of Millon's reagent to 2 mL of test solution. Mix; look for a white precipitate. Warm the solution in a boiling water bath until a red color appears.

Disposal Information: Dispose of the chemicals from the Millon's test in the "Waste Mercury Salts" container.

---

4. **Hopkins-Cole Test.** Add 2 mL of Hopkins-Cole reagent to 2 mL of test solution and mix. Hold the test tube at an angle. *Slowly* and *carefully* add 2 mL of conc H_2SO_4 (**Caution!!**) along its side so that in underlays the aqueous layer (Figure 40.1). Note any color change at the interface of the aqueous and sulfuric acid layers. If no change occurs, very gently agitate the tube causing a slight mix at the interface.

Disposal Information: Dispose of the test solutions (exclusive of those from the Millon's test) in the "Waste Organics" container.

Notes and Observations

CARBOHYDRATES AND PROTEINS

Date _____ Name _____ Lab Sec. _____ Desk No. _____

1. Sucrose is a disaccharide of glucose and fructose. Honey is *not* a disaccharide but rather a 1:1 ratio of glucose to fructose. Honey is a quicker energy source than sucrose. Explain.

2. What is a reducing sugar?

3. Candy bars, high in sucrose content, are quicker energy sources than bread even though both foods are high in carbohydrates. Explain.

4. Calculate the energy released in the complete combustion of 1.0 g of sucrose.

5. Alcohol is used as a disinfectant to kill bacteria before an injection or surgery. Explain its chemical effect on bacteria.

6. a. To avoid the dangers of metal poisoning, milk or egg white is used as an antidote for patients who have swallowed copper, lead, or mercury salts. Explain why it works.

    b. Why must an emetic (a substance that induces vomiting) follow such treatment?

7. Mercury salts have been used as germicides in the treatment of seed grain (corn, beans, sorghum, wheat, etc.). Explain their chemical effectiveness.

8. Write the formula of the peptide linkage.

CARBOHYDRATES AND PROTEINS

Date _____ Name _____ Lab Sec. _____ Desk No. _____

A/B. Carbohydrates. Molisch Test and Fehling's Test

	Molisch Test—Observation	Fehling's Test—Observation
glucose		
fructose		
sucrose		
maltose		
lactose		
starch		

C. Carbohydrates. Hydrolysis of Sucrose and Starch

	Result of Test
1% sucrose, Fehling's test	
1% starch, Fehling's test	
1% sucrose/I_2	
1% starch/I_2	

D. Proteins. Solubility and pH

Testing Reagent	Effect on Albumin—Observation
2 M NaOH	
1 M HCl	
1 M HNO_3	
0.05 M Na_2CO_3	
water	

E. Proteins. Denaturation of a Protein

Denaturing Agent	Effect on Albumin—Observation
heat	
conc HNO_3	
ethanol, 95%	

APPENDIX A

GLASSWORKING

 Custom-made glass rods and tubing are used for the handling of laboratory chemicals and for the construction of laboratory apparatus. Chemists must have a working knowledge of the simple manipulations involved in cutting, bending, and fire polishing glass. Hot glass and cold glass have the same appearance; therefore **never** place hot glass tubing on anything combustible and **never** touch glass that has recently been in a flame. Above all, do not hand hot glass to anyone, especially the laboratory instructor, unless you are trying to end an friendship (or not improve your grade).

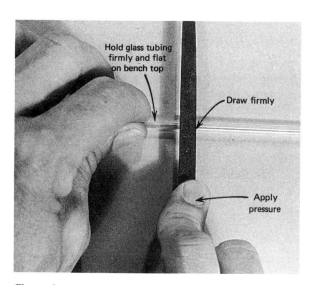

Figure A.1
Scratch the glass tubing with a triangular file

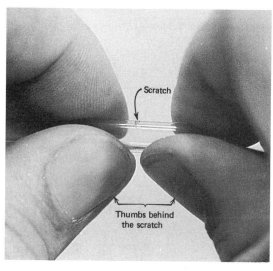

Figure A.2
Position the thumbs *behind* the scratch

Cutting Glass

 The procedure for cutting glass tubing or glass rod is to place a piece of the glass tubing or rod on the lab bench. Use a triangular file or glass scorer to make a deep scratch. Draw the file *once* across the glass; this requires some pressure and should not be a sawing action (Figure A.1). Place a drop of water or saliva on the scratch. Place the thumbs, almost touching, on each side and *behind* the scratch; hold the tubing at the waist with the scratch facing *outward* from the body (Figure A.2). Bend the tube backward as you pull it away from center (Figure A.3). With practice, the glass breaks evenly. If not, try again making the scratch a bit deeper.

Fire-Polishing

Any cut piece of glass (tubing or rod) has a rough, sharp edge. Fire-polishing smooths the edge to guard against cuts and the scratching of glassware. Hold and rotate the cut glass in the flame until the edge is fire-polished (Figure A.4). Do not overheat; if glass tubing is heated too much, the end closes.

Figure A.3
Pull and bend—*simultaneously*—
to break the tubing

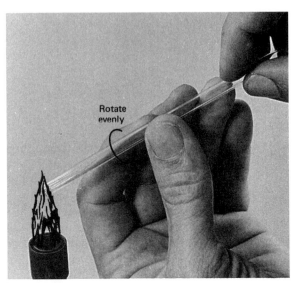

Figure A.4
Fire polish to *just* round the
edges

Bending Glass

The secret to bending glass tubing properly is to heat it sufficiently over the length of the bend. Heat and rotate the glass tubing (cut and fire-polish as before) in the hottest part of the flame (using a wing tip on the burner) until you feel a slight sag (Figure A.5). Remove the softened glass, hold it for *several seconds* to give the temperatures inside and outside the tube time to equilibrate. Slowly and smoothly bend the glass upward to the desired angle (Figure A.6). A "good" bend has the same diameter throughout with no constriction at the bend; poor bends are shown in Figure A.7.

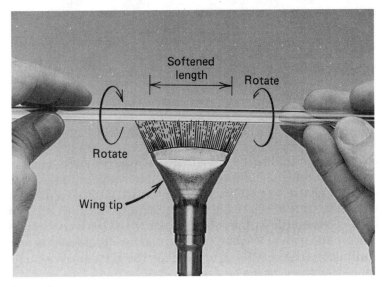

Figure A.5
Slow rotation in the hottest part of
the flame

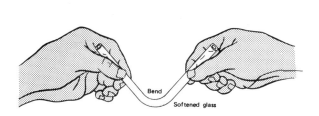

Figure A.6

Bend the softened glass tubing slowly upward

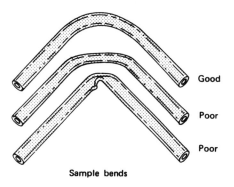

Good

Poor

Poor

Sample bends

Figure A.7

Good and poor glass bends

APPENDIX B

CONVERSION FACTORS

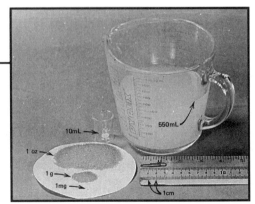

SI Prefixes

Prefix	Meaning (*Power of Ten*)	Abbreviation	Example using "grams"
femto-	10^{-15}	f	10^{-15} g = fg
pico-	10^{-12}	p	10^{-12} g = pg
nano-	10^{-9}	n	10^{-9} g = ng
micro-	10^{-6}	μ	10^{-6} g = μg
milli-	10^{-3}	m	10^{-3} g = mg
centi-	10^{-2}	c	10^{-2} g = cg
deci-	10^{-1}	d	10^{-1} g = dg
kilo-	10^{3}	k	10^{3} g = kg
mega-	10^{6}	M	10^{6} g = Mg
giga-	10^{9}	G	10^{9} g = Gg

SI and English Conversions

Physical Quantity	SI unit	Conversion Factors
Length	meter (*m*)	1 km = 0.6214 mi 1 m = 39.37 in 1 in = 0.02540 m = 2.540 cm
Volume	cubic meter (m^3) [liter (*L*)]	$1\ L = 10^{-3}\ m^3 = 1\ dm^3 = 10^3\ mL$ 1 L = 1.057 qt 1 oz (fluid) = 29.57 mL
Mass	kilogram (*kg*) [gram (*g*)]	1 lb = 453.6 g 1 kg = 2.205 lb $1\ Pa = 1\ N/m^2$
Pressure	pascal (*Pa*) [atmosphere (*atm*)]	1 atm = 101.325 kPa = 760 torr $1\ atm = 14.7\ lb/in^2$ (psi)
Temperature	kelvins (*K*) [degrees Celsius (°C)]	K = 273 + °C
Energy	joule (*J*) [calorie (*cal*)]	1 cal = 4.184 J 1 Btu = 1054 J

Length	
	1 meter (m) = 39.37 in = distance light travels in $\frac{1}{299\ 792\ 548}$ of a second
	1 inch (in) = 2.540 cm = 0.02540 m
	1 kilometer (km) = 0.6214 (statute) mile
	1 angstrom (Å) = 1×10^{-10} m = 0.1 nm
	1 micron (μm) = 1×10^{-6} m

Mass

1 gram (g) = 0.03527 oz = 15.43 grains
1 kilogram (kg) = 2.205 lb = 35.27 oz
1 metric ton = 1×10^6 g = 1.102 short ton = 0.9843 long ton
1 pound (lb) = 453.6 g = 7000 grains
1 ounce (oz) = 28.35 g

Volume

1 liter (L) = 1.057 fl qt = 1×10^3 mL = 1×10^3 cm^3 = 61.02 in^3
1 fluid quart $(fl\ qt)$ = 946.4 mL = 0.250 gal
1 fluid ounce $(fl\ oz)$ = 29.57 mL
1 cubic foot (ft^3) = 28.32 L = 0.02832 m^3

Pressure

1 atmosphere (atm) = 760.0 torr = 760.0 mm Hg = 29.92 in Hg
  = 14.696 lb/in^2 = 1.013 bars = 101.325 kPa
1 kilopascal (kPa) = 1 N/m^2
1 torr = 1mm Hg = 133.3 N/m^2

Energy

1 joule (J) = 0.2389 cal = 9.48×10^{-4} Btu = 1×10^7 ergs
1 calorie (cal) = 4.184 J = 3.087 ft lb
1 British thermal unit (Btu) = 252.0 cal = 1054 J = 3.93×10^{-4} hp hr
  = 2.93×10^{-4} kw hr
1 liter atmosphere $(L\ atm)$ = 24.2 cal = 101.3 J
1 electron volt (eV) = 1.602×10^{-19} J

Constants and Other Conversion Data

velocity of light (c) = 2.998×10^8 m/s = 186,272 mi/s
gas constant (R) = 0.08205 L atm/(mol K) = 8.314 J/(mol K)
  = 1.986 cal/(mol K) = 62.36 L torr/(mol K)
Avogadro's number (N_0) = 6.023×10^{23} /mol
$^\circ$F = 1.8°C + 32
K = $^\circ$C + 273.2
Planck's constant (h) = 6.625×10^{-34} J s/photon

VAPOR PRESSURE OF WATER

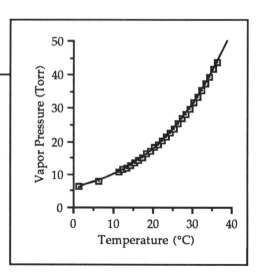

Temperature (°C)	Pressure (torr)
0	4.6
5	6.5
10	9.2
11	9.8
12	10.5
13	11.2
14	12.0
15	12.5
16	13.6
17	14.5
18	15.5
19	16.5
20	17.5
21	18.6
22	19.8
23	21.0
24	22.3
25	23.8
26	25.2
27	26.7
28	28.3
29	30.0
30	31.8
31	33.7
32	35.7
33	37.7
34	39.9
35	42.2
–	–
100	760.0

APPENDIX D

CONCENTRATIONS OF ACIDS AND BASES

Concentrated Reagent	Approximate Molarity	Approximate Mass %	Specific Gravity	mL to dilute to 1 L for a 1.0 M Solution
acetic acid	17.4	99.7	1.05	57.5
hydrochloric acid	12.1	37.0	1.19	82.6
nitric acid	15.7	70.0	1.41	63.7
phosphoric acid	14.7	85.0	1.69	68.1
sulfuric acid	17.8	95.0	1.84	56.2
ammonia (*aq*) (ammonium hydroxide)	14.8	29% (NH_3)	0.90	67.6

Caution: *When diluting reagents, add the more concentrated reagent to the more dilute reagent (or solvent). Never add water to a concentrated acid!*

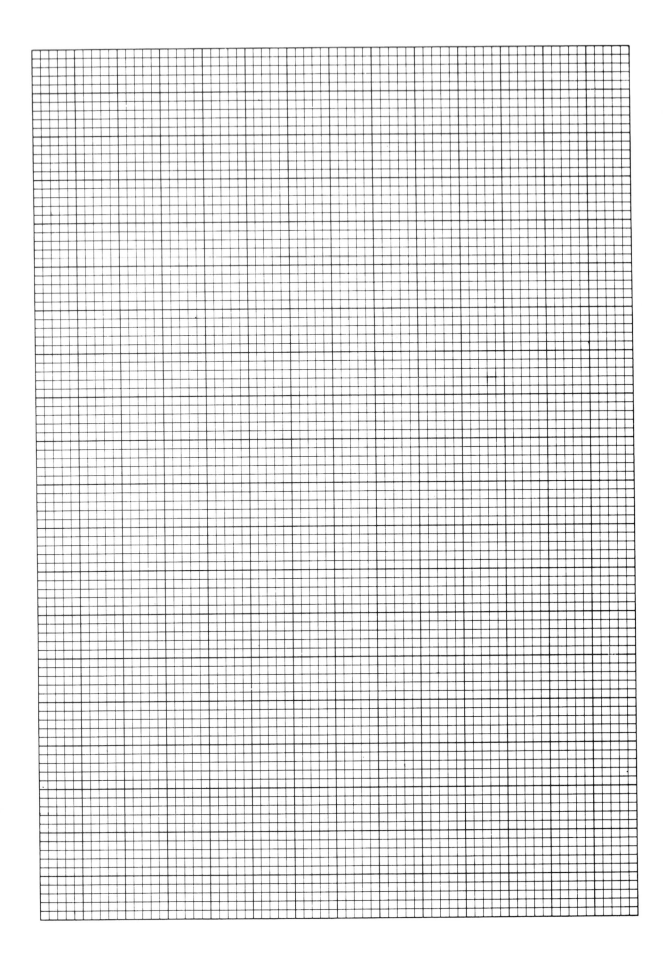

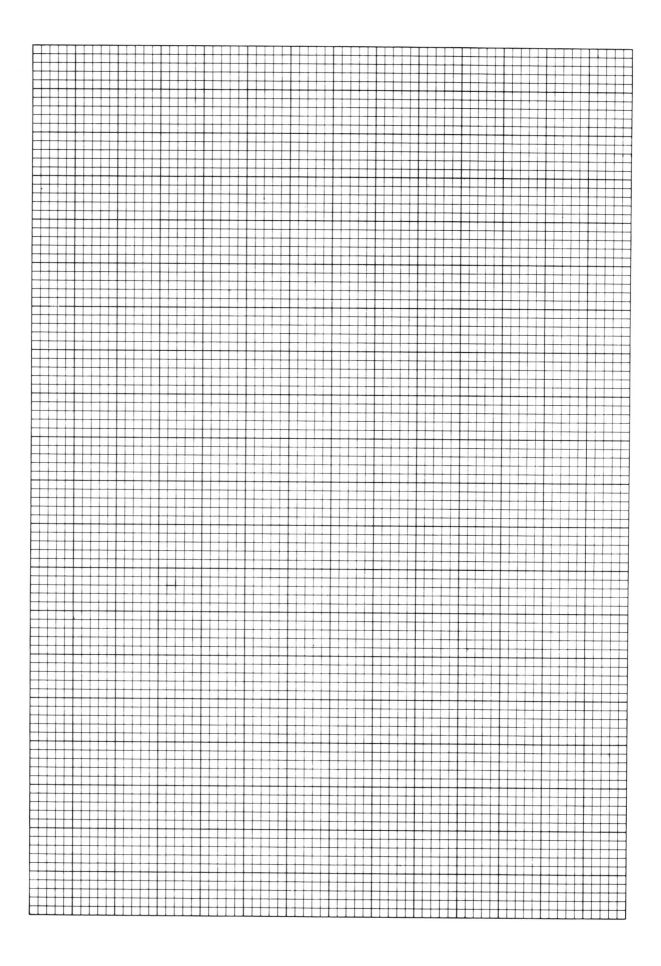

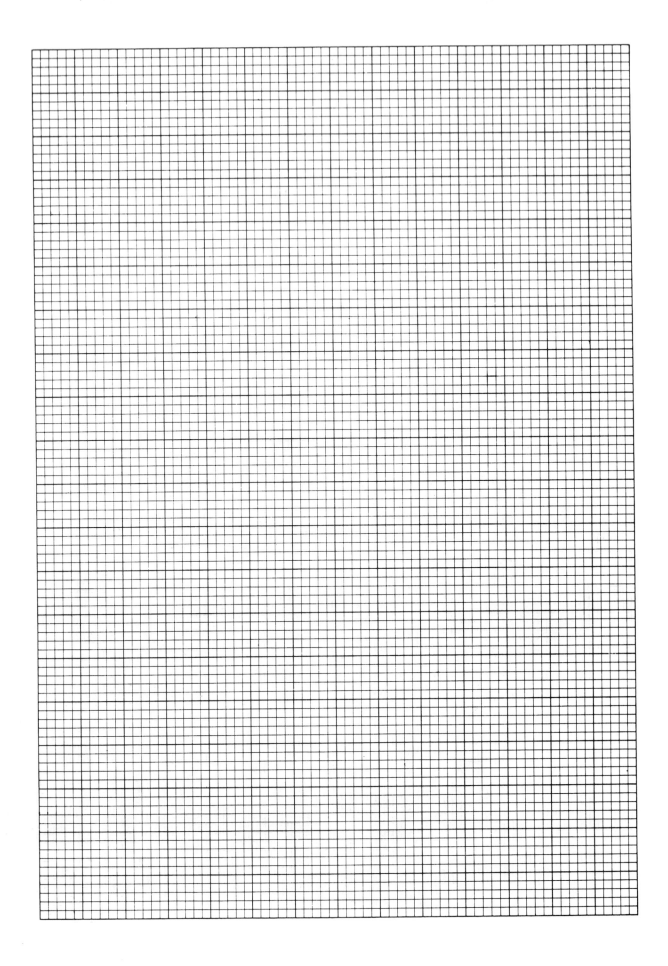

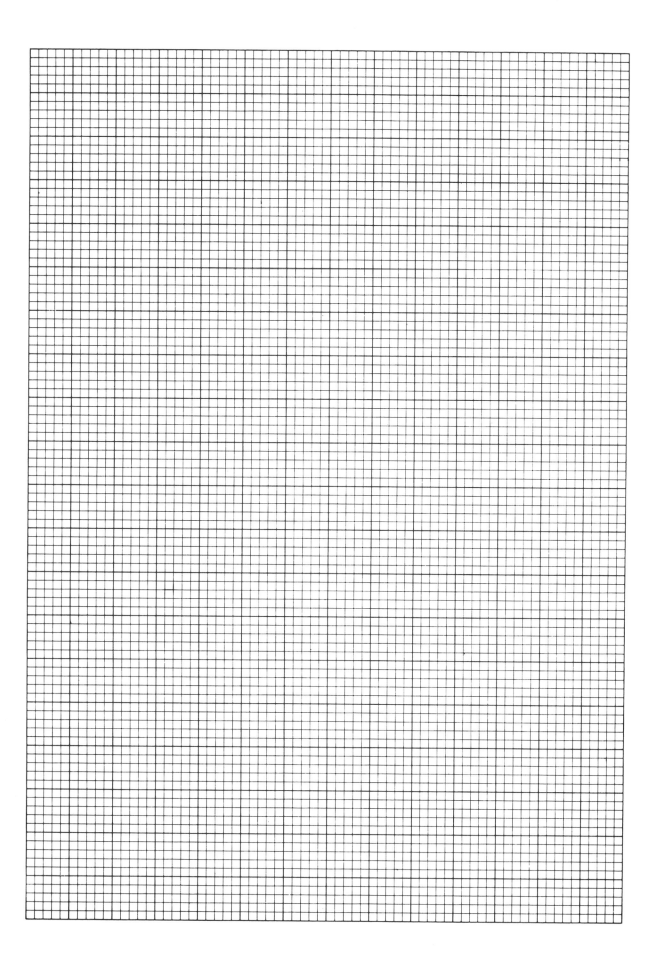

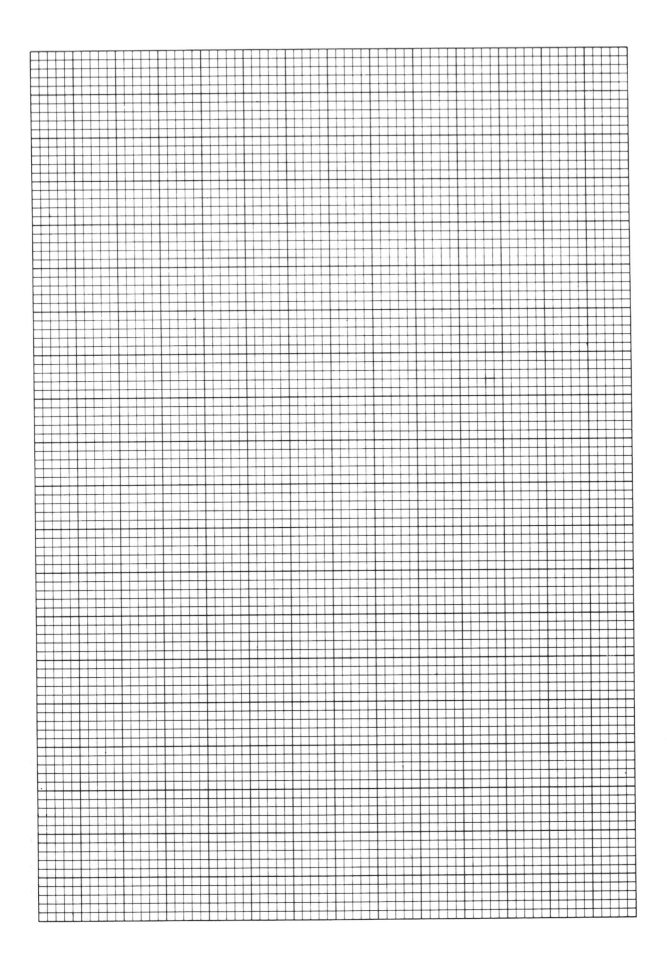

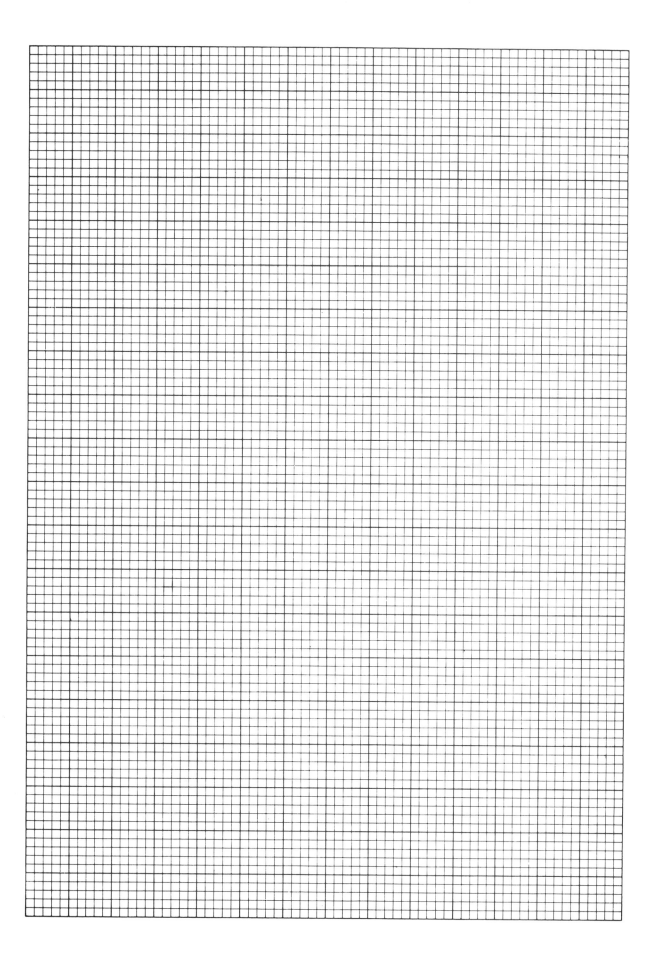

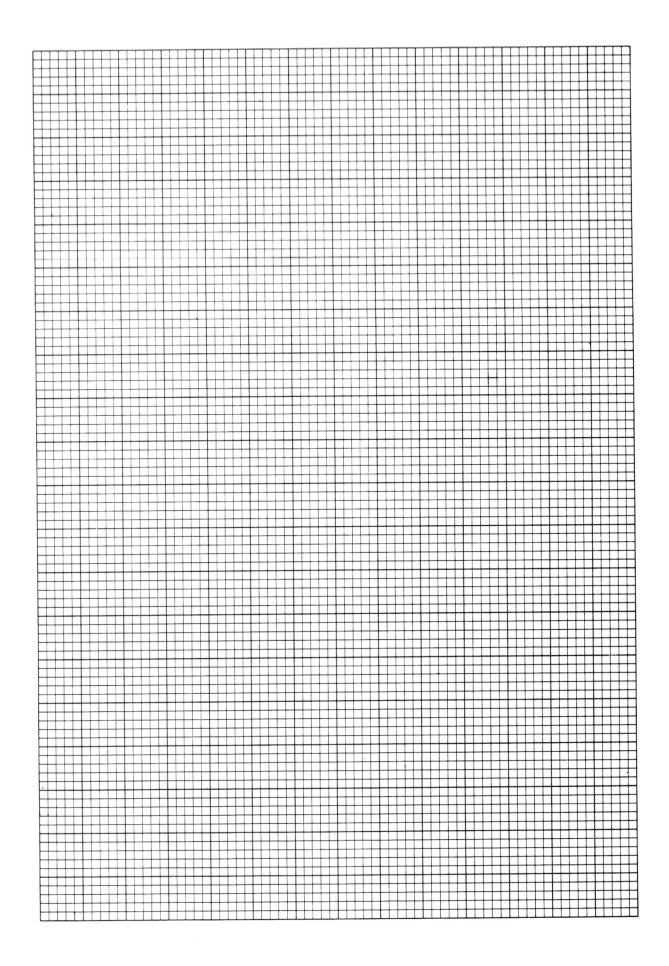

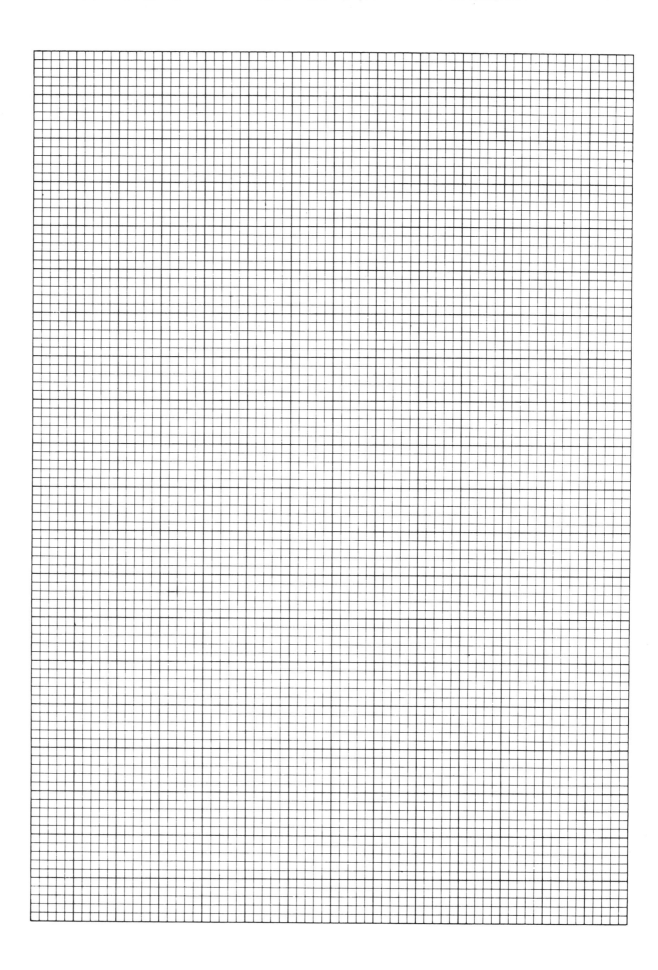

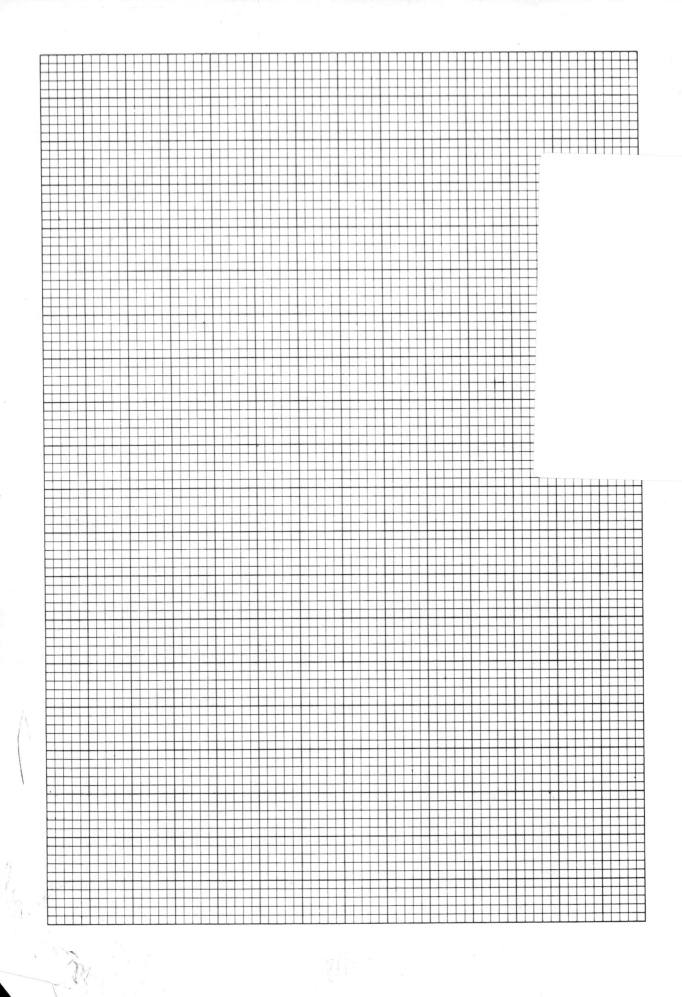